AF248558

Handbook of North Dakota Plants

ORIN ALVA STEVENS

North Dakota Institute for Regional Studies
Fargo, 1963

PHOTOLITHOPRINTED BY CUSHING - MALLOY, INC.
ANN ARBOR, MICHIGAN, UNITED STATES OF AMERICA
1963

North Dakota Institute for Regional Studies

The Institute, established March, 1950, is dedicated to the furtherance of scientific study of the resources—plant, animal, and mineral—of North Dakota and the Northern Plains and Prairies, and of the cultures which they have supported. The Institute is fortunate to sponsor as its first publication the present volume on the plants of the State. Further studies will be published as they are completed and as financial considerations permit, either as individual papers or as journal articles.

The present volume had its beginnings in 1891 with the establishment of the Agricultural College. It represents our present knowledge, based upon contributions of faculty of the School of Applied Arts and Sciences, and the Agricultural Experiment Station staff. Workers in the Extension Division, in other schools of the college and many others have had a part in it. It is our hope that it will be welcomed by the public for whom it has been prepared.

G. E. GIESECKE,
Dean, School of Applied Arts and Sciences

CONTENTS

INTRODUCTION

This book is intended to present general information on the wild plants of the State in a way that will be useful to the largest number of people. It probably will be most useful to teachers or others who have considerable knowledge of plant life. The descriptions are brief and are intended to give the most evident features which would help persons not familiar with botanical descriptions to recognize the plants. Comments on weeds, poisonous plants and those useful as ornamentals are included. Some cultivated or economic plants are mentioned for general information.

Complicated botanical terminology has been avoided as much as possible. Some terms are used in a popular rather than in a technical sense. Thus, "pod" is used for different sorts of fruits which are technically given other names. The word "cluster" is used for various groupings of flowers for which special terms are used in scientific descriptions.

Measurements of leaves, flowers and height of plant are intended to give an idea of size of plant. Sizes of leaves vary greatly and may even be larger or smaller than extremes listed here. Measurements have been checked from actual specimens and in some cases it has seemed desirable to modify the figures given in current manuals. For small flowers and seeds the metric system is more convenient but may not be familiar to most readers. A compromise has been attempted by giving the English equivalents often enough so they may be found on each page. In order to avoid cumbersome fractions, these equivalents are often only approximate. Small metric rules can be had from the larger dealers in school supplies. Several are printed in the back of this book. They can be cut out and pasted on a firm strip of cardboard if desired.

The common names best known or considered most suitable have been given, with addition of others which are often used. Those in quotation marks are local, often misleading or undesirable. Choice and spelling of names may seem inconsistent. Separation and compounding varies in different books. The tendency here is to eliminate hyphens either by combining or separating the two words.

The treatment of genera is somewhat unusual and may seem inconsistent. Since a genus is a group of related species, the genus is not described here unless it is represented by about three or more species. This saves space and for the purpose of this book, should not detract from its usefulness. Species are numbered consecutively for each family.

All plants known to occur in the State in wild condition, that is, native or introduced from other countries, are included. It is often a question whether or not to include certain plants which have been found only once or twice over a period of years and cultivated plants which appear occasionally from scattered seeds. In the present case, most of these have been included. In some groups, as in grasses, sedges and asters, distinctions between species are necessarily technical. However, these plants have been included to make the work complete and useful to those who may be interested in them.

Common species are indicated as generally distributed, or at least in a part of the State. For those of more limited or incompletely known distribution, counties are indicated. In some cases the plant may be present in many more counties. In other cases it probably is found in very few counties. Further discussion of distribution will be found on pages 17-22.

The book has had no editor. Some details which may seem unusual are intentional, others are oversights or exigencies of the printer. Many friends and associates have furnished suggestions for which the author is grateful. Several have read portions of the manuscript. Special acknowledgement should be made to former President Frank L. Eversull who secured funds for publication and to Miss Edith Russum who patiently typed and re-typed copy, assisted in proof reading and checking. Steyermark's **Spring Flora of Missouri** has been especially helpful in planning style of presentation.

Many of the illustrations have previously appeared in bulletins of the agricultural experiment station. Figure 302 is reproduced from Robbins, **Botany of Crop Plants,** by permission of the publishers, P. Blakiston's Son & Co. Most of the new drawings in the text and about one-half of those in the key to families were made by Alice Davis Smith. The others in the key and Nos. 208-11, 248-50, 257-60 and 274-277 are by Leo Hall. Figures 238, 247, 253 and 301 are by Don Hoag. 138-40 and 158-65 by Loren Potter. W. W. Moberg prepared the drawing for Figure 2. The colored frontispiece is by Alvina Halgrimson of McGregor, North Dakota.

Of the illustrations previously used, most of the grasses, figures 154-57 and 317 were by Aldeth Pinkham Wongness; 221 and 222 by Dorothea Gerbracht McCullough; 308-316 by Anita Blake Waldron. A few were from other publications: 128, 233 Colo. Exp. Sta.; 181 Kans. Bd. Agr.; 246 Manitoba Agr. Coll.; 182, 242, Mich. Exp. Sta.; 283 Mont. Exp. Sta.; 146-151, 212, 213 U. S. Dept. of Agr.

The photo for 273 was from Pearl Heath Fraser; 200, 235, 251, 252, 268, 286, 307 from Russell Reid. The others were taken for the author.

Fargo, North Dakota
February, 1950

The Appendix of the third printing of the *Handbook of North Dakota Plants* includes plants added up to the end of 1962 (marked*), more important corrections and changes in names. (See also *Rhodora* 63:39-46). The total number of species is now just over 1200.

Fargo, N. D.
May 1963

Orin Alva Stevens, D.Sc.
Emeritus Professor of Botany
North Dakota State University

Plant Collections In North Dakota

Since North Dakota is in the interior of the continent, its plants did not come to the attention of the first explorers, though many species native here also occur in the eastern states. A summary of our approximately 1000 species shows that about one-third of them are widely distributed. Very few, if any, are peculiar to North Dakota.

North Dakota has a less diversified topography than that of many other states. Hence, our list of species is only one-half to one-fifth that found in some other states. New plants are continually being introduced and no doubt a few native ones remain to be discovered. Probably some may have been destroyed entirely by changes due to human occupation. The total is probably not much larger than the number now known.

Early Plant Collectors

One of the first collections of plants in this region was that of the Lewis and Clark Expedition of 1804-06. They reached the present limits of North Dakota in late fall, wintered at Ft. Mandan and proceeded up the river early in the spring. Thus they had little opportunity to collect plants then and spent very little time here on their return trip. Many of our species, however, were first described from specimens which they collected while crossing the plains of Montana.

The largest early contribution was made by John Bradbury and Thomas Nuttall who came up the Missouri River in 1811 and spent some weeks in this area. Both collected plants for other botanists and some of our prairie species were first described from plants grown in England from seeds collected by these two men.

In 1839 the Nicollet Expedition came up the Missouri River, traveled across to Devils Lake and returned south. Many plants were collected and a list of these was published in the report of the expedition but few new ones were discovered. These three expeditions include most of the early work upon plants of this area.

Recent Plant Collectors

The North Dakota Agricultural College was established in 1890 and C. B. Waldron, the first member on the staff, began collecting specimens for the present herbarium. During the first few years, active work was carried on by C. B. Waldron, H. L. Bolley and assistants. From 1900-04, L. R. Waldron was actively engaged in collecting and identifying specimens. In 1900, Bulletin No. 46 was published, which listed 775 species of plants which had been collected in the State. This is approximately three fourths of the number now known to occur here.

In 1906-09, W. B. Bell collected in Williams (now Divide and Williams), McKenzie, Morton (area now in Grant), Ransom and Richland Counties as part of the North Dakota Agricultural College Soil and Geological Survey. H. F. Bergman continued this work in Barnes County in 1911-12 and collected extensively in other parts of the State in preparation for a new publication on the flora which appeared in 1918 as part of the sixth biennial report of the survey. This gave complete keys for identification of specimens and included 960 species.

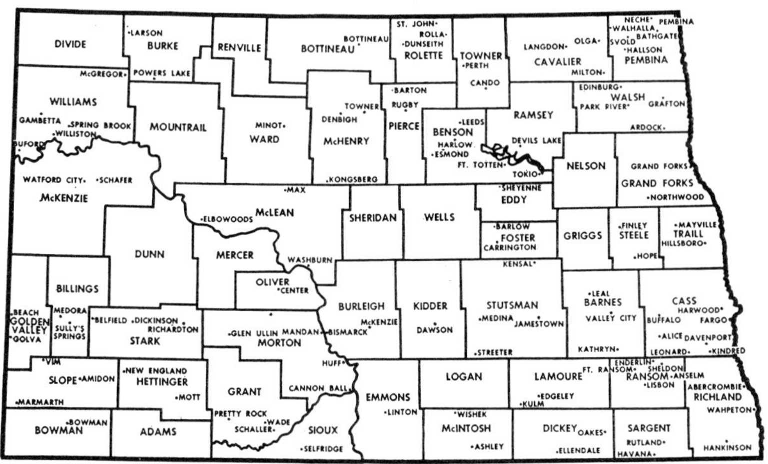

1. Sketch of North Dakota, showing counties and towns mentioned in text.

Dr. J. Lunell, a physician who lived at Leeds, Benson County, from 1894 to 1920, made an intensive study of the plants in that area and published a list of plants of the State (Am. Midl. Nat., 1915-18.). He described many new species, most of which are regarded by other botanists as minor variations of older species. His collections are now in the herbarium of the University of Minnesota. Dr. J. F. Brenckle, a physician at Kulm, LaMoure County, from 1899 to 1917, collected intensively in that area.

The present writer made occasional collections in various parts of the State in 1909-33, giving chief attention to weeds. Since 1933, he has given more attention to completing the records of distribution and to further study of the plants. This work was greatly facilitated by assistance from C.W.A., P.W.A., F.E.R.A. and N.Y.A. projects. Many specimens were mounted, repaired and rearranged. A set of file card maps was prepared showing distribution of specimens of each species. This has made it easier to secure additional specimens for counties not before represented. Although about 20,000 specimens of North Dakota plants are now in the collection, our knowledge of the distribution of many species is incomplete. The central part of the State is least well represented (fig. 16). For many introduced plants the specimens are quite insufficient to show the rate of spread. For some of those which have come in more recently or which are of especial interest, the records are better. In the last few years an attempt has been made to secure a specimen of each species from blocks of about four counties rather than one from each county.

Many specimens are received and identified each year from farmers, teachers, county extension agents and other people. Most of these specimens are not suitable for the herbarium or do not represent new records. However, some of them are of especial interest and are preserved each year. Several species have first come to our attention from such specimens. These records will be found in accounts of various plants in this book.

Changes in names of towns and counties may cause some confusion in using records of older publications. Some of the old, inland postoffices have been discontinued. Lunell referred frequently to "Butte", an elevated area a few miles southwest of Leeds. Later, that name was used for a town in McLean County. Billings, Morton, Ward and Williams Counties were divided in 1910-16 to form Golden Valley, Grant, Sioux, Burke, Mountrail, Renville and Divide Counties. Figure 1 shows locations of counties and towns mentioned. Sometimes collections made near a certain town were from an adjoining county, e.g., most of those from Leonard are from either Richland or Ransom County.

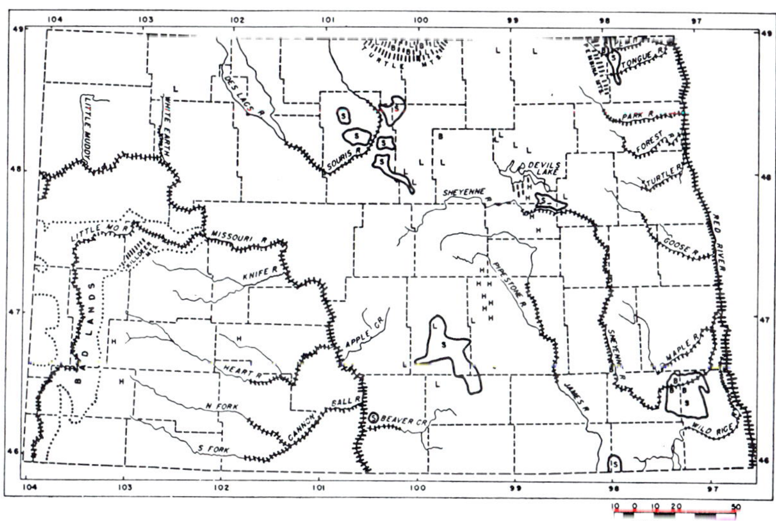

2. Map of North Dakota showing chief physical features (approximate) Wooded portions indicated by cross-bars. B, small, boggy areas, H, somewhat isolated, high hills; L, the larger lakes; S, areas with wind-blown sand.

NAMES OF PLANTS

Names of plants are part of our language and have developed in much the same way as names of other objects. Scientific names have been retained from the time of their use in the Latin language when it was the chief language of science. They have been developed according to systematic rules and in general are standard and uniform for all languages.

Common names have followed general usage and have had little control in method of formation or use. Some confusion arises because a particular plant is known by several different names and because one name is used for different plants in different countries or parts of the country. "Foxtail" is used by most people in North Dakota for *Hordeum jubatum,* which we prefer to call "wild barley"

3. View in Turtle Mts. Much of the trees and brush has been cleared for fields. Soil Conservation Service, photo by Adrian C. Fox.

because it is a species of barley, or "foxtail barley", combining the two names. In eastern United States, this plant is known as "squirreltail" while "foxtail" is used for what we call pigeongrass (*Setaria*). In Europe "foxtail" is a still different grass (*Alopecurus*) which resembles timothy and is grown for hay. Many common names are quite local. They have been suggested by some trifling or accidental reason and used without any thought of the convenience of most people. Sometimes these names become widely used or prove desirable and come into general use.

Much of the confusion in names is due to natural limitations on the number of words which a person can use. Many of our plants do not have common names in general use because the plants are recognized by so few people there is little occasion to refer to them. The common thistle in South Africa may be an entirely different thing from the common thistle in North Dakota, but it is easier to call both by a short, simple name than to use names which will insure that there is no duplication. If one wishes names for all plants, the scientific names should be used.

In the present book, no single authority has been followed for common names. Those which seem most desirable for North Dakota are preferred, and names which are misleading are avoided as far as possible. Common names have not been assigned for all species. For example *Stachys palustris* is called hedge nettle and no name is given for *S. aspera*, which may also be called hedge nettle, not distinguishing the two kinds. Qualifying adjectives would need to be added to separate them and this increase in length of names is not of sufficient interest nor value to justify writing them. Additional common names frequently used are given. A few of the local names are enclosed in quotation marks, as "honeysuckle" to indicate that these are undesirable and likely to cause confusion.

Scientific Names

The scientific name of a plant or an animal consists of a genus name and a species name. In general, we say that the individual plants of one kind make up a **species** and that closely related species make up a **genus**. Likewise, related **genera** (plural of genus) make up a family. Different kinds of oaks, violets or goldenrods are species (plural same as singular), while all of the many species of violets belong to one genus.

4. Large butte in Bad Lands area with characteristic erosion patterns. North slope has red cedar, other shrubs and grasses. Other slopes are mostly bare.

These names are sometimes compared to personal names as written: Smith, James; Smith, George, etc. This is not quite a fair comparison since all members of the human race belong to the same species, but it does suggest the way in which the names are handled. The genus name is really the original one and many of these were in use hundreds of years before our present system developed.

Viola was used by the Romans. Our word violet comes from it as does also Veilchen, the German name. A great many of these old names are found in our present English, either slightly modified as in *Viola, Tulipa* and *Lilium,* or intact as in *Arnica, Asparagus* and *Aster.*

As long ago as 1620, in one of the best early descriptive books on plants, Caspar Bauhin had 7 species of violets, which he described in short phrases, such as, "Viola alpina, folio in plures partes dissecto," i.e., an alpine violet with leaves divided into many parts. This method continued until 1753 when Linnaeus reorganized the descriptions and applied a specific name to each. The violet mentioned above he called *Viola pinnata,* referring to the divided leaves.

5. Prairie pond and hill scene, Benson County. Coarse sedges at edge of pond in foreground; marsh plants on further side of pond due to seepage from higher ground. Small trees and shrubs at foot of hill; small coulees among the hills. Wild ducks on pond. Soil Conservation Service, photo by Hufnagle.

This binomial method, used by Linnaeus in the "Species Plantarum", wherein he described all of the known species of plants, proved so useful that the book was adopted as the starting point and his names are still used unless some essential reason for change has been found. Most of the plants which he listed had been described before. The fact that his personal name is now associated with the plant shows that he was the first writer to use the name regularly in that form.

The authority for the name, usually abbreviated, is added in technical publications: *Viola pinnata* L., *Anemone cylindrica* A. Gray, *Penstemon grandiflorus* Nutt. Most botanists quote two authors if a name has been changed. Linnaeus placed our false

Solomon's seal in the same genus as lily-of-the-valley as *Convallaria stellata*. Later botanists agree that these two plants are not so closely related so the name now appears as *Smilacina stellata* (L.) Desf., because the French botanist, Desfontaines, established the new genus *Smilacina* in which these plants were placed.

There are other reasons why names have been changed from time to time. The "International Code of Nomenclature," first prepared in 1867 and revised at intervals, governs such matters, but even with general rules provided, many problems of interpretation arise. New information may change our ideas of relationships. Frequently new names have been given by an author who did not know that the plants had already been named by someone else. In this case, the older name is retained. Sometimes these old publications have not been generally known and often we are uncertain of the identity of the plants described in them.

At the present time it is customary to designate a particular specimen as the **type** of each new species. Any question in the future regarding the name will be determined by a study of that specimen. The earlier writers did not indicate their type specimens and these are selected by later authors after careful consideration. Much confusion has arisen because old descriptions have been accepted without reference to the specimens from which they were described. Now, when the specimens are examined, they may prove to be different species.

In a general way, we understand what is meant by genus and species but final decision in a particular case is often difficult. It must be based upon a thorough consideration of much material of related plants. Some groups are more variable than others. Most authors recognize **varieties** of species. These are larger groups than the so-called varieties of wheat or corn which are better known as **strains.**

Our pembina or highbush cranberry has been called *Viburnum opulus*, which is a European plant, often grown as an ornamental. Many authors have called ours *V. opulus*, var. *americanum*. At present, leading botanists consider it to be a distinct species and as a result of this it takes the name *V. trilobum* which had been given in 1785 but could not be used as long as the plant was considered the same species as *V. opulus* of 1753. Bluebell, wild spirea and other plants have been considered identical with Old World species by some and as distinct species by other authors.

In writing scientific names, the genus name is always begun with a capital letter. Species names are usually begun with a small letter. The present International Rules recommend capitalization of species names which refer to persons (*Viola Nuttallii*), also names used as nouns (*Viburnum Opulus, Polygonum Convolvulus*), but not geographical names. Many botanists feel that no species names should be capitalized and that method is followed in this book.

Scientific names differing from those used here are found in older books and some of them which have been commonly used are listed. The names used in this book do not agree entirely with those of any other book because changes are still being made and the permanence of many of them is uncertain. In some cases there have been actual errors of identification in previous reports; in others a different name is now recognized. In some cases there are different opinions on the status of the plant or its names.

PHYSIOGRAPHY

The Landscape in Relation to Plant Distribution

Most plants show a rather defini e relation to conditions under which they grow: temperature, moisture, light, length of day, soil type, plant and animal neighbors. The plants found in the Red River Valley, or even those on the hills of the Sheyenne River, are quite different from those of the hills of western North Dakota. The plants growing in a low meadow are largely different from those on an adjacent hill. Even a difference of a single foot in level may make an obvious difference in the plant cover.

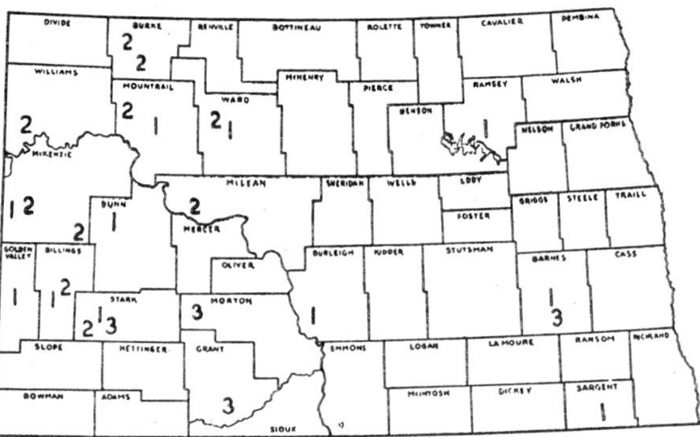

6. Selenium carrying milk-vetches. *Astragalus bisulcatus* (1) is widely distributed though local; *A. pectinatus* (2) is quite restricted to western part of State; *A. racemosus* (3) is local on dark clays.

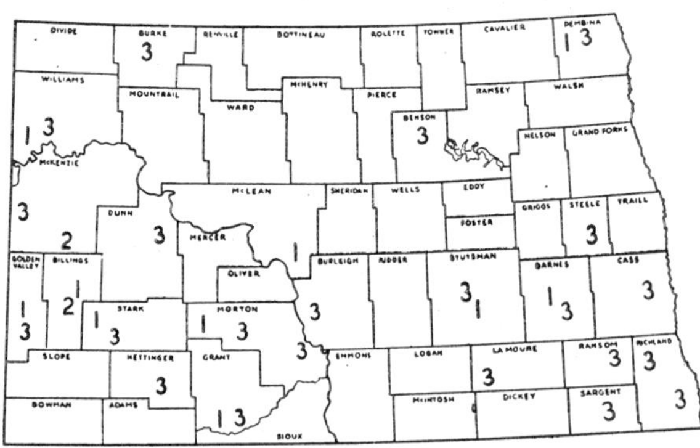

7. Dwarf Sagebrush, *Artemisia cana* (1) is frequent from Missouri River westward; Big Sagebrush, *A. tridentata* (2) is local in Bad Lands; White Sage, *A. ludoviciana* (3) is a common herbaceous species, undoubtedly in every county.

General Topography

The general features of the State can be described rather briefly.
It is a prairie region and the annual rainfall varies from about 22
inches in the east to 14 inches in the west.   The elevation varies from
800 feet in the northeast to 3000 feet in the southwest.   The highest
point is Black Butte in Slope County, 3468 feet.   The highest town
is Rhame in Bowman County, 3178 feet; the lowest, Pembina, 789
feet.

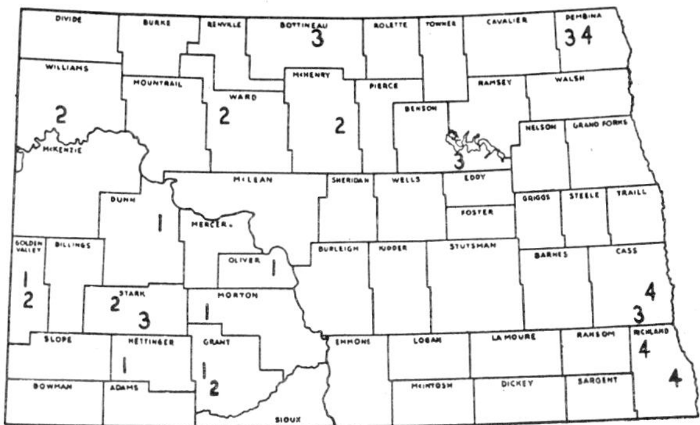

8. Some North Dakota ferns. Cliffbrake, *Pellaea glabella* (1) is limited to
exposed rocks; Little Clubmoss, *Selaginella densa* (2) is a common prairie
plant in the western part of the State; Grape Fern, *Botrychium virgini-*
anum (3) and Ostrich Fern, *Pteretis nodulosa* (4) are limited to moist
woods. Grape fern in Stark County is an instance of favorable local
conditions.

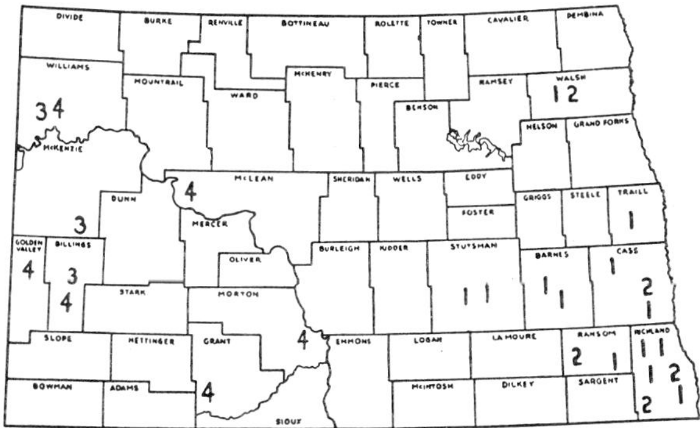

9. White Larkspur, *Aquilegia virescens* (1) and Eastern Virgin's Bower,
Clematis virginiana (2) are eastern plants; Blue Larkspur, *Delphinium*
bicolor (3) and Western Virgin's Bower, *Clematis ligusticifolia* (4) are
western species.

The chief streams are the Missouri River with its tributaries,
the Little Missouri, Knife, Heart and Cannon Ball from the west;
the Red River of the North with its tributaries, the Pembina,
Tongue, Park, Goose, Sheyenne, Maple and Wild Rice, also coming
from the west; the James River which roughly parallels the Sheyenne but continues southward to join the Missouri near Yankton,
South Dakota.

The Souris or Mouse River enters Burke County from Saskatchewan, continues in a loop through Ward, McHenry and Bottineau Counties back into Manitoba. Most of this loop and nearly an equal area east of the river was included in glacial Lake Souris, hence is a low plain. The Missouri and Little Missouri River Valleys usually have steep, high bluffs. Some of the other rivers have high bluffs in part: Sheyenne through Griggs, Barnes and most of Ransom County; James through Stutsman and northern LaMoure Counties; Des Lacs and Souris through Renville and Ward Counties; lower parts of White Earth Creek and western tributaries of the Missouri.

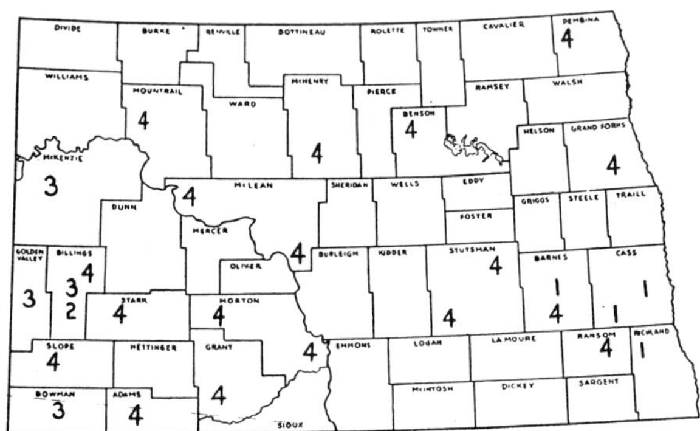

10. Bloodroot, *Sanguinaria canadensis* (1) is limited to woods in the southeast; Double Bladderpod, *Physaria brassicoides* (2) is quite local in the Bad Lands. One species of Bladderpod, *Lesquerella alpina* (3) is found only in the west, but another, *L. argentea* (4), is widely distributed on the prairie.

The Red River Valley is a flat area from 25 to 50 miles wide. The soils are largely black, developed under grass from clay, silts and sands deposited in Glacial Lake Agassiz basin. At the eastern edge of this valley in Minnesota, the ground rises rather rapidly into a hilly country which is largely covered with trees to within 40 miles of our border. On the North Dakota side, the rise is more gradual to gently rolling prairie without trees. The old lake beaches are gently sloping, often narrow and inconspicuous strips of sandy or gravelly soil. Between the successive beaches are slightly depressed strips which may be low meadow, marsh, slough or stream. These low areas contain some of the most varied flora to be found in the State.

From the Red River Valley westward to the Missouri River, the ground is mostly gently rolling prairie. A notable series of hills extends from Devils Lake south to Stutsman County. This part of the State, extending to about the middle, or a diagonal line from western Dickey to eastern Divide County, is covered with glacial drift to a depth of from 150 to 300 feet. The irregular depth of this cover has caused the formation of numerous ponds or small lakes, notably in Logan, Kidder, Benson, Pierce, Rolette, Ramsey and Towner Counties. The larger of these are a mile or more wide. They are shallow and many become dry during a series of dry years.

They are usually alkaline or saline, some of them excessively so. The borders of these saline lakes and other similar areas where water does not remain as a rule, provide a plant cover which is characteristic of strongly saline soils.

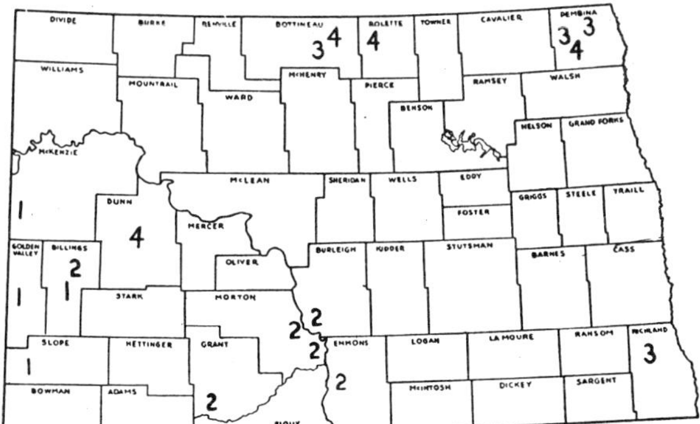

11. Mariposa Lily, *Calochortus nuttallii* (1) is restricted to the extreme west; Yucca, *Yucca glauca* (2) extends chiefly along the Missouri and Little Missouri Valleys; False Lily-of-the-valley, *Maianthemum can-adense* (3) and Disporum, *Disporum trachycarpum* (4) are eastern woodland plants. Disporum also grows in woods of the Killdeer Mts.

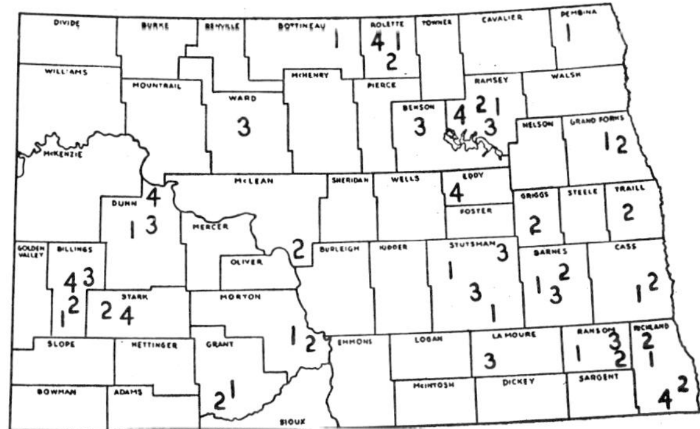

12. The different kinds of violets have different distributions. *V. rugulosa* (1) occurs in woods and thickets throughout the State; *V. pedatifida* (2) is a prairie plant, chiefly of the south and east; *V. nuttallii* (3) and *V. adunca* (4) are western prairie forms which extend well across the State.

At a distance of 25 to 50 miles east of the Missouri River lies the Altamont moraine, a belt with gravelly and stony soils. Here the elevation rises from 300 to 400 feet to the Missouri Plateau which occupies the southwestern third of the State. Along the eastern side of the Missouri River the bluffs are abrupt and about the only permanent streams are Apple Creek in Burleigh County, White Earth Creek in Mountrail County and Muddy Creek in Williams County.

The Missouri River marks a pronounced change in landscape and flora. There is little glaciation beyond this, the country is rougher and jagged hills become frequent, though there are extensive areas of gently rolling prairie. The Bad Lands form a strip along the Little Missouri River, only 3 or 4 miles wide at Marmarth in Slope County, but expanding over a much larger area in McKenzie County in which 11.9 per cent of the area is classed as "rough, broken land". Billings County, which is narrow and follows the Little Missouri River, has about 19 per cent "rough, broken land" and 6 per cent "scoria". Isolated buttes or small groups of them occur along the Missouri River and elsewhere, notably near Glen Ullin in western Morton County.

Devils Lake originally extended some 40 miles in length and drained into the Sheyenne River. Within historic time, it had no outlet and continually decreased in extent until very little of it remained after the drought years of 1934 and 1936. Since that, it has risen again and in 1947 was at about its 1938 level. On the south side of the lake are a few rugged hills including Sully's Hill, rising 300 feet above the lake.

The Turtle Mountains cover an oblong area about 40 miles long east to west on the northern boundary, lying in Manitoba, Rolette and eastern Bottineau Counties. They reach elevations of 400 to 600 feet above the surrounding plain and consist of countless small hills, lakes and ponds. The largest lake is Metigoshe, 12 miles north of the city of Bottineau. The Turtle Mountains were covered with trees except for a very few high points. Much of the area has been cleared for farm land, but it still has good recreational value.

13. Many native legumes are widely distributed, but False Lupine, *Thermopsis rhombifolia* (1) is western, and Silvery Lupine, *Lupinus argenteus* (2) is southwestern; Bushy Vetch, *Lathyrus venosus* (3) and Yellow Vetchling, *L. ochroleucous* (4) are limited to eastern wooded areas.

The Pembina Mountains include two escarpments extending southward through eastern Cavalier into Walsh County. These were thickly covered with brush and trees similar to the Turtle Mountain area. The Killdeer Mountains form a small, high area in Dunn County and their northeast slopes are wooded.

Soils

The soils of the Red River Valley and the sandy beaches have already been mentioned. In western Pembina County the Pembina delta comprises a sandy area several miles wide. The Elk Valley delta extends across western Grand Forks County into Steele and Traill Counties. The Sheyenne delta, mostly in Richland and Ransom Counties, is a still larger area of very sandy soil, including some large moving dunes and many which have become stabilized by vegetation.

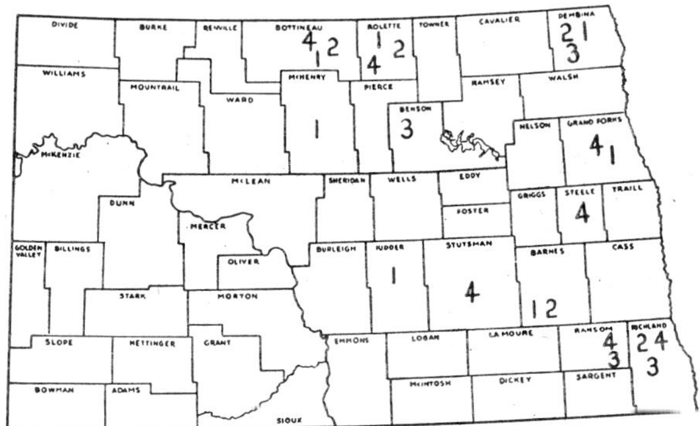

14. Swamp plants are localized and chiefly eastern as shown by: Joe-Pye Weed, *Eupatorium purpureum* (1); *Aster umbellatus* (2); Marsh Marigold, *Caltha palustris* (3); Pond Lily, *Nymphaea advena* (4).

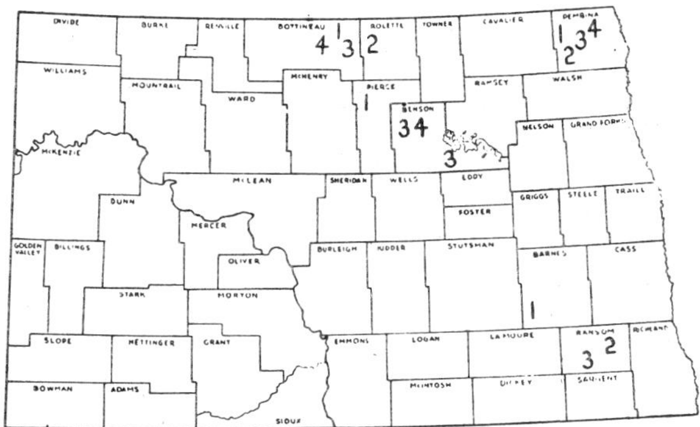

15. Cold, boggy springs in woods are rare and there are found: Small Enchanter's Nightshade, *Circaea alpina* (1); Swamp Buckthorn, *Rhamnus alnifolia* (2); Bugleweed, *Lycopus uniflorus* (3); Sweet Coltsfoot, *Petasites sagittata* (4).

McHenry County also has a dune area and sandy areas are found elsewhere, as in Kidder, Logan and Bowman Counties. The flood plains of the Missouri and Little Missouri Rivers, together with the lower parts of the other tributaries of the Missouri, have sandy areas.

The portion of the State between the Red River Valley and the Altamont moraine has in general, a fertile, black loam soil with interspersed areas of gravelly and sandy soils, poorly drained depressions and local alkali or saline spots. Farther west and on lighter soils, moisture becomes an increasingly limiting factor for plant growth. In the western half of the State, the soil is more brownish, the upper layer thin.

There are essentially no consolidated rock outcroppings except some soft sandstone layers in the Bad Lands where most of the exposures are Fort Union clays with frequent beds of lignite coal.

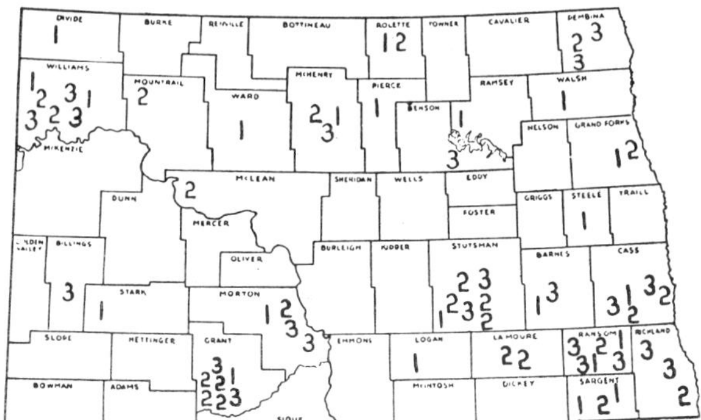

16. Example of extent of plant collections in North Dakota. Pasque-flower, *Anemone patens* (1), Canada Anemone, *A. canadensis* (2), and Cottonweed, *A. cylindrica* (3). These are probably present in every county but few collections have been made in the central part of the State and these plants happen not to have been collected in some other counties.

In many places and sometimes over a considerable area, the coal has burned, baking the clay into brick-like material locally called "scoria", which is usually in fragments, or fused into irregular clinkers, sometimes several feet thick.

Erosion of the marine and lacustrine clay beds has produced the bare clay buttes which sometimes almost completely lack vegetation, but often are well covered on the north slopes. The wash from these buttes produces flats or gentle slopes of strongly saline clays in the immediate vicinity. Limited exposures of the older deposits occur in two areas in eastern North Dakota, along the Sheyenne River below Valley City and on the Pembina River near Walhalla.

Vegetation Types

Plants respond readily to conditions for growth and thus we recognize distinct types of vegetation on different areas. Amount of water available is the chief factor in our region and this is modified greatly by the nature of the soil. Actual composition of soil is of less importance but in extreme cases this produces a marked change in the plants.

Water plants are widely distributed and remarkably alike over large areas of the earth. Cattails, bulrushes, pondweeds, bur-reed, various grasses and sedges grow in shallow water almost everywhere.   Often the same species will range entirely across North America, Europe and Asia, sometimes extending well into tropical regions.

Prairies develop in regions where rainfall is low, wind movement and evaporation high.   Forests develop under opposite conditions—abundant rainfall, low wind movement and evaporation. North Dakota lies in the northern part of the Great Plains, a large prairie area extending from Texas into Canada.   The eastern forest region covers the northern half of Minnesota and approaches quite closely to our northeastern border.   The southern edge of this forest runs northwestward, swinging across Manitoba and Saskatchewan some 100-200 miles north of us, but is represented in the Turtle and Pembina Mountains.

Distribution of Species Within the State

As a result of our position just beyond the edge of the eastern forest, areas which may be called forest are rare and occur only in certain places where conditions are more than ordinarily favorable.   The chief trees which characterize these areas are the aspen (*Populus tremuloides*) and balm of gilead (*Populus balsamifera*).

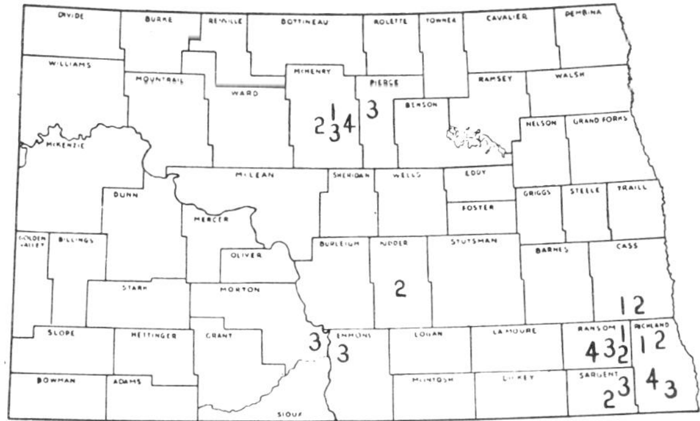

17. Some plants limited to sand dune areas are: Blowout Grass, *Redfieldia flexuosa* (1); Tumbleweed, *Cycloloma atriplicifolium* (2); Hairy Prairie-clover, *Petalostemum villosum* (3); Annual Skeletonweed, *Lygodesmia rostrata* (4).

The Turtle Mountains and Pembina Mountains were covered with trees and aspen is one of the chief species.   A narrow strip of trees along the Red River includes chiefly American elm, green ash and boxelder, with basswood on the lower levels and bur oak on ground which has been left higher than adjacent ground by stream cutting.   At intervals along the Red River small areas of aspen will be found.   These range from a few rods to several acres in size in some loops of the river.   Across the northern tier of counties prairie groves of aspen may be seen.   The Sheyenne River bluffs from Valley City to Kindred have frequent wooded slopes and gullies with considerable aspen.   Other small, isolated areas occur through the eastern and central part of the State and also on protected slopes deep gullies along the Missouri River.   The northern slopes and ˙es in the Killdeer Mountains are quite well covered with aspen.

In these wooded areas many herbaceous plants occur which are common in the eastern forest region but are rarely found elsewhere in North Dakota. False lily-of-the-valley (*Maianthemum canadense*) is perhaps one of the most characteristic of these species. It is quite definitely associated with the aspen groves and occurs in considerable abundance where conditions are suitable (fig. 11). The dog violet (*Viola conspersa*) has a similar distribution but is less common. Our first record of it was at Fargo in 1910. A specimen was not preserved and the station probably has been destroyed though the plant occurred in some abundance on the Minnesota side of the river in the same area. Later it was found near the Sheyenne River in northwestern Richland County and in the Killdeer Mountains.

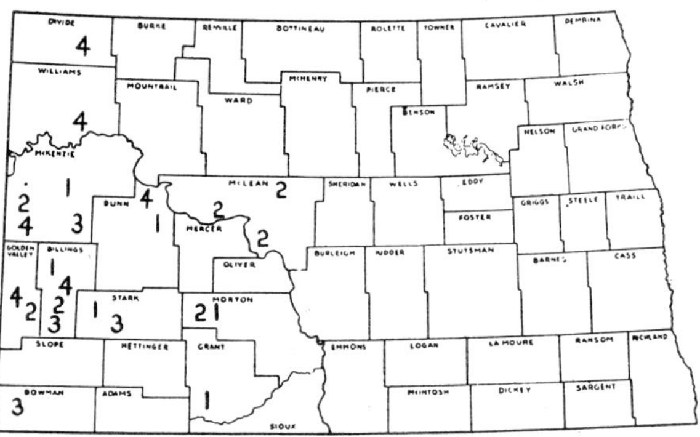

18. Some of the showy plants found only in the western part of the State are: Creasted Beardtongue, *Penstemon eriantherus* (1); Rabbit-brush, *Chrysothamnus graveolens* (2); Western Hawksbeard, *Crepis occidentalis* (3); Butte Aster, *Aster canescens* (4).

A number of plants common to the eastern forests occur here and there wherever a protected bank allows some tree growth to develop. Baneberry (*Actaea rubra*) was collected in such a place along the south shore of Girard Lake in Pierce County. Pink wood violet (*Viola rugulosa*) grows in hillside thickets even in western North Dakota. Mrs. Anna Meissner found the eastern grape fern (*Botrychium virginianum*) and an orchid (*Corallorhiza striata*) on a wooded bank of the Heart River in southeastern Stark County where the usual species are those of the Great Plains prairie.

Trees and shrubs occur also in western North Dakota, but the species and general appearance of the vegetation are quite different from those of the eastern part of the State. Cottonwood is the main tree of flood plains of the Missouri, Little Missouri and Cannonball Rivers. Shrubby willows are abundant in wet places, but the most characteristic shrub is buffaloberry (*Shepherdia argentea*). This grows on the drier parts of the flood plains and in the coulees between the hills. Dogwood, Juneberry and wild roses (*Rosa woodsii*) are also abundant. Big sagebrush (*Artemisia tridentata*) is limited to small areas along the Little Missouri and in the Bad Lands, usually growing only 2 or 3 feet high. Dwarf sage (*Artemisia cana*) is frequent from the Missouri River west, rarely occurring east of the Missouri River (fig. 7). Greasewood (*Sarcobatus vermiculatus*) grows only on some of the buttes, but rabbitbrush (*Chrysothamnus graveolens*) is quite common (fig. 18).

Creeping red cedar (*Juniperus horizontalis*) is very common on the hills, especially on north slopes. Rocky Mountain red cedar (*J. scopulorum*) is frequent on north slopes along the Little Missouri and sometimes reaches a diameter of a foot or more. Western yellow pine (*Pinus scopulorum*) extends into Bowman and Slope Counties from the southwest.

The plants of the wooded stream banks in the western part of the State are considerably different from those found in similar places in the east. Usually the higher ground is rather dry and the plants are those of the prairie. However, on slightly lower places which are moist and well shaded, species which are common to the eastern part of the State occur as, blue violet (*Viola papilionacea*), fringed loosestrife (*Lysimachia ciliatum*) and pink dogbane (*Apocynum androsaemifolium*)[1].

As previously mentioned, the prairie vegetation varies greatly with relatively slight differences in elevation and soil texture which affect the amount of moisture retained in the soil. A shallow drainage-way on a hillside will have different grasses and a better green color than the slopes only a few inches higher. The coulees, winding among the hills (fig. 5), have a dense, tall growth of grass and many more species of plants than occur in the sparse, short grass on the hills. *Anemone canadensis*[2], *Lilium philadelphicum*, *Agoseris glauca*, *Crepis runcinata*, *Hypoxis hirsuta*, *Lobelia spicata* and *Zygadenus elegans* are found in these slightly depressed areas with such a variety of grasses as *Agropyron trachycaulum*, *Elymus macounii*, *Andropogon furcatus*, *Muhlenbergia racemosa*, *Poa pratensis* and *Poa palustris*.

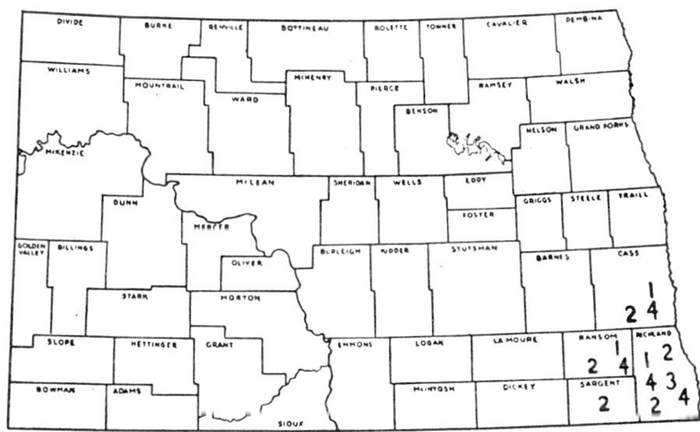

19. A few southern plants which barely reach North Dakota are: Silky Aster, *Aster sericeus* (1); Tall Blazing Star, *Liatris pycnostachya* (2); Tall Loosestrife, *Lythrum alatum* (3); Cup-plant, *Silphium perfoliatum* (4).

By contrast, the grasses of the ridges are *Bouteloua gracilis*, *Stipa comata*, *Andropogon scoparius*, *Koeleria cristata;* other plants, *Aster ericoides*, *Artemisia frigida*, *Castilleja sessiliflora*, *Oxytropis lambertii*, *Petalostemum purpureum*, *Chrysopsis villosa*, *Gaura coccinea*, *Linum rigidum*, *Sphaeralcea coccinea*, *Penstemon albidus* and *Gaillardia aristata*.

[1] Stevens, O. A. North Roosevelt Park. North Dakota. Am. Bot. 49:104-110. 1943.
[2] Common names are omitted here in order to present a more detailed list.

On the driest, most exposed slopes or tops of hills one finds *Polygala alba, Carex filifolia, Gutierrezia sarothrae, Solidago nemoralis, Liatris punctata, Astragalus triphyllus, Cryptantha bradburiana, Phlox hoodii* and *Aster oblongifolius.*

The grass which first invades dune sand is *Redfieldia flexuosa* but the one most characteristic of sandy soils all over the prairie is *Calamovilfa longifolia.* Some other plants in dune sand are *Petalostemum villosum, Corispermum* spp. *Penstemon grandiflorus, Rumex venosus* and *Psoralea lanceolata* (fig. 17). Characteristic also of sandy soils are *Tradescantia occidentalis, Oenothera nuttallii* and *Helianthus petiolaris.*

Highly saline areas are frequent though usually small in extent, and occur especially on shores of shallow ponds or edges of coulees where there is no drainage for water coming over exposed clay strata. Plant growth in such places is very limited. *Salicornia rubra* tolerates the greatest degree of salinity and frequently grows alone on these dried pond margins. Next come *Suaeda depressa, Chenopodium rubrum, C. glaucum, Atriplex argentea, Glaux maritima, Heliotropium curassavicum, Plantago eriopoda, Triglochin martima* and *Aster pauciflorus.*

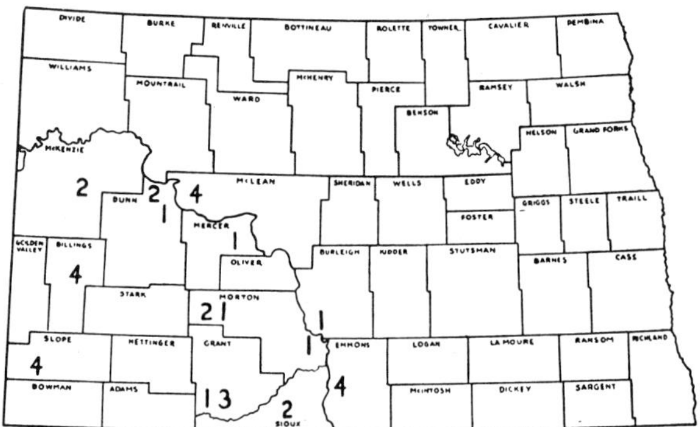

20. Some plants which seem to come into North Dakota from the south along the Missouri River are: Fetid Marigold, *Dyssodia papposa* (1); Venus' Lookingglass, *Specularia leptocarpa* (2); Dwarf Fleabane, *Erigeron divaricatum* (3); False Buffalograss, *Munroa squarrosa* (4).

Of the grasses, *Distichlis stricta* often forms a solid turf on saline flats or is mixed with *Agropyron smithii. Puccinellia nuttalliana* is frequent in wet soil and *Hordeum jubatum* often abundant. *Poa arida* may be common on those places where a greater variety of plants are found.

Somewhat similar conditions prevail on the exposed clay beds, chiefly in Bad Land areas. *Iva axillaris, Atriplex dioica, Suaeda fruticosa, Penstemon erianthera, Oenothera caespitosa* and *Aster canescens* may make a scanty cover on such ground.

General Distribution of Plants over the State

When one crosses the Missouri River to Mandan, he sets his watch back an hour and sees signs which read, "Where the West begins". This is the approximate point which geographers designate as the beginning of the semi-arid section of the plains and the change in flora is quite pronounced. In another paper[1] I discussed the plant distribution somewhat more fully, but the general features will be repeated here.

[1]Am. Jour. Bot. 7:231-242. 1920.

Until the publication of Rydberg's manual[2], there was no book which covered adequately our region. The long-standard eastern manual[3] covered the area as far west as the Mississippi Valley, interpreted as including all of Minnesota, and the comparable western book[4] included the Rocky Mountain area and more or less of the Great Plains. The region between the Red River Valley and the Missouri River was thus a sort of no man's land, and has in fact, an intermingling of plants from the two great regions.

I had calculated[1] that about one-fourth of our species belonged to the eastern region, one-tenth were western, while another two-tenths were about equally divided between eastern and western, overlapping in the middle zone. The rest are widely distributed or introduced from other countries. In the eastern group are columbine, baneberry, virgin's-bower, bloodroot, moonseed, wild leek, prickly ash and others. In the western group are low larkspur, double bladderpod, greasewood, mariposa lily and western red cedar (fig. 9, 10).

Some of them barely cross our boundary, *Leucocrinum, Astragalus longifolius*, and others only into the Bad Lands, *Stanleya, Delphinium bicolor, Artemisia tridentata*. In the middle ground we have *Viola nuttallii, Mertensia lanceolata, Chamaerhodos, Astragalus racemosus, A. bisulcatus* and *Carex filifolia* from the west. A small but distinct group from the northern forest appears in the Turtle Mountains, Pembina Mountains and locally in cold bogs along the Sheyenne River. These include *Mitella nuda, Ribes triste, Pyrola asarifolia, Circaea alpina, Shepherdia canadensis, Cornus canadensis, Erigeron lonchophyllus, Achillea multiflora* and *Cinna latifolia* (fig. 21).

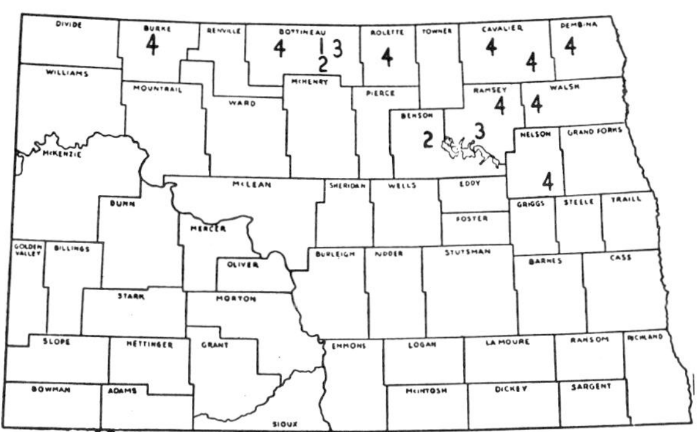

21. A few plants found only in the north are: Northern Yarrow, *Achillea multiflora* (1); Stickseed, *Hackelia floribunda* (2), Fleabane, *Erigeron lonchophyllus* (3) chiefly in wooded areas. Showy Loco, *Oxytropis splendens* (4) extends across the State on the prairie.

Some species from the south barely enter the State and we find these especially in the sandy area in Richland and Ransom Counties: *Bouteloua hirsuta, Cyperus rivularis, Lythrum alatum, Verbena stricta, Euonymus atropurpureus* and *Lobelia syphilitica* (fig. 19). From the southwest, by the Missouri and Little Missouri Valleys come *Rumex venosus, Lupinus pusillus, Abronia micrantha, Psoralea lanceolata, Specularia perfoliata, S. leptocarpa, Eriogonum annuum* and *Erigeron divaricatum*.

[2]Rydberg, P. A. Flora of the Prairies and Plains of Central North America. 1932.
[3]Fernald, M. L. and B. L. Robinson. Gray's New Manual of Botany. 1908.
[4]Coulter, J. M. and A. Nelson. New Manual of Rocky Mountain Botany. 1909.

Introduced Plants

At least 15 per cent of our plants are not native, but have been introduced from other countries, mainly from Europe. Dandelion and Frenchweed, now so common, began to come in about 1890, soon after settlement of the State began. The list has grown steadily. Nearly every year, one or two more are added. Rarely do we know just how or when they came. Some species are so common and widely distributed that it is uncertain whether or not they are native to America. Kentucky bluegrass (*Poa pratensis*) and lamb's quarters (*Chenopodium album*) are examples. Quite a number are widely distributed, native to both eastern and western hemispheres.

Thus, we have all degrees of naturalization. It is customary to apply the term "naturalized" or "introduced" to those species which are firmly established. *Bromus inermis* is a striking example of a plant which was introduced as a crop plant and has become widely established on roadsides and in waste ground. More often, cultivated species will be found for a season or two "escaped from cultivation" but do not become really established as wild plants. The term "adventive" is often applied to weeds or other plants which appear for a season but do not become established.

PLANT CHARACTERS

A knowledge of the parts of a plant is needed just the same as a knowledge of the parts of a tractor or auto is needed for the best use of those machines. One may plant and harvest crops with little knowledge of the plants, and one may drive an auto with no knowledge of the engine. When difficulties arise, one needs more information about the inner workings.

Roots and Stems

Underground parts are frequently important, though ordinary roots are not much used in descriptions. Many perennials send up new stems or flower stalks and leaves from rhizomes, bulbs, tubers, corms or thickened, tuber-like roots. These parts should be noted carefully. Some examples of each and the distinguishing features are:

Rhizome—a shallow, horizontal, underground stem; uniform in thickness, shows joints at which branches, roots and upright stems arise (fig. 22). Examples: quackgrass, many other grasses and sedges, Solomon's seal, hedge nettle, rough sunflower, large bindweed.

Bulb—a short, upright underground stem, covered with scales which make most of the bulk (fig. 23). Examples: onion, camas, pink woodsorrel.

Tuber—a thickened end of a rhizome. Examples: potato, Jerusalem artichoke, nutgrass.

Corm—like a bulb, but the stem is thick and scales much reduced (fig. 24). Examples: gladiolus, Jack-in-the-pulpit, blazing star.

Tuberous roots—thickened parts of true roots, not stems. Examples: sweet potato, Indian breadroot, water hemlock.

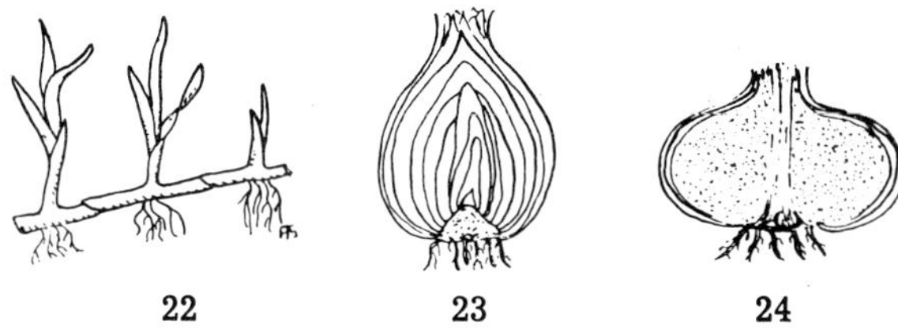

Many plants spread by means of long, sometimes thick roots. These are more irregular than rhizomes in both size and direction of growth. They do not show joints but may develop roots, branches or stems at any point. Examples are perennial sow thistle, leafy spurge, field bindweed, milkweed. The size and extent of both rhizomes and horizontal roots vary greatly in different plants. Some rhizomes grow many feet in a single year and may produce stems at only intervals of several inches. Other plants have short rhizomes which produce a thick clump of stems that increases only a few inches each year.

Stems are ordinarily erect, but many may be creeping, climbing or supported more or less by other plants. Some plants develop runners, above ground, horizontal stems, which take root at the joints as in strawberry, silverweed and buffalograss. Sometimes the lower parts of branch stems lie on the ground and root at a few joints, but do not form true runners.

Frequently the only true stem is underground and leaves and flower stalks arise from it. The common blue violet (*Viola papilionacea*) has a short, upright stem below ground. Bloodroot has a rhizome from which the leaves and flowers come. Plants with bulbs often have no other stem, the leaves and flower stalks coming from the bulb. Plants such as dandelion and plantain are often called "stemless" because they lack an elongated, leafy stem.

Leaves

Leaves have two common types of arrangement and a third which is much less common.

Alternate—one at a place, a series of them forming a spiral around the stem (fig. 25).

Opposite—in pairs, one opposite another, the next pair at right angles to the first (fig. 26).

Whorled—three or more in a circle around the stem, as in bedstraw (fig. 27), Joe-pyeweed and culver's root.

The type of arrangement is quite constant for any particular kind of plant though the lowest or uppermost leaves may be a little different from the others. Leaves often appear crowded together without any definite arrangement. This is because the stem is so short that the "joints" are crowded close together. Dandelion leaves are actually in a spiral on a very short stem. Many woody plants develop short flowering branches, such as apple "fruit spurs" or the second year flowering branches of Caragana.

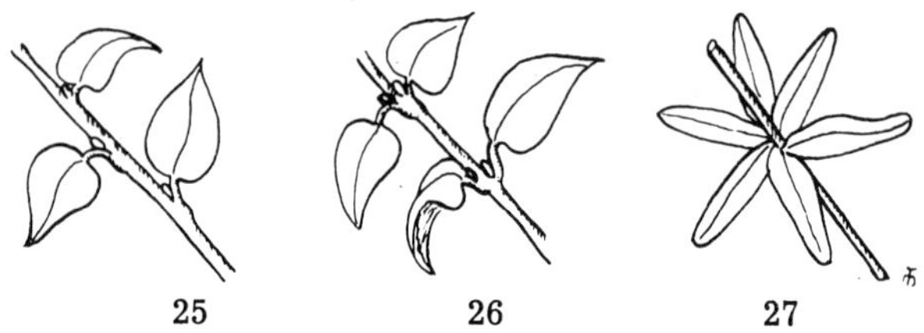

The leaf is usually composed of two parts, the broad, thin portion **(blade)** and the slender stalk **(petiole)** which joins the blade to the stem. The petiole may be very short or entirely absent, or it may be as long or longer than the blade. Usually it varies with height of the leaf on the stem, but its relative length is fairly constant for a particular species.

Stipules are sometimes found at the base of the petiole. They are small, often leaf-like parts, one on each side of the petiole. Sometimes they are modified into a spiny form and occasionally they are quite large. They are characteristic of several groups, particularly the buckwheat, rose and pea families (fig. 222).

Leaf types are divided into simple and compound. **Simple** leaves are those which have a single blade, though it may be partly divided into lobes. **Compound** leaves are divided into several distinct portions, called leaflets. Familiar examples are clovers with 3 leaflets; rose and ash, which are **pinnately compound,** with several pairs of leaflets (fig. 28); wood sorrel (fig. 29) lupine and Virginia creeper, which are **palmately compound,** with 3, 5 or more leaflets radiating from one point.

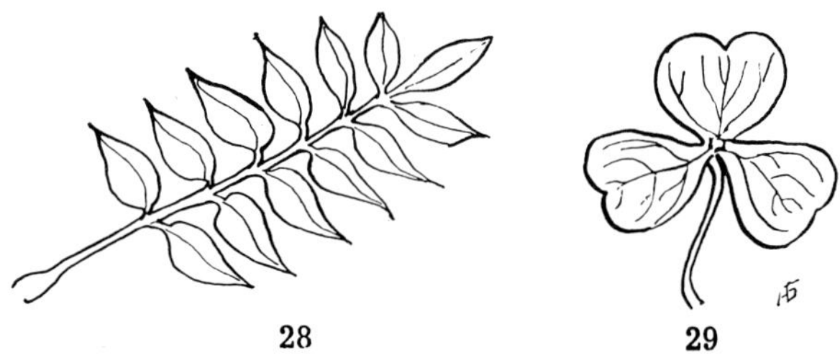

In some cases, leaves may be twice- or thrice-compound (decompound). This is more common in tropical plants with large leaves, but some members of the carrot family have repeatedly divided leaves. In meadowrue and columbine, the leaves are said to be **ternately** decompound. The leaf stalk divides into three parts, these again and again divide before the leaflets are reached (fig. 198).

Leaf shapes are divided into several main types but there are all sorts of intermediate forms, so that the description for any particular plant may range from one type to another, or the prevailing type may be between two of the usual forms. Ovate (fig. 30), lanceolate (fig. 31), oblong (fig. 32), elliptic (fig. 33), and rounded (fig. 34) are the ones most used here.

Leaf margins are often variously toothed, lobed, parted or divided, so that quite a long list of terms is used in describing them. In the present work, we have avoided these details so far as possible. The number, shape, size or slope of teeth may be distinctive.

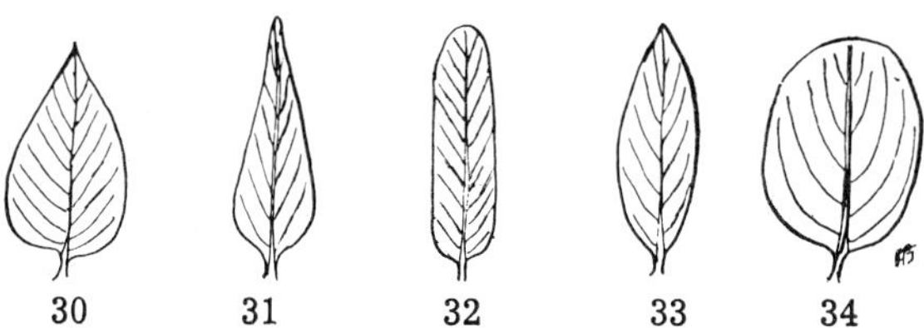

Leaf surfaces also receive attention. A dark or light green color may be due to the amount of coloring matter or to other internal characters not readily observed. The color may be caused by a layer of waxy material **(cuticle)** on the surface or by small hairs. We have avoided numerous terms used to describe hairs. If they are far apart and stiff, the leaf feels rough. If they are fine and close together, it may feel quite soft.

Flowers

The way in which flowers are placed on the plant is an important feature, but in many cases, we have used the general term "clustered" instead of technical terms. Such groups as **umbel** (fig. 35) in carrot and milkweed, **head** in clover and sunflower (fig. 36) and **spike** in plantain (fig. 37) and wheat, are distinctive.

In many cases, a single flower or several flowers come out just above a leaf base **(axillary;** in leaf axil—fig. 38). The flower may have a distinct stalk **(pedicel),** or if none the flower is **sessile.** If

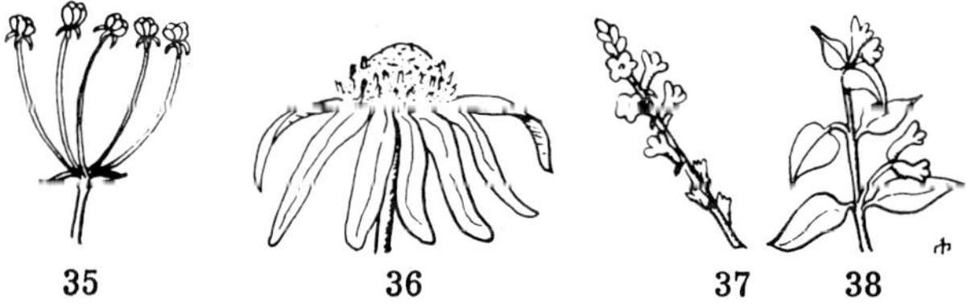

the stalk branches and bears several flowers, we have an axillary cluster, but if there is only one flower, we say the flowers are "solitary in the axils". In general, flowers are either axillary or terminal, but frequently they appear in the upper axils as well as forming a terminal cluster.

Parts of the flower. We usually think of a flower as having four sets of parts (fig. 39), green outer parts **(sepals** comprising **calyx),** colored parts **(petals** comprising **corolla),** pollen bearing **(stamens)** and seed bearing **(pistils)** parts. In many flowers one or more parts are missing and the simplest possible flower would be one stamen or one pistil. Sometimes there is little or no distinction

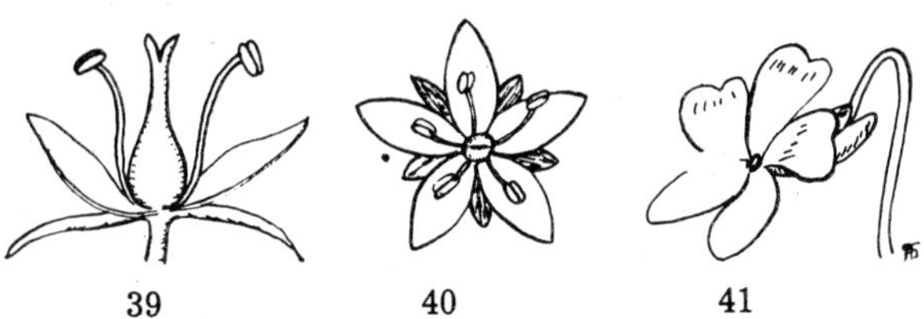

39   40   41

between sepals and petals. Frequently petals are absent and sepals are colored (anemone, clematis). Often small leaves or scales **(bracts)** are closely associated with a flower or flower cluster and sometimes these bracts are colored (poinsettia, dogwood). If petals are united, we do not speak of petals, but of corolla and its lobes.

The number of each part is quite important except in case of large numbers. Five is the most common number; four occurs in several families. Stamens are sometimes twice or more as many as petals or corolla lobes, sometimes less. Pistils are most often less in number than petals and often are grown together to form a **compound ovary** or **pistil,** as indicated by number of cavities in the ovary (fig. 44) or by number of stigmas.

Shape of corolla is important. If all petals are the same size and shape, we say the corolla is **regular** or **symmetrical** (fig. 40). If some are larger than others or differently shaped, it is **irregular,** (fig. 41), but this is not a satisfactory term because the right and left sides are similar, though upper and lower are not. We have tried here to avoid use of technical terms and use saucer-shaped for a nearly flat type, cup-shaped, bell-shaped, etc., for deeper sorts. Description of colors is complicated and we have tended to use "pink", "purple", etc., in a broad sense rather than to attempt finer distinctions.

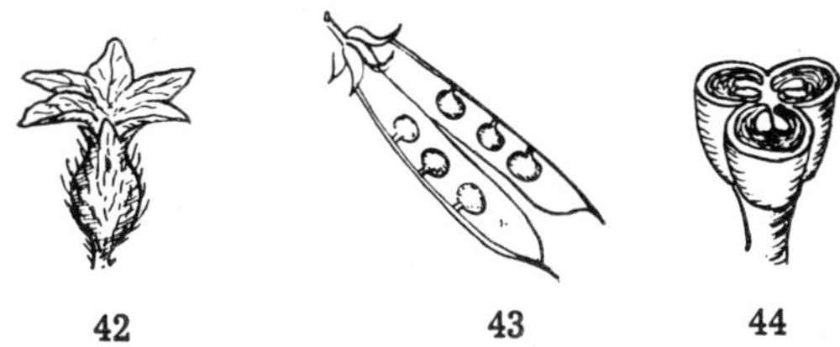

42   43   44

The ovary is called **superior** when stamens, petals and sepals are attached to the tip of the stem just below the ovary (fig. 39). In flowers with **inferior** ovaries, calyx and corolla seem attached to the tip of the ovary, as in a cucumber (fig. 42) or honeysuckle.

Fruit

Fruit is defined in botany as the ripened ovary with seeds enclosed, together with any other closely attached parts. Fruits are either fleshy (as a berry) or dry. The many-seeded, dry fruits usually split open (fig. 43), releasing the seeds. This kind is usually a capsule of which various forms are found. **Pod** is an indefinite term applied to any of these, and it is used here as the simplest way of referring to them. The one-seeded dry fruits do not split open and are commonly not distinguished from seeds. The most common of this type is the **achene** of sunflower, smartweed, dock and sedge.

HOW TO USE THE KEYS

A "key" is much used in various branches of biology and in some other studies, to aid in quick identification of specimens. In brief, it contrasts characters of plants and divides into smaller and smaller groups until a particular one is singled out. It may be compared to a road which forks repeatedly and has at each fork a sign indicating what is to be found along each branch. Usually, only two branches are given but occasionally three or more may be used.

The "short key" is offered in the hope that it will help the beginner to learn the use of a key. It does not include all plants listed in the book but it does include nearly one-half of the families, particularly the larger groups and some others with conspicuous flowers.

Suppose one has a specimen of ordinary mustard. Flowers are practically always necessary and fairly mature fruits should also be at hand if possible. The first division of the key is given by the first and sixth lines beginning "a" and "aa". Here all plants are divided into two groups, those (a) which have grass-like leaves and dry, chaffy flower parts, and those (aa) which have different characters. Mustard has broad leaves and soft flower parts, so it belongs in the second group.

The second group is divided into "d" and "dd" on the basis of milky sap. This may be a bit troublesome for sap does not show as well when plants have wilted. Mustard does not have milky sap, so it belongs under "dd". Here the division is on size and arrangement of flowers. Those of mustard are fairly large and each stands by itself, not grouped in compact heads. Likewise under "jj", the flowers are colored and conspicuous and we continue through "mm", "nn", "o" and "pp" to "q". Here the line which agrees with our specimen is followed by the name of the group to which it belongs, the mustard family, Brassicaceae.

For final identification of the plant, one now turns to the page where a further key is given to all kinds of mustards found wild in North Dakota. This key, used in the same way, leads to the name of the genus or species. In most of the keys, the paired lines are indicated only by indentation and similar wording. This method is usually regarded as the simplest to follow and one learns to go through the key quite rapidly.

The student will soon learn the characteristics of many of the chief families and new members of these groups will be readily recognized. In using keys, one must follow the divisions carefully and not jump hit or miss into some other place. If our plant fits under "aa", we have nothing to do with b, bb, c or cc lines. If it agrees with "d", we have nothing to do with any lines after "dd". In this key, we have tried to avoid this difficulty by not using the same letters again in later divisions.

Sometimes one finds that nothing seems to fit. This indicates that an error has been made farther back. Sometimes the key calls for features not shown by the specimen because of immaturity or lack of part of the plant. In such cases one can try one branch and if it does not work out satisfactorily, try the other branch. This method tends to lead to aimless wandering and one should make sure of each step whenever possible.

When the family is reached, the brief summary of its characteristics and mention of familiar plants which belong to it will help make sure of results so far. If these characters do not agree, probably an error has been made at some point, or perhaps the plant is not covered by the "Short Key". The regular key to families is intended to cover all of the plants, including many which for some reason do not fit readily into the general characters of their families.

The "Short Key" is something of an experiment, though it has been used by students for several years. It should include 75 to 90 per cent of the plants which a person might happen to pick up. No key is complete nor perfect. Comparisons often involve features which happen not to be present in the specimens at hand or which may be difficult to see or understand. Often the difference between two things is hard to recognize unless material of both is at hand.

The long key is intended to include all species described in this book. It should serve also for adjoining states or provinces where conditions and plants are similar to those of North Dakota. In more distant areas different kinds of plants are found which would not fit into the keys. New plants are added to the state list each year and one of these might not fit though the chances are that the key will place it approximately. Specimens can be sent to the Botany Department, Agricultural College, Fargo, N. D., for verification or identification.

SHORT KEY TO PRINCIPAL PLANT FAMILIES

a. Grass-like plants, with long, narrow leaves and chaffy flower parts.
  b. Leaves rounded, not flat; flower parts 6; fruit a many-seeded capsule.
         Rushes (Juncaceae)
  bb. Leaves flat; stamens 3, in 1 or 2 chaffy scales; fruit 1-seeded.
    c. Stems usually hollow, leaf sheaths open (except Bromus); leaves in 2 rows.
         Grasses (Poaceae)
    cc. Stems not hollow, leaf sheaths closed; leaves in 3 rows.
         Sedge (Cyperaceae)
aa. Not grass-like; leaves usually wider and shorter.
  d. Plants with milky sap.
    e. Flowers in dense, involucrate heads as in a dandelion.
         Lettuce group (in Asteraceae)
    ee. Flowers not in involucrate heads.
      f. Fruit 1 or 2 long pods; seeds with hair tufts.
        g. Flowers in dense umbels, stamens and pistil not evident.
         Milkweed (Asclepiadaceae)
        gg. Flowers in clusters but not in umbels, structure ordinary.
         Dogbanes (Apocynaceae)
      ff. Fruit short, a 3-seeded capsule; seeds not tufted.
         Spurges (Euphorbiaceae)
  dd. Plants without milky sap.
    h. Flowers small, in dense involucrate heads (as in thistle or aster).
      i. Flowers greenish, some heads or flowers producing pollen only. *Ragweed* and *Wormwood* groups (in Asteraceae)
      ii. Flowers usually colored, all heads bearing seeds.
         Asters (Asteraceae)
    hh. Flowers not in involucrate heads.
      j. Flowers small, green, whitish or pink; fruit 1-seeded.
        k. Leaf with a membranous sheath around the base; seed-like fruits 3-angled or flattened. *Knotweeds.* (Polygonaceae)
        kk. Leaves without sheathing bases; fruit usually flattened, not 3-angled.
          l. Flower clusters spiny; seeds black, smooth, shining.
         Pigweeds (Amaranthaceae)
          ll. Flower clusters not spiny; seeds not usually so smooth.
         Goosefoots (Chenopodiaceae)
      jj. Flowers usually larger, colored, more conspicuous.
        m. Flowers shaped as in pea and clover; leaves compound.
         Legumes (Fabaceae)
        mm. Flowers not so shaped; leaves simple or compound.
          n. Petals 6, leaves parallel veined; plants with bulbs or rhizomes. *Lilies* (Liliaceae)
          nn. Petals 4 or 5, sometimes more, or united; leaves usually net veined.
           o. Petals 4.
            p. Flowers 2—3 mm. wide; leaves in whorls of 4, 6 or 8; fruit 2 burs. *Bedstraw* (Rubiaceae)
           pp. Flowers usually larger; leaves single; fruit a capsule.
            q. Flowers 3—10 mm. wide; stamens 6.
         Mustards (Brassicaceae)
            qq. Flowers 10—75 mm. wide; stamens 8.
         Evening Primroses (Onagraceae)
         oo. Petals 5, or variously united.
          r. Petals separate or mostly so.
           s. Flowers 2—3 mm. wide, usually in large umbels; fruit dry, splitting into two halves.
         Parsleys (Apiaceae)
          ss. Flowers usually larger, not in umbels.
           t. Flowers somewhat irregular in shape; fruit a capsule, seeds attached to sides.
         Violets (Violaceae)

    tt. Flowers regular; fruit various, seeds centrally attached.

      u. Stamens many, united into a tube.
          Mallows (Malvaceae)

     uu. Stamens 5 or 10, or if many, not united.

       v. Stamens many; pistils usually many.

        w. Leaves often pinnately compound, stipules usually present.
          Rose family (Rosaceae)

        ww. Leaves often ternately compound, no stipules.
          Buttercups (Ranunculaceae)

      vv. Stamens 5 or 10; pistil 1.

        x. Petals separate, slender below, often notched at tip.
          Pinks (Caryophyllaceae)

        xx. Petals slightly united at base, not notched.

         y. Leaves of 3 broad leaflets.
          Wood-sorrels (Oxalidaceae)

         yy. Leaves simple, small.
          Flax (Linaceae)

rr. Petals partly or entirely united.

  z. Twining vines, yellow and parasitic or green with large flowers.
      Bindweeds (Convolvulaceae)

 zz. Erect, spreading or creeping, not twining plants.

  a. Corolla on top of ovary.

   b. Fruit fleshy; mostly woody plants.
      Honeysuckles (Caprifoliaceae)

   bb. Fruit a dry capsule; not woody plants.

    c. Corolla 2-lipped, anthers of stamens united.   *Lobelias* (Lobeliaceae)

    cc. Corolla symmetrical, anthers not united.   *Bellflowers* (Campanulaceae)

 aa. Corolla attached below the ovary, enclosing it.

  d. Fruit consisting of 4 nutlets which separate when ripe.

   e. Corolla regular or nearly so; stamens 5.

    f. Corolla slightly irregular; leaves opposite. *Vervains* (Verbenaceae)

    ff. Corolla regular; leaves alternate.
      Borages (Boraginaceae)

   ee. Corolla usually 2-lipped, stamens 2 or 4.   *Mints* (Lamiaceae)

  dd. Fruit a capsule or berry.

   g. Corolla 2-lipped or only slightly irregular; stamens 2 or 4.
      Figworts (Scrophulariaceae)

   gg. Corolla regular; stamens 5.

    h. Ovary and fruit 1-celled.

     i. Seeds attached to outer wall of capsule. *Gentians* (Gentianaceae)

     ii. Seeds attached centrally.
      Primroses (Primulaceae)

    hh. Ovary and fruit 2 or 3-celled.

     j. Ovary and fruit 3-celled.
      Phloxes (Polemoniaceae)

     jj. Ovary and fruit 2-celled.

      k. Fruit 4-seeded; style 2-branched.
      Waterleaf (Hydrophyllaceae)

      kk. Fruit many seeded; style single. *Nightshades* (Solanaceae)

GENERAL KEY TO FAMILIES

Page

A. Trees, shrubs or woody vines. (If herbaceous
   go to AA, page 33)
  B. Woody vines, climbing.
    Leaves compound.
      Leaflets 5; fruit a blue berry. *Parthenocissus* 204
      Leaflets 3 (fig. 45); fruits many, feathery.
                          Clematis 145
    Leaves not compound.
      Climbing by tendrils; leaves broad, usually
        lobed (fig 46).   *Vitis* 204
      Climbing by twining; leaves finely
        toothed.   *Celastrus* 201
  BB. Trees or shrubs.
    C. Trees: central stem 5—20 m. (17—65 ft.) high.
      (If shrubs go to CC.)
    Leaves opposite, compound; leaflets 3—7;
      fruits winged.
      Fruits in pairs, wing on one side only
        (fig. 47).   *Acer negundo* 202
      Fruits single, wing on both sides (fig. 48).
                      Fraxinus 224

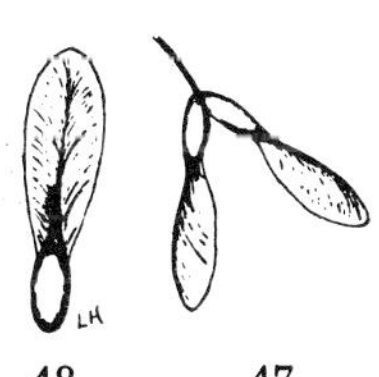

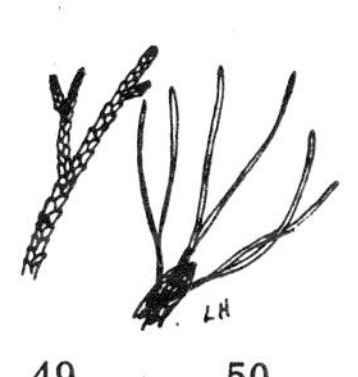

    Leaves alternate, simple.
      Leaves needle or scale-like.
        Leaves very small, scale like (fig. 49).
                    Juniperus 45
        Leaves slender, stiff, needle-like (fig. 50).
                    Pinus 45
      Leaves not needle nor scale-like.
        Branches thorny; flowers white, showy;
          fruit fleshy.
          Thorns stout; fruit with 3—5 seeds.
                   Crataegus 175
          Thorns slender; fruit with 1 large seed.
            Prunus americana 175
        Branches not thorny.
         Flowers in summer, showy; leaves large,
          rounded.   *Tilia* 204
        Flowers in spring, not showy.
          Leaves narrow with very fine teeth.
                   Salix 111

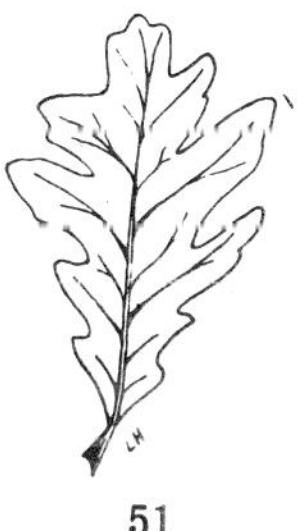

        Leaves broad with coarse teeth or lobes.
         Leaves obovate, 10—15 cm. (4—6
          in.) long, pinnately lobed
          (fig. 51).   *Quercus* 115
         Leaves ovate to oblong, not deeply
          lobed.
          Teeth of leaves sharp.
            Leaves smooth; seeds in slen-
              der catkins (fig. 52) *Betula* 115
           Leaves rough; seeds not in
              catkins (fig. 53). *Ulmus* 116
         Teeth of leaves rounded.
            Leaves smooth; seeds hair-
              tufted in pods. *Populus* 110
           Leaves soft hairy; fruit single,
              fleshy   *Celtis* 116

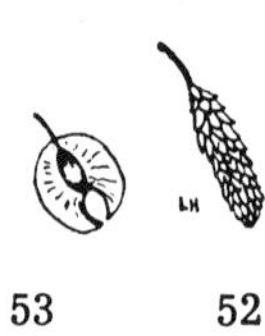

53   52

54

55

56   57

58

CC. Shrubs; not over 5 m. (17 ft.) high; stems
   several or much branched.
   Low creeping plants.
      Leaves scale or needle-like.
         Flowers yellow, 6 mm. (¼ in.) wide; rare
            sandhill plant.   *Hudsonia* 207
         No true flowers; "cedars" in western N. D.
                                       Juniperus 45
      Leaves rounded; flowers white, bell shaped
         (fig. 54).   *Arctostaphylos* 221
   Upright plants, not creeping.
      Leaves opposite.
         Leaves silvery with small scales.
                                    Elaeagnaceae 211
         Leaves not silvery; fruit fleshy.
            Flowers 3—4 mm. wide, greenish;
               leaves finely toothed.   *Rhamnus* 203
            Flowers 5—8 mm. wide, white, pinkish
               or yellow.
               Flowers clustered at leaf base; fruit
                  3—6-seeded.
                  Flowers pinkish white; fruit white.
                                    Symphoricarpos 262
                  Flowers yellow; fruit yellow or red.
                                       Lonicera 263
               Flowers in terminal, flat clusters;
                  fruit 1-seeded.
                  Petals 4; leaves entire.   *Cornus* 220
                  Corolla lobes 5; leaves toothed or
                     lobed.   *Viburnum* 261
      Leaves alternate.
         Leaves compound.
            Stems spiny or prickly.
               Stems stout; fruit splitting open (fig. 55).
                                    Zanthoxylum 195
               Stems slender; fruit fleshy.
                  Flowers 3—8 cm. wide; fruit smooth,
                     pear-shaped (fig. 56).   *Rosa* 174
                  Flowers 1 cm. or less; fruit rounded
                     (fig. 57).   *Rubus* 173
            Stems not spiny.
               Flowers greenish, not showy; fruit with thin
                  flesh.   *Rhus* 200
               Flowers yellow or purple; fruit not fleshy.
                  Flowers purplish in a long spike.
                                       Amorpha 187
                  Flowers yellow, few in a loose cluster.
                                 Potentilla fruticosa 168
         Leaves simple.
            Stems thorny or spiny.
               Thorns stout, 2—5 cm. long, on or ending
                  branches.
                  Leaves narrow, cylindrical; Bad Lands
                     shrub.   *Sarcobatus* 134
                  Leaves broad and flat.
                     Fruit 3—5 seeded, sepals persistent
                        on it.   *Crataegus* 175
                     Fruit 1-seeded, sepals not persistent.
                                         Prunus 175
               Spines small, 1 cm. or less, at leaf base.
                  Young stems quite spiny; leaves 3—5-
                     lobed (fig. 58).   *Ribes* 165

59

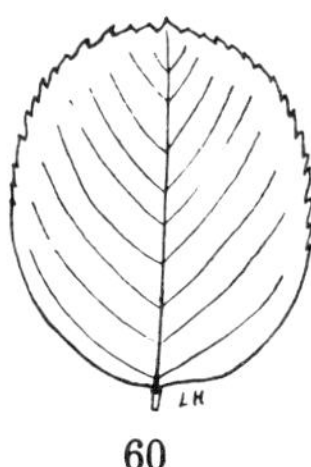

60

61

62

Young stems not spiny; leaves not lobed.
Stem yellowish; flowers purple.
Lycium 249
Stem brown; flowers yellow.
Berberis 148
Stems not thorny nor spiny.
Leaves narrow, very gray.
Western, dry land shrubs.
Leaves 3-toothed at tip (fig. 59).
Artemisia tridentata 291
Leaves entire. *Artemisia cana* 291
Bog or sand hills plants. *Salix* 111
Leaves broad or short and rounded.
Fruit fleshy.
Fruit bitter; rare, low, swamp plant.
Rhamnus alnifolia 203
Fruit edible; common shrubs.
Leaves not lobed.
Leaves rounded with few teeth
toward tip (fig. 60).
Amelanchier 174
Leaves elongated, finely toothed.
Prunus 175
Leaves palmately 3—5-lobed
(fig. 58). *Ribes* 165
Fruit dry or splitting open.
Fruit 4-lobed, splitting to show red
seeds (fig. 61). *Evonymous* 201
Fruit not so.
Fruit a nut enclosed by cup or tube.
Corylus 114
Fruit in cone-like clusters.
Fruit clusters erect, persistent.
Alnus 115
Fruit clusters drooping, falling
in spring. *Betula* 115
AA. Herbs; main stems not living over, unless at
very base.
B. Water plants; growing under water or
mostly, or floating, sometimes left on mud
banks. (If ordinary land plants go to BB.)
Small floating leaves, 2—10 mm. (usually
3—6 mm., ⅛—¼ in.) wide (fig. 62).
Lemnaceae 99
Not entirely floating, rooting in ground under
water.
Flowers conspicuous, colored.
Flowers white.
Stems and leaves entirely in water, stems
chiefly horizontal.
Ranunculus aquatilis 146
Stems usually upright, mostly out of
water if flowering.
Leaves large, 5—20 cm. (2—8 in.)
long; petals 3, pistils many.
Alismaceae 50
Leaves small, 1—3 cm. long; petals
4; pistil 1. *Cardamine* 159
Flowers yellow or pink.
Flowers pink, leaves floating.
Polygonum natans 124

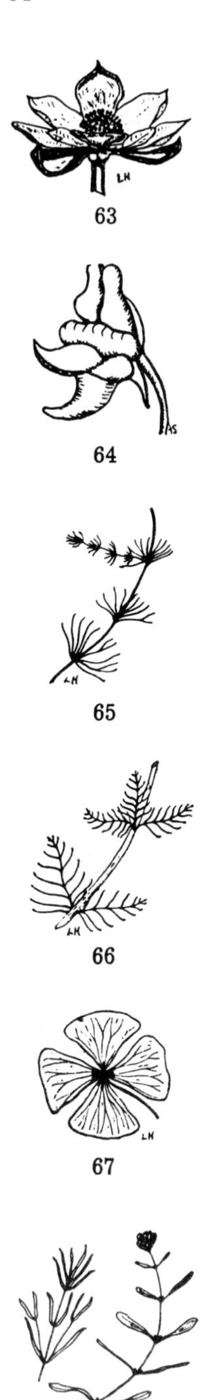

Flowers yellow.
  Leaves finely divided, often mostly
    under water.
    Petals separate (fig. 63), pistils
      many.   *Ranunculus flabellaris* 146
    Petals united (fig. 64), pistil 1.
            Utricularia 258
  Leaves rounded, not divided.
    Leaves 2—5 cm. wide, stem erect,
      2—4 dm. (8—16 in.) high. *Caltha* 141
    Leaves 5—20 cm. wide, floating.
           Nymphaeaceae 140
Flowers inconspicuous unless in mass,
  usually greenish.
  Leaves finely divided, mostly under water.
    Leaf segments forked; leaves
      crowded at branch tips (fig. 65).
           Ceratophyllum 140
    Leaf segments comb-like; leaves not
      crowded (fig 66).   *Myriophyllum* 216
  Leaves broad or narrow but not finely
    divided.
    Leaves relatively wide and short.
      Leaves circular, 1—3 cm. wide, of 4
        leaflets (fig. 67).   *Marsilea* 43
      Leaves ovate or oblong.
        Flowers purple or whitish; seeds
          many, very small.
          Leaves 2—5 mm. long; small
            creeping plant.   *Elatine* 207
          Leaves 10—20 mm. long; stems
            upright or floating.   *Veronica* 256
        Flowers greenish; fruits 1-seeded.
          Leaves 2—5 mm. (¼ in. or less)
            long, floating.
            Callitriche palustris 199
          Leaves 3—10 cm. (1½—4 in.) long,
            floating or submerged.
            Potamogeton 49
    Leaves narrow or elongated or both.
      Leaves in whorls of 6 or more; stem
        usually coming above water.
           Hippuris 216
      Leaves alternate, opposite or in
        whorls of 3—4.
      Leaves very narrow, 1—3 cm. long,
        mostly in water.
        Leaves opposite or in whorls of
          3—4.
          Leaves with fine, sharp teeth;
           stems short, densely leafy
           (fig. 68).   *Najas* 47
          Leaves not sharply toothed;
           stems slender, elongated.
           Flowers solitary at leaf
            bases. (fig. 69).
            Callitriche hermaphroditica 199
          Flowers on long, slender
           terminal stalk.   *Anacharis* 51

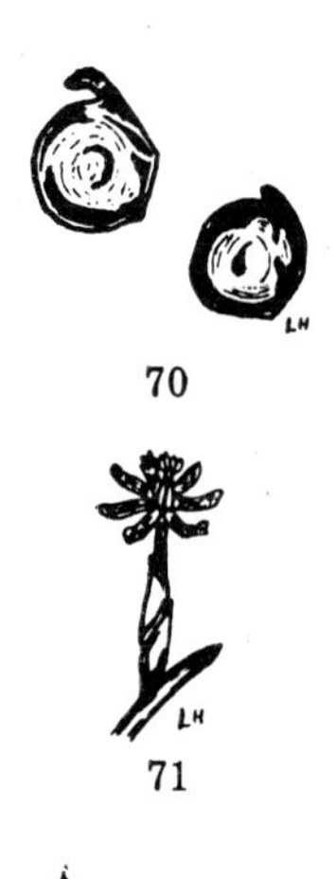

70

71

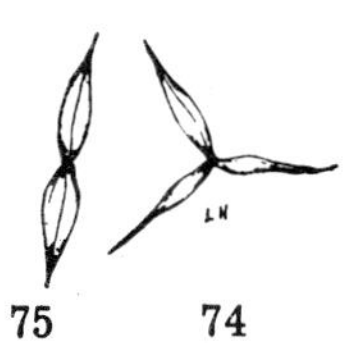

73   72

75   74

76

77

Leaves alternate.
   Flowers greenish without
      calyx or córolla; fruit 1-
      seeded (fig. 70).
                  Potamogetonaceae 46
   Flowers yellowish with 6 seg-
      ments (fig. 71); fruit many-
      seeded.   *Heteranthera* 100
Leaves elongate or stems leaf-like,
   usually out of water.
   Stems mostly leafless or leaves
      stem-like.
      Flowers with 6 chaffy seg-
         ments (fig. 72); fruit many
         seeded.   *Juncaceae* 100
      Flowers without sepals or pe-
         tals (fig. 73); fruit 1-
         seeded.   *Cyperaceae* 84
Leaves present, elongate, 2—10
   dm. long.
   Leaves entirely under water,
      erect.   *Vallisneria* 52
   Leaves mostly above water.
      Flowers in conspicuous clus-
         ters; not enclosed by scales.
         Flowers in cylindrical spikes;
            leaves 1—2 m. (3—6 ft.)
            long.   *Typhaceae* 46
         Flowers in rounded heads;
            leaves 2—4 dm. (8—16
            in.) long. *Sparganiaceae* 46
      Flowers enclosed by scales, in
         various sorts of clusters.
         Leaves in 3 vertical rows
            (fig. 74).   *Cyperaceae* 84
         Leaves in 2 vertical rows
            (fig. 75).   *Poaceae* 52
BB. Not water plants; sometimes in water but
leaves mostly above surface; ordinary plants.
Plants nearly leafless, not green in color.
   Plant body a slender, yellow vine (fig 76);
      flowers white.   *Cuscuta* 230
   Plants with erect, purplish or white stems
      and flowers.
      Flowers and stems pure white.   *Monotropa* 221
      Flowers and stems purplish or whitish.
         Flower tubular 1 cm. long (fig. 77); mostly
            prairie plants.   *Orobanche* 258
         Flower not tubular, spreading; rare woods
            plants.   *Corallorrhiza* 110
Ordinary plants with green leaves.
   C. Flowers not much colored, greenish or whit-
      ish, less than 5 mm. (⅕ in.) wide. (If dis-
      tinctly white or colored go to CC, p. 38.)
      Leaves long, slender; plants more or less
         grass-like.
         Flowers chaffy.
            Flower parts 6, pointed (fig. 73),
               seeds many, very small. *Juncaceae* 100

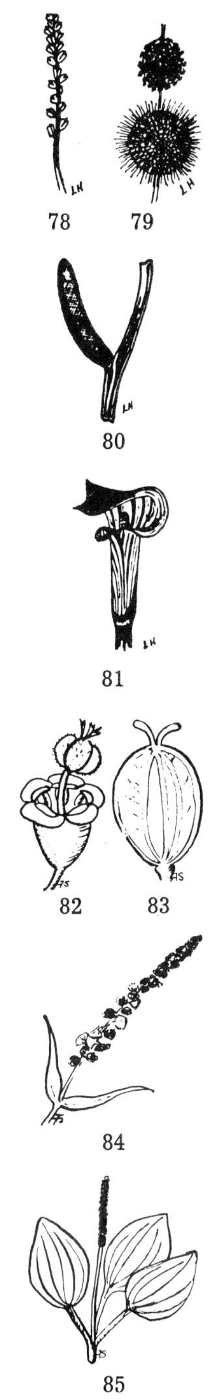

Flower parts 2—3 scales (fig. 73);
   seed 1.
   Leaves in 3 vertical rows (fig. 74);
      stems solid, often 3-angled.
                  Cyperaceae 84
   Leaves in 2 vertical rows (fig. 75);
      stems hollow, rounded.   *Poaceae* 52
Flowers not chaffy; leaves elongated but
   less grass-like.
   Flowers in a slender spike 5—20 cm.
      (2—8 in.) long (fig. 78); leaves
      thick.   *Triglochin* 50
   Flowers in rounded or short, thick
      clusters 2—3 cm. long.
      Flowers in several rounded heads
         (fig. 79), 1—2 cm. wide. *Sparganium* 46
      Flowers in a cylindrical cluster 2—3
         cm. long (fig. 80), apparently on a
         leaf.   *Acorus* 99
Leaves not long and slender.
   Leaves of 3 broad leaflets; flowers en-
      closed in a purplish spathe (fig. 81).
                      Arisaema 98
   Leaves otherwise; flowers not in a spathe.
      Plant tufted, moss-like; leaves fine,
         sharp pointed.   *Paronychia* 139
      Plant not moss-like.
         Plants with milky juice; pods 3-
            seeded (fig. 82).   *Euphorbia* 197
         Plants without milky juice; pods
            usually not 3-seeded.
         Leaves compound or deeply divided.
            Leaves compound.
               Petals and sepals 0, stamens
                 conspicuous; fruit 1-
                 seeded.   *Thalictrum* 148
              Petals present, small, white.
                 Fruit a purple berry.   *Aralia* 216
                 Fruit dry.
                    Fruit splitting into 2 dry
                      halves each 1-seeded
                      (fig. 83).   *Ammiaceae* 217
                 Fruit a several seeded,
                    curved pod.
                       Desmanthus 178
         Leaves deeply lobed or divided;
            fruit 1-seeded.
            Staminate flowers in heads
               (fig. 84) in slender, ter-
               minal clusters.   *Ambrosia* 281
            No slender, terminal flower
              clusters.
              Leaves very rough; fruits
                 enclosed in papery clus-
                 ters.   *Humulus* 116
              Leaves soft; fruits not so
                 enclosed.   *Cannabis* 117
        Leaves not compound nor deep-
            ly divided.
            Leaves all basal (fig. 85);
               flowers in slender spikes
               2—30 cm. long.   *Plantago* 259

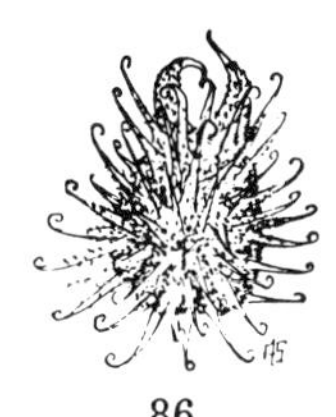

86

87

88

89

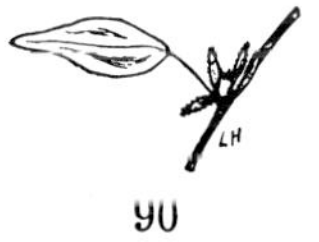

90

Stem leafy; flowers usually
not in slender spikes.
Leaves about as wide as long.
Stems weak, mostly creeping;
leaves rounded, 5—10 mm.
(⅕—⅖ in.) long.
Stellaria media 139
Stems erect, stout or fairly so.
Fruit a spiny bur 1—2.5 cm.
long (fig. 86). *Xanthium* 282
Fruit not a spiny bur.
Flowers in a terminal, flat
cluster; plant with sting-
ing hairs. *Laportea* 118
Flowers in small, scattered
heads; leaves soft hairy.
Iva xanthifolia 281
Leaves longer than wide, often
quite narrow.
Leaves opposite or whorled.
Leaves whorled, 4—6 at one
place, narrow, 1—3 cm.
long.
Flowers clustered on a slen-
der stalk (fig. 87); pod
3-seeded.
Polygala verticillata 196
Flowers on separate stalks
(fig 88); pod many
seeded. *Mollugo* 135
Leaves opposite.
Plants 1—2 m. (3—7 ft.) high
with stinging hairs.
Urtica 118
Plants 1—4 dm. (4—16 in.)
high without stinging
hairs.
Flowers in whitish heads
3—6 mm. (⅛—¼ in.)
wide. *Iva* 281
Flowers in branching clusters, not in heads.
Pilea 118
Leaves alternate.
Leaf base enclosed by membranous sheath (fig. 89).
Polygonaceae 119
Leaf base not enclosed
by sheath.
Flowers clustered at leaf
base and surrounded by
small leaf about 1 cm.
long (fig. 90).
Parietaria 119
Flowers not so clustered or
small leaves slender
and spiny.
Fruit 1-seeded, the coat
usually thick and
black.

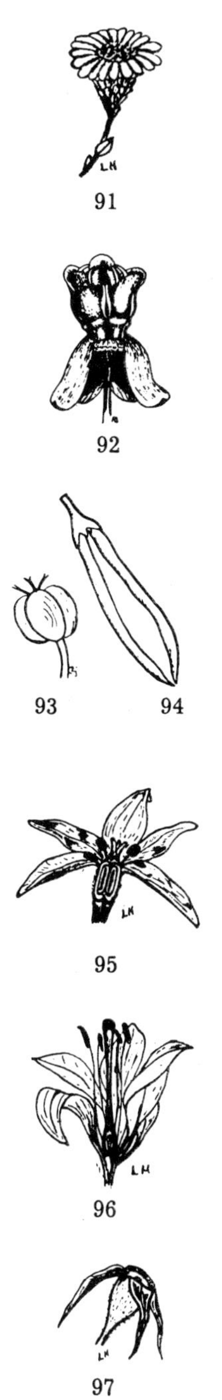

Spines at base of
        flowers.
                Amaranthaceae 134
                    and *Salsola* 133
    No spines below
        flowers.
                Chenopodiaceae 126
            Fruit about 6-seeded;
                rare plant. *Lechea* 207
CC. Flowers distinctly white (sometimes greenish)
    or otherwise colored.
    Flowers in involucrate heads (dense clusters of
        various sizes, surrounded by 1 or more rows of
        scale-like leaves, fig. 91).   *Asteraceae* 267
    Flowers not in such heads.
        Plants with milky juice.
            Flowers in dense umbels, structure not evi-
                dent (fig. 92).   *Asclepiadaceae* 227
            Flowers not in dense umbels, but sometimes
                closely clustered.
                Flowers not bell shaped, greenish or yellow-
                    ish or with small, rounded petal-like
                    tips; fruit a short, 3-seeded pod (fig. 93).
                                Euphorbiaceae 196
                Flowers bell shaped, white or pink; fruit
                    2 slender pods (fig. 94).   *Apocynaceae* 226
        Plants without milky juice.
            D. Petals absent or not different from sepals.
                (If present go to DD.)
                Stamens many, pistils several or many.
                                Ranunculaceae 140
                Stamens 3—10, pistils 1—5.
                    Perianth parts (petals, sepals) 6, usually
                        showy; stamens 3 or 6.
                    Stamens 6.
                        Perianth attached to top of ovary (fig. 95).
                                Hypoxis 106
                        Perianth parts attached to stalk below
                            ovary (fig. 96).   *Liliaceae* 102
                    Stamens 3; flowers blue to white; leaves
                        grass-like.   *Sisyrinchium* 107
                    Perianth parts 3—5, often showy or else ab-
                        sent.
                        Perianth parts 3 or 4, usually colored.
                            Perianth parts 3 (fig. 97), purple; low
                                plant in woods.   *Asarum* 119
                            Perianth parts 4 (fig. 98), white or blue.
                                Clematis 145
                        Perianth parts 5 (rarely 4 or 6).
                            Sepals greenish; petals usually 0;
                                stamens 10, pistils 5.
                                Penthorum 163
                            Perianth white, greenish or colored;
                                stamens 5—8, pistil 1.
                                Leaf base enclosed by membran-
                                    ous sheath (fig. 89).
                                Polygonum 122
                                Leaf base without sheath.
                                Santalaceae 119
            DD. Petals present and distinct from sepals.
                Fleshy, spiny plants without leaves; petals
                    many.   *Cactaceae* 210

Ordinary leafy plants, sometimes somewhat fleshy.
  E. Petals separate (sometimes slightly
      united at base; if partly or entirely
      united, go to EE, p. 40a)
    Flower pouch-like, hanging horizontally (fig.
      99); tall plants in woods.   *Impatiens* 203
    Flowers not pouch-like.
      Petals and sepals 3 each.
        Petals pink, blue or lavender. *Tradescantia* 99
        Petals white.
          Stamens and pistils many; water plants.
                          Alismaceae 50
          Stamens 6; pistil 1; woodland plant.
                          Trillium 106
      Petals and sepals, 2, 4, or 5, rarely 6—10.
        Petals 10, large, white (fig. 100); plant
          coarse, rough.   *Mentzelia* 210
        Petals 2—6.
          Sepals 2, sometimes very small.
            Petals 2, deeply 2-lobed (fig. 101);
              fruit a sticky bur.   *Circaea* 216
            Petals 4—8; fruit a pod.
              Petals 8, white; sepals similar, soon
                falling.   *Sanguinaria* 149
              Petals 4 or 6; sepals green, persistent.
                Petals long and narrow; sepals
                  very small, scale-like (fig. 102).
                          Fumariaceae 150
                Petals broad; sepals evident.
                  Stamens many.   *Papaver* 150
                  Stamens 5 or less, sometimes
                    more.   *Portulacaceae* 136
          Sepals 4—6; sometimes very small or
            partly united.
            Sepals 6; leaves of many oblong, 3-
              toothed leaflets.   *Caulophyllum* 148
            Sepals 4—5.
               Petals 4; stamens 4—8.
                Petals attached on top of ovary.
                  Flowers 2—5 mm. ($\frac{1}{12}$—⅕ in.)
                    wide; leaves whorled (fig.
                    103); fruit a small bur.
                          Galium   260
                Flowers 2—8 cm. (¾—3 in.)
                    wide (fig. 104); leaves al-
                  ternate; fruit a pod.
                        Onagraceae 212
              Petals attached to stem below
                ovary or to calyx.
                Petals on edge of calyx cup.
                        Ammannia 212
                Petals on stem below ovary.
                  Pod 2-celled; petals usually
                  yellow or white (fig. 105).
                        Brassicaceae 151
                Pod 1-celled; petals pink (fig.
                  106).   *Capparidaceae* 163
            Petals usually 5; stamens 5, 10 or
             more.
              Leaves rounded, 5 mm. wide,
                covered with sticky hairs which
                bend down to capture insects;
               rare bog plant (fig. 107).
                        Drosera 163

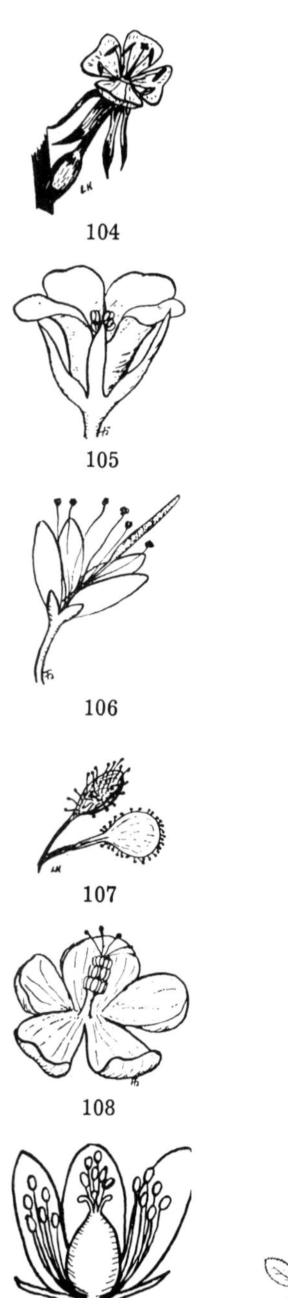

104

105

106

107

108

109

Leaves without such hairs.
  Petals attached on top of ovary;
    sepals often very small.
    Fruit dry, splitting into 2
      halves (fig. 83); leaves
      compound.   *Ammiaceae* 217
    Fruit a many seeded pod; plant
      very rough (fig. 100).
                    Loasaceae 210
  Petals attached to stem below
    ovary or to calyx.
    Stamens 15 or more.
      Filaments of stamens more
        or less united; pistil 1.
        Filaments united to form a
          slender tube (fig. 108).
                    Malvaceae 204
        Filaments united at bases
          into several groups
          (fig. 109).
                    Hypericum 206
      Stamens separate; pistils of-
        ten many; leaves usually
        compound.
        Leaves usually with stip-
          ules (fig. 110); pistils
          separate or compound.
                    Rosaceae 166
        Leaves without stipules
          (fig. 111); pistils 3—
          many, always simple.
                    Ranunculaceae 140
    Stamens 5—10.
      Leaves compound.
        Fruit splitting into 4 or 5
          spiny parts (fig. 112);
          plant creeping.
                    Tribulus 195
        Fruit not so.
          Pod 5-celled; leaves of 3,
            heart shaped leaflets
            (fig. 113).   *Oxalis* 194
          Pod 1-celled (rarely 2);
            leaves various.
                    Fabaceae 176
      Leaves simple.
        Stamens 10 or approxi-
          mately.
          Leaves opposite; petals
            often 2-cleft.
                    Caryophyllaceae 136

110

111

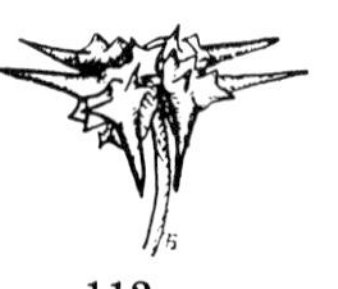

112

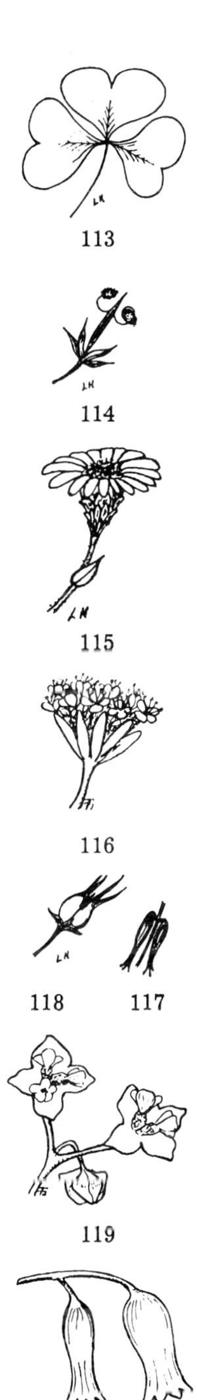

Leaves alternate; petals
rounded at tip.
Style 1.
Seeds many, angled.
Lythrum 212
Seeds 3, rounded.
Polygala 196
Styles 5; seeds 10, flat.
Linum 195
Stamens 5 (10—15 false
stamens in *Parnassia*).
Fruit long beaked, split-
ting into 5, 1-seeded
parts (fig. 114).
Geranium 193
Fruit a many seeded
pod.
Pistil 2, united at least
below.
Saxifragaceae 164
Pistil 1.
Violaceae 207
EE. Petals united, either partly or completely.
Corolla attached on top of ovary.
Flowers in heads.
Anthers of stamens united at edges (fig. 115).
Asteraceae 267
Anthers not united; rare weed (fig. 116).
Scabiosa 263
Flowers not in heads but sometimes rather close-
ly clustered.
Lobes of corolla 4; leaves opposite or
whorled (fig. 103).   *Rubiaceae* 259
Lobes of corolla 5; leaves alternate.
Vines; corolla saucer shaped, regular.
Cucurbitaceae 263
Erect herbs; corolla tubular, 2-lipped.
Lobelia 266
Base of corolla enclosing ovary.
Corolla irregular, usually more or less 2-lipped;
stamens 2 or 4.
Fruit splitting into 4 nutlets.   *Lamiaceae* 239
Fruit not so splitting.
Fruit 1-seeded (fig. 117).   *Phryma* 258
Fruit with many small seeds (fig. 118).
Scrophulariaceae 250
Corolla regular or nearly so.
Several pink flowers in a flat, calyx-like cup
(fig. 119); fruit 1-seeded.   *Nyctaginaceae* 135
Flowers not so grouped; fruit 4—many-seeded.
Sepals and petals similar, 6 partly united in
tubular form (fig. 120).   *Polygonatum* 106
Sepals and petals not alike, usually 5 each.
Stamens 10; woodland plants with basal,
rounded leaves.   *Pyrola* 221
Stamens 2, 4 or 5.
Stamens 2 or 4; flowers small, closely
clustered at bases of opposite leaves
(fig. 121). *Mentha* 246 and *Lycopus* 244

121

122

123

124

125

126

Stamens 5.
  Fruit splitting into 4 nutlets (fig. 122).
    Leaves alternate.　　*Boraginaceae* 234
    Leaves opposite.　　*Verbenaceae* 238
  Fruit a several to many (or 1—3)
    seeded pod or berry.
    Fruit 4-seeded; seeds large, rounded.
      Flowers 2.5—5 cm. (1—2 in.) wide
        (fig. 123); leaves triangular.
          Convolvulaceae 228
      Flowers 5—15 mm. (⅕—⅗ in.)
        wide (fig. 124); leaves divided.
          Hydrophyllaceae 233
    Fruit many seeded; seeds usually
      small.
      Stamens opposite corolla lobes
        (fig. 125).　　*Primulaceae* 221
      Stamens alternate with corolla
        lobes (fig. 126).
        Leaves opposite.
          Fruit 1 or 2-celled; seeds
            very numerous.
              Gentianaceae 224
          Fruit 3-celled; seeds 6—15.
              Polemoniaceae 231
        Leaves alternate; seeds numerous.
              Solanaceae 246

FERNS AND FERN ALLIES—Pteridophyta

These are commonly included in descriptive manuals on seed plants, though they constitute a separate group. Most of them have true leaves, delicate in texture. Reproduction is by means of spores borne either on the lower side of ordinary leaves or on special leaves, parts of leaves or stems, usually in early summer. All are perennial and some spread by rhizomes. There are no leafy, branching stems.

Ferns can often be distinguished from flowering plants when neither flowers nor spores are present, by the fact that leaves of true ferns come up from the ground in a coil. Veins of the leaf usually fork repeatedly, being neither straight and unbranched as in grasses and lilies, nor finely branched as in broad-leaved plants. Leaves of milfoil or other flowering plants are sometimes called "ferns" because they are finely divided.

Key to Species or Genera

Leafless plants with jointed, simple or branched stems; spores in
    small cones at ends of stems (fig. 128). 13—18. *Equisetum*
Plants with leaves.
  Low, moss-like prairie plant with very small leaves.
      19. *Selaginella densa*
  Larger plants, leaves medium to large, usually compound.
    Leaves circular, with 4 leaflets; water and mud plants.
      12. *Marsilea vestita*
    Leaves more or less elongated, not circular, not water plants.
      Leaf stalks solitary, stout and bare below.
        Spores on special leaves, the ordinary leaves short, broad,
          coarse. 10. *Onoclea sensibilis*
        Spores on ordinary leaves or parts of leaves.
          Stalk 3 dm. (1 ft.) high, only upper part of blade spore
            bearing. 1. *Botrychium virginianum*
          Stalk 3—10 dm. high; spores borne along edges of leaf
            blade. 8. *Pteridium aquilinum*
      Leaves clustered, stalks often slender, scaly or leafy to near base.
        Leaflets few, oblong, 1—2 cm. (½ in.) long, the edges rolled
          over the spores (fig. 127). 9. *Pellaea glabella*
        Leaves much divided or cut into fine segments.
          Spores on special leaves which have no flat blades.
            11. *Pteritis pennsylvanica*
          Spores on under side of ordinary leaves.
           Leaves 1.5—3 dm. (6—12 in.) long; spore clusters at first
            covered by a thin membrane which soon breaks up.
            Membrane attached all around the edges of spore cluster.
              7. *Woodsia obtusa*
            Membrane attached only at one side. 6. *Cystopteris fragilis*
          Leaves 4—10 dm. long; spore clusters with a more persistent
            membrane.
            Leaves once pinnate, 5—7 cm. (2—3 in.) wide.
              4. *Dryopteris thelypteris*
            Leaves twice pinnate, 7—20 cm. wide.
             Spore clusters oblong, membrane attached along one
             side. 5. *Athyrium angustum*
             Spore clusters rounded, membrane attached at one
             point.
              Leaf-segments finely and sharply cut.
                3. *Dryopteris spinulosa*
             Leaf-segments coarser, the final lobes about 5—7 mm.
              (¼ in.) wide and more rounded.
                2. *Dryopteris cristata*

1. *Botrychium virginianum* (L.) Sw. GRAPE FERN. Leaf single, blade 1—2 dm. (4—8 in.) long, finely divided, on a stout stalk 2—4 dm. high; upper part of leaf spore bearing, not flat. Occasional in woods; Richland, Cass, Pembina, Benson, Bottineau and Stark Counties (fig. 8).

2. *Dryopteris cristata* (L.) A. Gray. SHIELD FERN. Leaves 2—6 dm. long, primary segments lanceolate, 5—10 cm. (2—4 in.) long, deeply divided into several pairs of oblong segments. Local in moist woods along Pembina and Sheyenne Rivers in Pembina, Richland and Ransom Counties.

3. *Dryopteris spinulosa* (O. F. Müll.) Watt. WOOD FERN. Similar to last except for more sharply cut leaflets. Collected by Albert H. Shunk in 1916 along Sheyenne River near Sheldon, Ransom County.

4. *Dryopteris thelypteris* (L.) A. Gray. MARSH FERN. Leaves about 10 cm. wide, hardly tapered toward tip, primary segments deeply divided into oblong segments 5 mm. (⅕ in.) long, often arranged zigzag. Found by Albert H. Shunk in 1916.

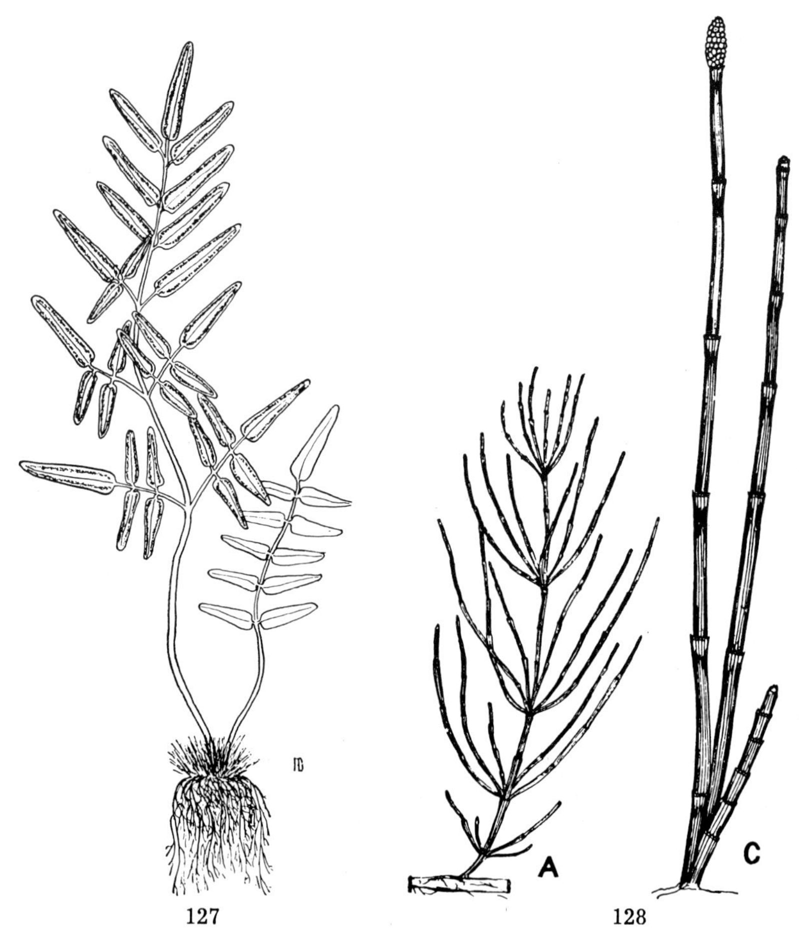

127   128

127. Cliffbrake (*Pellaea*)
128. Horsetails: A. *Equisetum arvense;* C. *E. kansanum.*

5. *Athyrium angustum* (Willd.) Presl. LADY FERN. Similar to No. 4 but leaves more finely and sharply cut. In a wooded swamp along Sheyenne River in Richland County near Leonard and at Walhalla.

6. *Cystopteris fragilis* (L.) Bernh. BLADDER FERN. Leaves 1—3 dm. (4—12 in.) long, 2—7 cm. (1—3 in.) wide, in tufts from a short crown. Local on moist, wooded banks throughout the State. One of the most widely distributed of all ferns, growing from 79° N. in Greenland to 55° S. in South Georgia (east of Cape Horn, S. Am.).

7. *Woodsia obtusa* (Spreng.) Torr. Closely resembles No. 6 but occurs in western part of State in drier locations.

8. *Pteridium aquilinum* (L.) Kuhn. COMMON BRAKE. BRAKEN. Leaves stout, 6—10 dm. (2—3 ft.) high, the complete leaf 2—4 dm. long and about as wide, with many oblong leaflets 1—2 cm. (⅜—¾ in.) long. One record from Svold, Pembina County, in 1937 by H. A. Graves and myself. This is also a very widely distributed fern. Spores are borne under the rolled edges of the leaflets.

9. *Pellaea glabella* Mett. CLIFFBRAKE. (Fig. 127). Leaves 1—2 dm. long, several from a brown, scaly crown; leaflets 5—9 (lowest sometimes divided), rather thick, oblong, 5—15 mm. long; leaf stalks wiry, dark purple. Mostly in cavities of large rocks on hillsides; Dunn, Oliver, Grant, Hettinger and Golden Valley Counties (fig. 8). Previously reported as *P. atropurpurea.*

10. *Onoclea sensibilis* L. SENSITIVE FERN. A coarse plant with broad, thick, coarsely lobed leaves 1—3 dm. long; spores on a separate stalk in a grape-like cluster 1—2 dm. long. Known from only two or three locations near Sheyenne River in Richland and Ransom Counties.

11. *Pteretis pennsylvanica* (Willd.) Fern. OSTRICH FERN. Our largest fern, bearing several leaves 6—12 dm. (2—4 ft.) long from a thick crown. The separate spore-bearing leaves appear in July from the center of the crown. They suggest an ostrich feather but are not finely branched. Abundant, at least formerly, in many places along the Red, Sheyenne and Pembina Rivers (fig. 8). It does well along the north and east sides of buildings. Small plants should be chosen for transplanting.

12. *Marsilea vestita* Hook. and Grev. WATER FERN. (Fig. 67). Stem slender, on surface of ground; leaves circular, 1—3 cm. wide, of 4, wedge-shaped leaflets; spores in oblong cases 4—6 mm. long, on short branches from stem. Occasional in low spots, central and western part of State. The leaves float on shallow water or form bright green patches on wet ground.

Equisetum HORSETAIL. SCOURING RUSH

Leafless plants from rhizomes; stems simple or branched, hollow, the joints easily separating, some small scales at top of joint; spores in small cones at end of stem.

Key to Species

Stems usually with many branches.
Spore cones on early, special, brownish stems which soon wither; sheaths of sterile stem joints 3-toothed. 13. *Equisetum arvense*
Spore cone stems producing green branches; sheaths 4-toothed.
14. *Equisetum pratense*
Stems without or with few branches; cones later, on tips of green stems.
Stems soft, dying at end of summer, branched irregularly.
15. *Equisetum fluviatile*
Stems hard, remaining green into the next year, usually not branched.
Stems with conspicuous gray bands at joints. 18. *Equisetum praealtum*
Stems without gray bands.
Spore cones with a very small, sharp tip. 17. *Equisetum laevigatum*
Spore cones rounded at tip. 16. *Equisetum kansanum*

13. *Equisetum arvense* L. FIELD HORSETAIL. (Fig. 128A). Green stems 3—6 dm. (1—2 ft.) high, densely branched; spore-bearing stems 1—2 dm. high, not branched, appearing in May. Common, especially on wet, sandy soils.

14. *Equisetum pratense* Ehrh. Resembles the last. Specimens from Cass, Barnes and Pembina Counties but I have not observed the plant in the field.

15. *Equisetum fluviatile* L. Stems 1—1.5 m. (3—5 ft.) high, weak, with few branches. Swampy places, Richland, Benson, McHenry and Pembina Counties.

16. *Equisetum kansanum* J. H. Schaffner. (Fig. 128B). Stems 3—6 dm. high, usually not branched; spore cones 1—1.5 cm. (½ in.) long, with a rounded tip. The commonest species in sandy soil, moist or quite dry. Spore cones appear in June.

17. *Equisetum laevigatum* A. Br. Similar to last, apparently less common. Cones with a small point 1 mm. long. Dr. Schaffner examined a number of our specimens.

18. *Equisetum praealtum* Raf. Stems stout, 3—8 dm. high, the joint sheaths 4—10 mm. (⅙—⅖ in.) long, mostly white, with black base and tip. Frequent but less common than No. 16. Often referred to *E. hyemale* L.

19. *Selaginella densa* Rydb. SMALL CLUBMOSS. Low, moss-like plant; stems creeping, 5—15 cm. long with upright branches; spore bearing cones 1—2 cm. long at branch tips. Abundant on dry prairie, east to McHenry County (fig. 8). Spores produced in May.
The report of *Lycopodium complanatum* at Fargo was undoubtedly based upon a mistaken impression of the collector.

SEED PLANTS Spermatophyta

CONIFERS — Gymnospermae

PINE FAMILY—Pinaceae

The word gymnosperm means "naked seed", referring to the fact that the seeds are borne on the **surface** of scales, not **enclosed** by the scale. When the scales of the cone separate at maturity the seeds fall out. This feature is not so evident in cedars and some other members of the group, in which the scales become fleshy and stick together forming a berry. The members of this group do not bear flowers but have two sorts of cones composed of scales. The scales of pine cones become quite hard and woody at maturity. The cones which produce pollen are delicate and drop off after the pollen is shed.

Slender, evergreen leaves ("needles") are another well known feature of the group. The leaves of larch or tamarack are slender but soft and drop in the fall. Those of red cedar and arbor vitae ("white cedar") are very small and scale-like. The wood is usually soft and uniform in texture, because it is composed of only one type of cell (tracheids) and does not have vessels or fibers which occur in the so-called "hard woods". However, the late summer wood in some species is quite hard so that alternating hard and soft rings appear in the lumber, as in fir and yellow pine.

Key to North Dakota Species

Leaves 1—1.5 dm. (4—6 in.) long; large trees with woody cones.
                                                    1. *Pinus ponderosa*
Leaves 2—12 mm. (1/12—½ in.) long; shrubs or trees with berries.
  Leaves 5—12 mm. long; usually a low, almost creeping shrub.
                                                    2. *Juniperus communis*
  Leaves scale-like, 1—3 mm. long.
    Small to medium sized tree.   3. *Juniperus scopulorum*
    Low, creeping shrub.   4. *Juniperus horizontalis*

1. *Pinus ponderosa* Dougl. WESTERN YELLOW PINE. Tree up to 20 m.
   (65 ft.) high; leaves in threes; cones woody, 8—10 cm. (3—4 in.) long.
   Local·in Bad Lands area, chiefly in Slope County, occasional north to
   McKenzie County. Our plant is var. *scopulorum* Engelm. This is the
   only pine in the State.[1] Red or Norway pine (*Pinus resinosa*), tamarack
   (*Larix laricina*), white and black spruce (*Picea alba* and *P. mariana*)
   occur in Minnesota up to within about 40 miles of our eastern boundary.
   Blue spruce (*P. pungens*) is commonly planted as a dooryard or wind-
   break tree. Black Hills spruce, a form of *P. alba,* is also planted.

2. *Juniperus communis*, var. *depressa* Pursh. DWARF JUNIPER. Widely
   spreading shrub, usually less than 1 m. (3 ft.) high. Berries blue-black,
   5—6 mm. (¼ in.) wide, ripening in 2 or 3 years. Occasional on slopes
   in Bad Lands and Pembina Mountains.

3. *Juniperus scopulorum* Sarg. ROCKY MOUNTAIN RED CEDAR. Shrub or
   tree, trunk up to 3 dm. in diameter; berries 5—8 mm. wide, blue-black,
   often quite gray with waxy coating, requiring 2 years to mature.
   Abundant in Little Missouri area, occasional in some gulches along
   Missouri River. This has been extensively used locally for fence posts
   and timbers. The eastern red cedar (*J. virginiana*) is not known to
   occur naturally in North Dakota. Its berries ripen in one year. Planted
   specimens of it at Fargo shed pollen Apr. 29, compared to May 8 for
   J. scopulorum (15 yr. av.).

4. *Juniperus horizontalis* Moench. CREEPING CEDAR. Main stems usually
   creeping on ground, forming dense mats on buttes, hills and slopes.
   Common in western part of State east to Missouri River; occasional in
   Pembina Mts.

FLOWERING PLANTS Angiospermae

The name angiosperm means "enclosed seeds", from the fact
that the ovules of the flower are enclosed within the cavity of the
ovary. When the seeds mature, this outer covering either splits to
release them or remains to be broken down by decay or to be eaten
by animals. While this whole group is known as "flowering plants",
the flower in many cases is no more conspicuous than the smallest
cones of conifers.

The two main divisions are:

  Monocotydedons—Leaves parallel veined; flower parts
  mostly in 3's or 6's.

  Dicotyledons—Leaves netted veined; flower parts mostly
  in 5's or 4's.

The moncotyledons include 17 families in our State, cattails to
orchids, inclusive. Lilies, grasses and grass-like plants are the most
common members. The term "monocotyledon" refers to the seed
which has but one cotyledon, but this feature is not readily recognized from ordinary examination. The seedlings do not come up

----

[1] *Pinus flexilis* James, **Limber Pine,** was collected in Slope County by R. P. Williams in
1949.

with the pair of "seed leaves" (cotyledons) which are so characteristic of dicotyledons like mustard, beans, cucumbers and sunflowers. In smilax, trillium and a few other monocotyledonous plants, the leaves are broad. In some plantains, Russian thistle, skeleton weed and other dicotyledons, the leaves are narrow. However, most monocotyledons can be readily recognized by leaf and flower characters given above.

CATTAIL FAMILY Typhaceae

Stout water plants growing from rhizomes. Leaves dark green, upright, 1—2 m. (3—7ft.) long, 2—5 cm. (1—2 in.) wide; flowers very small, closely crowded in an oblong spike on a stout stalk arising between the leaves; staminate flowers forming upper part of spike; stamens 2—7, pistil 1; petals and sepals represented only by bristles; tiny fruits 1-seeded, scattering in fluffy masses in late summer.

Key to Species

Leaves 20—35 mm. (¾—1½ in.) wide; staminate flowers close above
　　the pistillate.　　　　　　　　　　　　　　　1.　*Typha latifolia*
Leaves usually not over 15 mm. wide; staminate flowers 2—10 cm.
　　above the pistillate.　　　　　　　　　　　2.　*Typha angustifolia*

1. *Typha latifolia* L. BROAD-LEAVED CATTAIL. A very common plant in ditches, edges of ponds, etc. Flowers in early June.
2. *Typha angustifolia* L. NARROW-LEAVED CATTAIL. This was first recognized by Neil Hotchkiss of the Fish and Wildlife Service, U. S. Dept. Int., at Long Lake in Burleigh County in 1942. Specimens were sent us later by M. C. Hammond. At the Lower Souris Wildlife Refuge in McHenry County, Mr. Hammond first recognized the plant in 1943. It apparently has been only recently introduced because Mr. Hammond had spent several years at Lower Souris and was a keen student of plants.

BUR-REED FAMILY Sparganiaceae

Marsh plants somewhat resembling cattails, but leaves pale green, stiffer, 3—10 dm. (1—3 ft.) long, 7—12 mm. (¼—½ in.) wide; flowers in several rounded heads on a stalk between the leaves; fruiting heads 2—3 cm. (¾—1½ in.) wide, staminate ones smaller, above the pistillate heads; stamens 5; pistil 1; fruit obovoid, 5—7 mm. long, brown, the outer coating spongy. Flowers in June, fruit ripe in August.

We have recognized only one species, *Sparganium eurycarpum* Engelm. (fig. 79), but one or two others, which are quite similar, may be found. The plants are fairly common and grow in the edges of shallow ponds which usually dry up during summer. They are generally considered of little importance as duck food plants.

PONDWEED FAMILY Potamogetonaceae

Aquatic plants with leaves under water or floating; flowers small, greenish, in axillary spikes, usually under water, sometimes projecting just above surface; 4—6 small, green sepals in some species; pistils and stamens 1—6; fruit a 1-seeded achene.

This is the most important group of plants for duck food. The birds eat large quantities of seeds and in some cases tubers or other parts of the plant. The species are difficult to identify. Another plant likely to be looked for here is *Anacharis occidentalis* (p. 51).

Key to Species

Leaves opposite; staminate and pistillate flowers separate.
  Dioecious; stamen 1, enclosed in a membrane; pistil 1, not enclosed.
                                                    1. *Najas flexilis*
  Monoecious; pistils 2—5, stamen 1, both usually enclosed in same
    sheath.   2. *Zannichellia palustris*
Leaves alternate; flowers perfect, in spikes or other clusters.
  Flowers 4—7 in a loose cluster on a long, slender stalk which be-
    comes coiled at the lower part; stamens 2.   3. *Ruppia maritima*
  Flowers not stalked, several to many in short or elongated spikes;
    stamens 4.
    Plants with both floating and submerged leaves.
      Submerged leaves without blades, floating ones oblong, 2—5
        cm. wide.   4. *Potamogeton natans*
      Submerged leaves with narrow blades.
        Submerged leaves lanceolate, long petioled.
                                          5. *Potamogeton nodosus*
        Submerged leaves linear, sessile.   6. *Potamogeton gramineus*
    Plants with only submerged leaves.
      Leaves ovate or lanceolate, 2—5 cm. (1—2 in.) wide (fig. 129C).
                                          7. *Potamogeton perfoliatus*
      Leaves linear, 2—5 mm. (1/12—⅕ in.) wide.
        Stipules united along their edges to leaf blades.
          Leaves tapering to a sharp point; stipules hardly wider than
            leaf base. (fig. 129A).   8. *Potamogeton pectinatus*
          Leaves not tapering, rounded at tip; lower stipules 2—5
            times as wide as leaf.   9. *Potamogeton vaginatus*
        Stipules not united to leaf blade.
          Stipules less than 1 cm. long, early falling off.
                                          10. *Potamogeton pusillus*
          Stipules 1—5 cm. long, persistent.
            Plants long and "stringy"; leaves 5—20 cm. long, 2—4
              mm. wide.   11. *Potamogeton zosteriformis*
            Plants more bushy; leaves 3—5 cm. long, 1—2 mm. wide.
                                          12. *Potamogeton foliosus*

1. *Najas flexilis* (Willd.) Rostk. & Schmidt. NAIAD. (Fig. 68). Stems long and slender or densely tufted; leaves enlarged at base, 8—15 mm. (¼—½ in.) long, 1—2 mm. wide, with 20—30 fine teeth; fruit 2—3 mm. long, ovoid, smooth. One specimen from Fargo (Lee 1338). It may be much more common than this would indicate. *N. marina* L. has been reported for North Dakota but we have no specimens.

2. *Zannichellia palustris* L. HORNED PONDWEED. Slender branches 1—4 dm. (4—16 in.) long from a rhizome; leaves 3—5 cm. long, hardly 1 mm. wide; fruits 2—4 mm. long, oblong but flattened, slightly curved with 3 or 4 short, blunt teeth on the back. Specimens from various parts of State.

3. *Ruppia maritima* L. WIGEON GRASS. Very slender plants; leaves 2—10 cm. long, less than 1 mm. wide. This is mainly a sea coast plant. We had for a long time only some material collected by Bolley in 1904 at Island Lake near Dawson, Kidder County. In 1939, R. N. Bach sent specimens from Lake George, Kidder County, near Streeter. We have also a luxuriant specimen with stems 4 dm. long, collected by H. S. Telford in 1941 at Spiritwood Lake, Stutsman County.

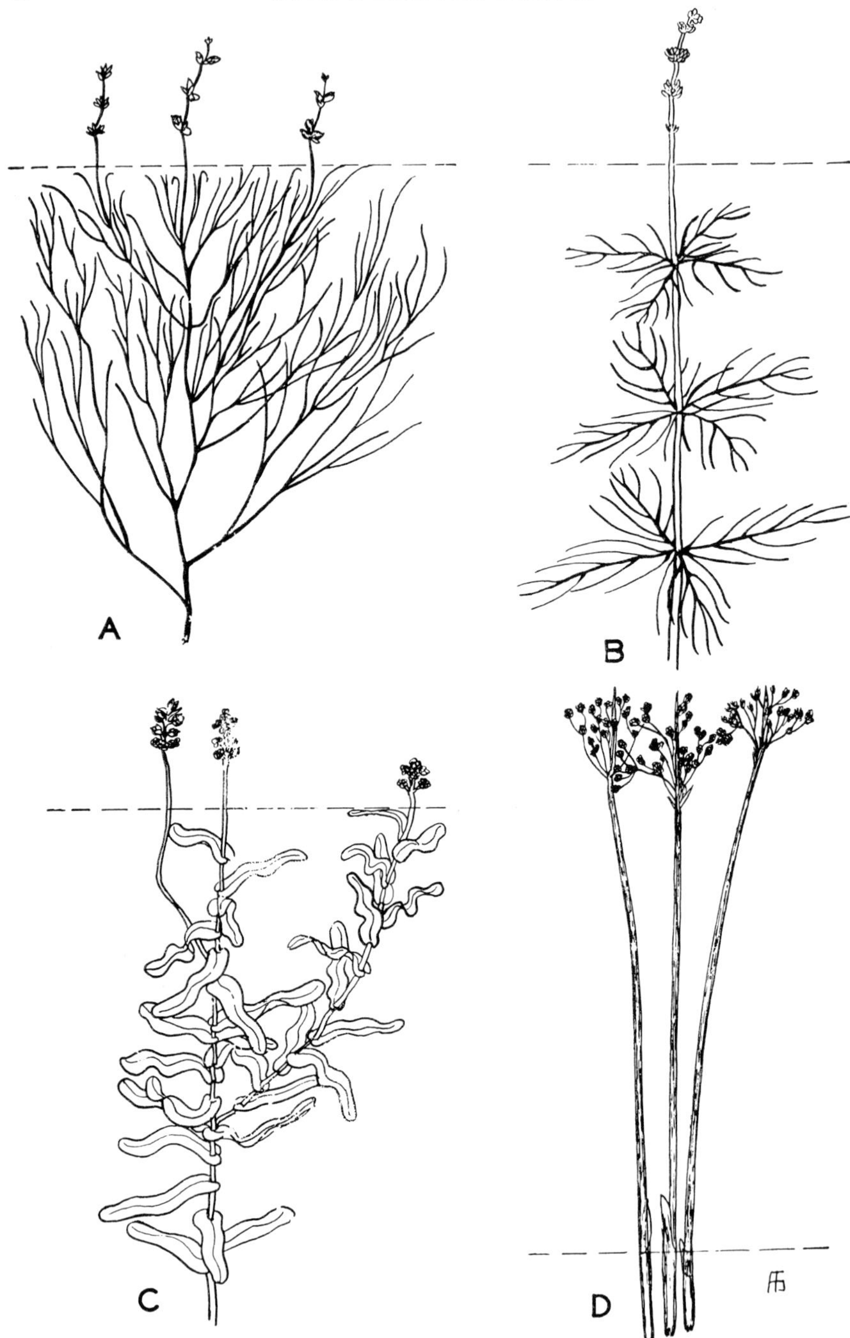

129. Water plants: A. Sago Pondweed (*Potamogeton pectinatus*); B. Water Milfoil (*Myriophyllum*); C. Clasping-leaved Pondweed (*Potomogeton perfoliatus*); D. Bulrush (*Scirpus validus*).

4. *Potamogeton natans* L. FLOATING-LEAVED PONDWEED. Stems 1—2 m. (3—7 feet) long, usually not branched; floating leaves oblong, 5—10 cm. long, 2—5 cm. wide; stipules free from petiole, lanceolate, 2—5 cm. long; flower spikes dense, cylindric, 2—3 cm. long, projecting above water. This is one of the common "weeds" which so often form a thick growth in water 1—2 m. deep in the Minnesota lakes but North Dakota ponds are usually too shallow or saline for it to grow well. Specimens from Turtle Mts. only.

5. *Potamogeton nodosus* Poir. LONG-LEAVED PONDWEED. Somewhat like No. 4, but leaves 1—2 dm. long, tapering. We have it only from the Red River at Fargo and Wahpeton. Also known as *P. americanus* Cham. & Schlect.

6. *Potamogeton gramineus* L. VARIABLE PONDWEED. Stems slender, much branched, said to become as much as 3 m. long; submerged leaves narrow, 1—5 cm. long, 2—15 mm. wide; floating leaves oblong-lanceolate, 1.5—6 cm. long, 8—25 mm. wide; flowers in dense spikes, 1—2 cm. long on short, axillary peduncles. We have it from Emmons, Barnes, Stutsman and Williams Counties, but it should be fairly common. The floating leaves are smaller than those of No. 4, and the fine, branching stems should readily distinguish it. Often recorded as *P. heterophyllus* Schreb.

7. *Potamogeton perfoliatus* L. CLASPING-LEAVED PONDWEED. (Fig. 129C). Stems moderately stout, usually erect in water 1—2 m. (3—7 ft.) deep; leaves ovate or lanceolate, clasping the stem, 5—10 cm. (2—4 in.) long, 1—2 cm. wide at base, somewhat crisped; flower spikes terminal at about surface of water, 1 cm. long. We have it from Richland to Steele County and from Bottineau County. It is one of the common species in suitable locations and is easily recognized by the large leaves which thickly cover the erect stems.

8. *Potamogeton pectinatus* L. SAGO PONDWEED. (Fig. 129A). Stems slender, much branched, very leafy; leaves very narrow, 3—15 cm. long, sharp pointed; flowers in interrupted spikes; fruits rounded, 3—4 mm. long. Probably the most common species because it will grow in shallow, saline water. It forms large masses of material just below the surface of the water. The horizontal stems in the mud form tubers which are much sought by ducks.

9. *Potamogeton vaginatus* Turcz. SHEATHED PONDWEED. Similar to No. 8 but growing in deeper water and producing more elongated stems. We have it only from Lake Ardock, Walsh County, but it may be fairly common.

10. *Potamogeton pusillus* L. SMALL PONDWEED. Stems very slender, much branched, 1—6 dm. long; leaves very slender, 2—8 cm. long; flowers 3—10, in rounded clusters on axillary peduncles. One specimen from Lake Tobiason, Steele County, seems to be this species.

11. *Potamogeton zosteriformis* F e r n a l d. EEL-GRASS. PONDWEED. FLAT-STEMMED PONDWEED. Stems flat, "winged", much branched, 1—2 m. long; leaves 5—30 cm. long, about 4 mm. wide; spikes axillary, 10—15 flowered. A rather common and striking species because of the long, parallel-sided leaves and wide, flat stems. It has been called *P. compressus* L. and *P. zosterifolius* Schum.

12. *Potamogeton foliosus* Raf. LEAFY PONDWEED. Somewhat like No. 8 and No. 10 in appearance and apparently a common species.

ARROWGRASS FAMILY Juncaginaceae

Plants of wet soil with long, slender, rounded or 3-angled leaves and small, greenish flowers on a central stalk from base of plant. Sepals 6, stamens 6, pistils 3 or 6, partly united; fruit dry, separating into 3 or 6 parts.

Key to Species

Pistils 6; flower cluster stout, 2—6 dm. (8—24 in.) high, many-flowered.   1. *Triglochin maritima*
Pistils 3; flower cluster slender, 1—3 dm. high.   2. *Triglochin palustris*

1. *Triglochin maritima* L. Arrowgrass. (Fig. 130). Perennial from a thick base; leaves 2—6 dm. long, somewhat 3-angled, 4—6 mm. thick; flowers purplish when in bloom, 5 mm. wide; fruit oblong, 5—6 mm. long. June, July. Frequent to common in wet, saline places. This plant is poisonous because it often contains sufficient prussic acid to kill livestock if they eat it in quantity.

2. *Triglochin palustris* L. A much smaller, more slender plant. Local around springs or boggy places, all parts of State.

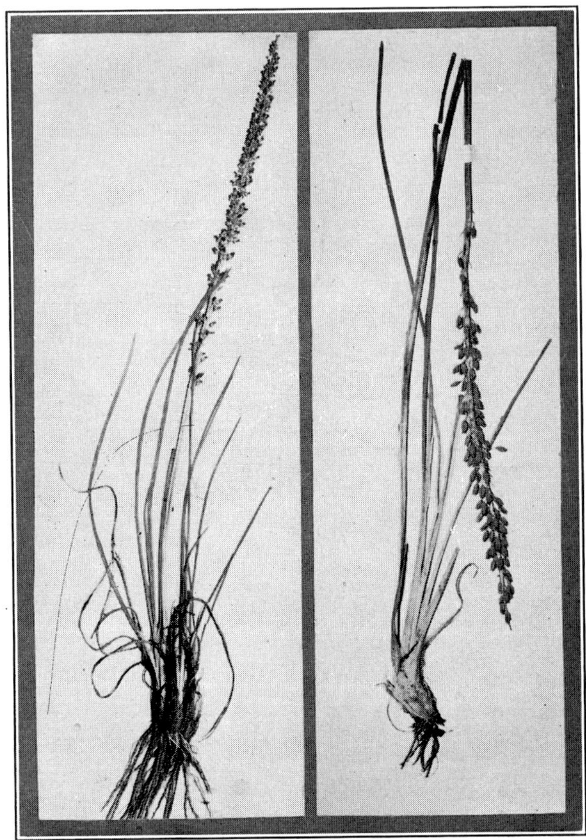

130. Arrowgrass (*Triglochin maritima*). Photo from pressed specimens.

WATER-PLANTAIN FAMILY Alismaceae

Marsh or water plants with usually broad leaves. Flowers perfect, in large clusters, often showy, white or greenish; petals 3, sepals 3, pistils and stamens many, each pistil maturing into a flat, 1-seeded achene.

Key to Species

Pistils about 25 in a single ring; flowers 5—7 mm. (¼ in.) wide.
Pedicels erect; achenes longer than wide, grooved on back.
1. *Alisma subcordatum*
Pedicels recurved; achenes as wide as long, ribbed on back.
2. *Alisma geyeri*
Pistils 75—100 in several series; flowers 12—25 mm. wide.
3. *Sagittaria cuneata*

1. *Alisma subcordatum* Raf. WATER-PLANTAIN. (Fig. 131). Perennial from a short crown; leaves on long petioles, erect or floating, cordate, ovate or oblong, 3—10 cm. (1—4 in.) long; flowers in a much branched cluster 2—6 dm. (8—24 in.) high from the crown. July, Aug. A common plant especially where shallow water dries up during summer.

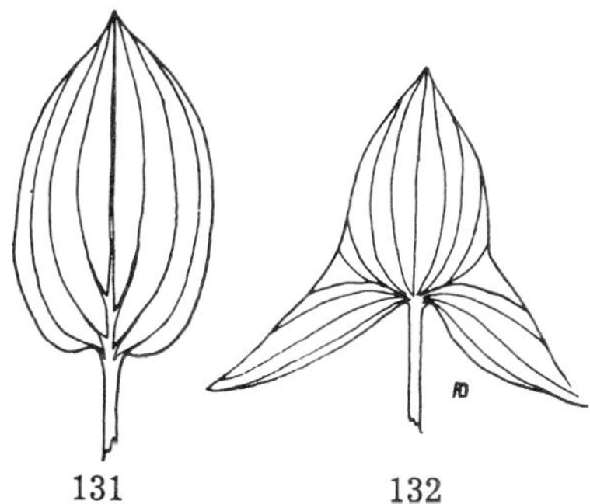

131 132

131. Water-plantain (*Alisma subcordatum*).
132. Arrowhead (*Sagittaria*).

2. *Alisma geyeri* Torr. GEYER'S WATER-PLANTAIN. A smaller plant, 2—4 dm. high; leaves 5—8 cm. long, narrowed at both ends. Rather local, though we have it from 8 counties in different parts of State. It seems restricted to edges of permanent ponds. Carl A. Geyer was botanist to Nicollet's Expedition and discovered this plant in our area in 1839.

3. *Sagittaria cuneata* Sheldon. ARROWHEAD. (Fig. 132). Perennial, similar to No. 1, but leaves sagittate, with two, long, tapering basal lobes; remainder of blade oblong or ovate, 1—2 dm. long, 3—6 cm. wide; flowers 1—2.5 cm. wide, conspicuous, in a branched cluster 2—6 dm. high. July, Aug. Less common than *Alisma* but frequent in similar places. One specimen from Pembina County has been referred to *S. latifolia* Willd. but this needs further study. The plants have slender rhizomes bearing tubers which are eaten by ducks and in some countries by people.

WATERWEED FAMILY Hydrocharitaceae

Aquatic or mud plants resembling pondweeds. Flowers dioecious or monoecious; sepals 3, petals 3 or 0, stamens 3—9, pistil 1; fruit a many seeded capsule.

Key to Species

Stamens 9; leaves 5—15 mm. (⅕—⅗ in.) long, in whorls on floating stems.
1. *Anacharis occidentalis*
Stamens 1—3; leaves 1—5 dm. (4—20 in.) long, arising from base of plant in water.
2. *Vallisneria americana*

1. *Anacharis occidentalis* (Pursh) Victorin. WATERWEED. Stems 2—10 dm. (8—40 in.) long, floating; leaves 5—15 mm. (⅕—⅗ in.) long, 1 mm. wide; flowers dioecious, staminate solitary on very slender stalks 5—15 cm. (2—6 in.) long from upper leaf bases; sepals and petals

1—1.5 mm. long, rounded; pistillate without stalks, but with a tubular portion 2—5 cm. long; capsules 2—5 cm. long. We have this from several counties in southern part of State and from Benson County. It is easily overlooked unless the staminate flowers are seen floating on the water, attached to their pink, thread-like stalks.

2. *Vallisneria americana* Michx. EELGRASS. "WILD CELERY". Leaves ribbon-like from base of plant, 1—5 dm. long; flowers similar to those of No. 1 but staminate break off and open on surface of water; pistillate on slender stems coming to surface, then submerging by coiling the stalk. This is the "wild celery" so famous as duck food. Our only record is from Wood Lake, Benson County (H. S. Telford in 1941). Formerly referred to *V. spiralis* L.

GRASS FAMILY Poaceae (Gramineae)

Annual or perennial herbs with upright, leafy stems, long, narrow leaves and chaffy flowers (fig. 133). Petals and sepals 0, stamens usually 3, pistil 1 with 2 feathery stigmas; fruit a caryopsis (1-seeded, seed coat and ovary wall combined in a single, membran-

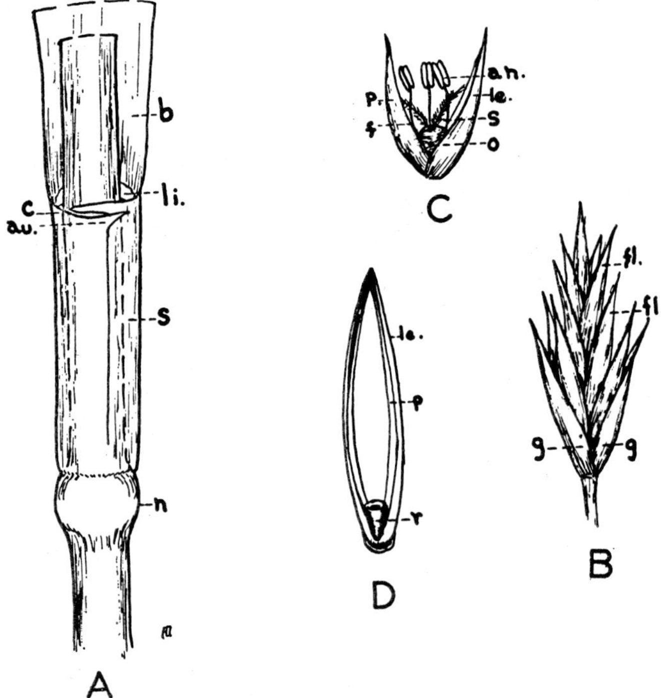

133. Grass stem and flowers. A. Portion of stem and leaf base: n, node or joint; s, leaf sheath; b, base of blade; au, auricle; c, collar; li, ligule. B. spikelet: g, glumes; fl, flowers. C. flower: le, lemma; p, palea; an, anther and f, filament of stamen; o, ovary; s, stigmas. D. Mature grain enclosed by chaff: le, lemma; p. palea; r, rachilla.

ous covering); stamens and pistil enclosed by 2 scales (lemma and palea); flowers (florets) grouped in 1—many-flowered spikelets, each more or less enclosed by 2 basal scales (glumes); spikelets arranged in spikes or panicles (often called "heads").

One of the most important families, including native and cultivated forage grasses as well as corn, rice, wheat, rye, barley and oats. Grasses are usually recognized as such, but sedges, rushes and a few others are often confused with them. It is a difficult group and distinctions are necessarily technical. The family is divided into several tribes. Usually an unknown grass can be placed readily in one of these by general features of flower arrangement. Further information on grasses is given in N. D. Exp. Sta. Bull. 300 and 301.

Key to Tribes

Glumes falling off with florets at maturity, or absent from the first.
  Glumes none or very small.
    Stems stout, 1—2 m. (3—7 ft.) high; spikelets cylindrical, 2—3 cm. (1 in.) long.   *Zizanieae* p. 79
    Stems slender, 3—10 dm. high; spikelets flat, 3—5 mm. (⅙ in.) long.   *Oryzeae* p. 79
  Glumes present, usually well developed.
    Spikelets in one-sided spikes.   *Chlorideae* p. 75
    Spikelets not in one-sided spikes.
      Palea absent, spikelet with apparently 3 scales.   *Alopecurus* p. 69
      Palea present, spikelets with 4 or 5 scales.
        Palea thick, hardened; spikelets not in pairs.   *Paniceae* p. 80
        Palea membranous; a fertile, sessile spikelet paired with a sterile, stalked spikelet.   *Andropogoneae* p. 83
Glumes persisting on stalk after florets have fallen.
  Spikelets sessile in spikes or dense, spike-like panicles.
    Spikelets in one-sided spikes, either dense or loose.   *Chlorideae* p. 75
    Spikelets evenly placed, spikes not one-sided.
      Spikelets 1-flowered; lemma thin membranous, not awned.   *Phleum* p. 69
      Spikelets several-flowered or 1-flowered; lemma thick and awned.   *Hordeae* p. 62
  Spikelets on slender stalks; panicles widely branched or sometimes dense and spike-like.
    Spikelets 1-flowered or additional flowers imperfect.
      Glumes 4, lemma firm or hardened.   *Phalarideae* p. 78
      Glumes 2, lemma thin membranous.   *Agrostideae* p. 68
    Spikelets 2—many-flowered, flowers usually all perfect.
      Lemma with a bent or twisted awn at or below middle.   *Aveneae* p. 67
      Lemma awnless or with a straight awn at tip.   *Festuceae* p. 53

FESCUE TRIBE Festuceae

One of the largest groups, including bromegrasses, bluegrasses, fescues and many others. Usually recognized by the broad, spreading panicles with spikelets of several or many flowers. In a few cases the panicles are dense and spike-like.

Key to Species or Genera

Lemma 1—3-nerved.
  Lemma irregularly truncate, plant usually in running water.   40. *Catabrosa aquatica*
  Lemma not truncate; plants usually not in running water.
    Plant 2—3 m. (6—10 ft.) high; leaves 2—3 cm. (1 in.) wide.   44. *Phragmites communis*
    Plant less than 1 m. high; leaves 1 cm. or less wide.
      Spikelets 1—3-flowered; low, long-leaved plant of sand dunes.   11. *Redfieldia flexuosa*
      Spikelets 12—40-flowered.
        Stems creeping on mud, branches 5—15 cm (2—6 in.) high.   39. *Eragrostis hypnoides*

Plants tufted, stems erect, 2—6 dm. (8—24 in.) high.
 Plant strong scented; spikelets 2.5—3 mm. (⅛ in.) wide.
                                          37. *Eragrostis cilianensis*
 Not strongly scented; spikelets 1—1.5 mm. wide.
                                          38. *Eragrostis pectinacea*
Lemma usually 5—9-nerved (3 in some species of *Poa* and *Festuca*).
 Spikelets strongly flattened, lemma with sharp keel.
  Lemma thick, hard, smooth; creeping, dioecious plant.
                                          42. *Distichlis stricta*
  Lemma thin; stems mostly upright and flowers perfect.
   Lemma short awned, finely ciliate on keel. 43. *Dactylis glomerata*
   Lemma obtuse or acute, not awned nor ciliate.   22-36. *Poa*
 Spikelets rounded, lemma not keeled.
  Lemma with basal tuft of hairs.
   Tall, leafy, marsh grass with rhizomes.   21. *Fluminea festucacea*
   Tufted, bare-stemmed grass in aspen woods.
                                          45. *Schizachne purpurascens*
  Lemma without basal tuft of hairs.
   Lemma obtuse, not awned at tip.
    Lemma faintly 5-nerved, 2—3 mm. long.
                                          17. *Puccinellia nuttalliana*
    Lemma strongly 7-nerved.
     Spikelets 10 mm. or more long, 7—13-flowered.
                                          19. *Glyceria borealis*
     Spikelets 6 mm. or less long, 4—6-flowered.
      Spikelets 3 mm. long; plant 3—6 dm. high in wet
       meadows.   20. *Glyceria striata*
      Spikelets 4—6 mm. long (fig. 138); plant 6—10 dm. high
       in ponds or ditches.   18. *Glyceria grandis*
   Lemma acute or awned.
    Lemma 7—12 mm. long, 5—9-nerved with an awn from a
     2-toothed tip.   1-11. *Bromus*
    Lemma 3—6 mm. long, 3—5-nerved.
     Lemma with a short awn at tip.   12-16. *Festuca*
     Lemma not awned.   22-36. *Poa*

Bromus   BROMEGRASS. CHESS. CHEAT

Medium sized to large grasses with fairly broad leaves, spreading panicles and moderately large grains. Leaves can be distinguished from other grasses by the closed sheaths (fig. 133).

Key to Species

Perennial from rhizomes or tufts.
 Forming dense sod from rhizomes; lemma scarcely awned.
                                          1. *Bromus inermis*
 Stems tufted or scattered; lemma short awned.
  Lemmas hairy only at base and along edges; tall plants in woods.
                                          3. *Bromus ciliatus*
  Lemmas hairy all over.
   First glume 3—5-nerved, second 5—9-nerved; prairie plant.
                                          2. *Bromus anomalus*
   First glume 1-nerved, second 3-nerved; woodland plant.
    Leaf sheaths extending beyond next leaf base above.
                                          4. *Bromus latiglumis*
    Leaf sheaths not reaching base of next sheath.
                                          5. *Bromus purgans*
Annual, weedy species.
 Lemmas, excluding awns, 10 times as long as wide; awn 10—15 mm.
  (½ in.) long.   6. *Bromus tectorum*
 Lemmas, excluding awns, 4—5 times as long as wide; awn 3—10
  mm. long.

134

135

134. Smooth Brome (*Bromus inermis*).
135. Downy Brome (*Bromus tectorum*).

Panicle dense, the branches erect or ascending.
    Lemmas glabrous.   7. *Bromus squarrosus*
    Lemmas covered with fine hairs.   8. *Bromus mollis*
Panicle open, the branches spreading.
    Panicle branches rather stiff, not flexuous.   9. *Bromus commutatus*
    Panicle branches slender, lax or flexuous.
        Palea shorter than lemma, awn bent in drying.
                                                   10. *Bromus japonicus*
        Palea as long as lemma, awn nearly straight. 11. *Bromus arvensis*

1. *Bromus inermis* Leyss. SMOOTH BROME. (Fig. 134). Perennial by rhizomes which form a dense sod; stems 6—15 dm. (2—5 ft.) high; panicle 1—2 dm. (4—8 in.) long, nearly as wide, the branches at first erect, drooping at maturity; lemmas thin, nearly flat, 9—12 mm. (⅜—½ in.) long. Flowers in mid-June, to some extent throughout summer. · Planted since about 1890, now thoroughly naturalized along roads, fence rows and waste ground. Our best pasture grass for eastern part of State.

2. *Bromus anomalus* Rupr. Tufted, prairie perennial, resembling scattered, dwarfed plants of Smooth Brome, but spikelets densely covered with hairs lying flat. Records from Bottineau, Ward, Williams and McKenzie Counties. *B. porteri* Nash is regarded as the same species.

3. *Bromus ciliatus* L. FRINGED BROME. Resembles Smooth Brome in general appearance, but grows in large clumps in or near woods. Hairs on lemma bases may be long and spreading. Specimens from Cass, Barnes, Benson, Rolette and Bottineau Counties.

4. *Bromus latiglumis* (Shear) Hitchc. Resembles No. 3 and apparently is the most common form. Specimens from Richland, Ransom, Cass, Barnes, McHenry and Burke Counties; one from Grant County (Bell 1287) has overlapping sheaths but only a few hairs toward base of lemma as in *B. ciliatus*.

5. *Bromus purgans* L. CANADA BROME. Similar to No. 4. Specimens from Pembina and McHenry Counties only.

6. *Bromus tectorum* L. DOWNY BROME. (Fig. 135). Annual, sometimes developing large tufts 3—6 dm. high; stems and leaves fine, soft hairy; panicle branches drooping. Frequent along roadsides in east, more common in pastures or waste ground westward. Matures in mid-summer and is troublesome to livestock because of the stiff awns. It furnishes some early pasture.

7. *Bromus squarrosus* L. A specimen from Esmond, Benson County, was collected in 1939 by J. R. Swallen of the U. S. Department of Agriculture. Specimens received in 1941 from Hal Stefansson and Lyle Currie, County Extension Agents in Bottineau and McKenzie Counties, respectively, seem to be the same. It resembles Nos. 9 and 10. Our material has fine stems 3 dm. high with conspicuous brown nodes when mature. The short panicles have large spikelets 15—25 mm. long.

8. *Bromus mollis* L. SOFT BROME. One specimen from Barton, Pierce County, collected by L. R. Waldron in 1903, was identified as this species by J. R. Swallen.

9. *Bromus commutatus* Schrad. HAIRY CHESS. Annual, 3—6 dm. high; panicles somewhat smaller, denser, or longer than those of Smooth Brome; lemmas boat-shaped, firm, with an awn 3—5 mm. long. Fields and waste ground. This and the next species seem to be our most common annual bromes except for Downy Brome. They appear erratically, sometimes in abundance locally for a year or two.

10. *Bromus japonicus* Thunb. JAPANESE CHESS. I think this is more common than the last, but species in this group are very similar and difficult to distinguish. Most of our specimens are from western half of State.

11. *Bromus arvensis* L. FIELD BROME. One specimen from Pembina County (Bergman 2067) was verified by J. R. Swallen.

    Bromus catharticus Vahl., RESCUE GRASS, has been reported, but was probably an escape from planted material. It has strongly flattened, pointed spikelets.

Festuca   FESCUE

Small to fairly large grasses of various habits, mostly uncommon in North Dakota. The species are numerous and difficult to identify. Ours are easily separated from each other but some are much like species of *Poa, Puccinellia* or other members of this tribe.

Key to Species

Annual, very slender, 1—3 dm. (4—12 in.) high, often in dense
  patches.   12. *Festuca octoflora*
Perennials, growing in tufts.
  Leaves very narrow, 1—3 mm. (1/25—⅛ in.) wide; lemma awned.
    Tufts low, dense, widely spreading, 1—2 dm. high.   13. *Festuca ovina*
    Tufts slender, upright, 3—6 dm. high.   14. *Festuca scabrella*
  Leaves moderately wide, 4—8 mm. wide; lemmas not awned.
  Woodland plants; spikelets 3—5-flowered, lemma 4 mm. long.
                                    15. *Festuca obtusa*
  Meadow plants; spikelets 5—9-flowered, lemma 5—6 mm. long.
                                    16. *Festuca elatior*

12. *Festuca octoflora* Walt. SIX-WEEKS FESCUE. Dry prairie slopes, quite common at least some years in western part of State. A recent record is for Richland County. Matures in July and has little forage value.

13. *Festuca ovina* L. SHEEP FESCUE. A widely distributed species with several varieties. Rare in North Dakota. Specimens from Pembina, Benson, Pierce, McHenry and LaMoure Counties. Personally, I have seen it but once in the Big Coulee near Esmond, Pierce County. The Walhalla specimen (Bergman 2010) grew near woods in sandy soil and is quite slender, 4 dm. high. Sheep fescue is often planted and appears in lawns as scattered, thick, dark green tufts with many hard, fine leaves.

14. *Festuca scabrella* Torr. ROUGH FESCUE. Tufted, quite stiff and upright; leaves 1—3 dm. (4—12 in.) long, with conspicuous old leaf bases from the year before persisting; stems 3—6 dm. high; panicle narrow, 5—15 cm. (2—6 in.) long; spikelets 4—6 flowered, 7—10 mm. (⅓ in) long. We have it from Benson, Rolette and Burke Counties. In Rolette County, I found it on the prairie tip of Butte St. Paul, one of the highest points in the Turtle Mts. Previously recorded as *F. hallii*, which is regarded as the same species.

15. *Festuca obtusa* Biehler. NODDING FESCUE. Stems 3—8 dm. high; leaves 1—3 dm. long, 4—8 mm. wide; panicle branches long, slender, drooping; lemma hard, smooth, not distinctly nerved. A tall, lax grass in woods. We have it only from Richland, Cass, Barnes and Pembina Counties but it probably occurs in favorable places in other eastern counties. Previously called *F. nutans* Willd.

16. *Festuca elatior* L. MEADOW FESCUE. Densely tufted, leafy; leaves 1—4 dm. long, 4—8 mm. wide; stems 4—8 dm. high; panicle dense, short branched, 2—3 dm. long; lemma smooth, hard, resembling an *Agropyron* but a little smaller. Native of Europe and Asia, much planted in more moist regions, an occasional escape here. *F. pratensis* Huds. is usually regarded as the same species.

17. *Puccinellia nuttalliana* (Schult.) Hitchc. SALT MEADOWGRASS. ALKALI-GRASS. Slender, tufted perennial; leaves 2—4 dm. long, 1—3 mm. wide; stems 3—6 dm. high, panicle 1—2 dm. long, with many short branches which usually bend downward at maturity; spikelets 4—7 mm. long, much like a *Poa* except for the rounded, open lemmas. Common, usually in scattered tufts at edges of ponds or on saline, wet flats. Hitchcock's "Manual" reported also *P. distans* (L.) Parl. in North Dakota. On later examination of all of our material, Mr. J. R. Swallen reported that Bergman 418 from Stutsman County closely resembled the ones which had been referred to *P. distans*, but he thought best to consider ours all *P. nuttalliana*. It was previously called *P. airoides* (Nutt.) Wats. & Coult.

18. *Glyceria grandis* S. Wats. TALL MANNAGRASS. (Fig. 138). Perennial, stems leafy, 1—1.5 m. (3—5 ft.) high; leaves soft, 2—6 dm. (8—24 in.) long, 6—12 mm. (¼—½ in.) wide; panicle branches widely spreading or drooping, 1—3 dm. long; spikelets oblong, 5—6 mm. long; grains plump, oblong, 1.3 mm. long, dark brown, readily separating from lemmas. Common in shallow water or wet places. The leaves are pale green, soft and palatable, but plants usually occur in limited quantities where they cannot be harvested. The spikelets fall apart easily at maturity, and floating on the water, gave rise to the name mannagrass.

19. *Glyceria borealis* (Nash) Batch. NORTHERN MANNAGRASS. Similar to last but more slender, panicle branches fewer, slender, upright; spikelets slender, 1—1.5 cm. long. Wet places; rare. We have it from Cass, Stutsman, Grant and Billings Counties.

20. *Glyceria striata* (Lam.) Hitchc. FOWL MANNAGRASS. Perennial, 2—6 dm. high; leaves 1—3 dm. long, 2—6 mm. wide; panicle 8—15 cm. long, branches spreading or drooping; spikelets smaller than No. 18, grains about 0.75 mm. long. Often abundant in wet meadows and seepage areas. It resembles Kentucky Bluegrass but can be distinguished by the plump, rounded grains and later maturity. The nearly black grains are a characteristic feature of mannagrasses. Previously called *G. nervata* Trin.

21. *Fluminea festucacea* (Willd.) Hitchc. HOLLOWSTEM. SPRANGLETOP. Perennial from long, hard rhizomes; leaves hard, stiff, the shorter ones often upright, 1—6 dm. (4—24 in.) long, 5—10 mm. (⅕—⅖ in.) wide; stems 1—2 m. (3—7 ft.) high, panicle 1—2 dm. long, with numerous branches; spikelets slender, 8 mm. long. One of our most common slough grasses, forming a solid stand on low flats which are often flooded. Sometimes mixed with *Carex atherodes* or other plants. Also occurs along ditches and in shallow water. Flowers in early June but requires abundant water and often does not bloom. It is one of the lowland grasses commonly used for hay because it is of fair quality and in most years the ground becomes dry enough so it can be harvested. The local name "hollowstem" refers to the unusually large opening in the stem. (See N. D. Exp. Sta. Bull. 349.)

Poa BLUEGRASS

Small to medium sized, fine-leaved grasses, growing in tufts or forming a sod by rhizomes; producing a good quality of hay and maturing early. The chief species are rather easily recognized but technical distinctions are difficult and identity of several native, western forms is still uncertain.

Key to Species

Spikelets distinctly flattened, lemma V-shaped.
  Annual, introduced plant, frequent in lawns.   22. *Poa annua*
  Perennials, native or naturalized in fields and meadows.
    Rhizomes present, plants forming a sod.
      Stems flattened; panicle narrow, upright.   23. *Poa compressa*
      Stems not flattened; panicle usually spreading.
        Lemma hairy on keel and edge, not on midnerve.
            24. *Poa pratensis*
        Lemma hairy on midnerve as well as on keel and edge.
          Panicle branches short, stiff, crowded.   25. *Poa arida*
          Panicle branches longer, not stiff nor crowded.
             26. *Poa glaucifolia*
    Rhizomes absent, plants tufted.
      Flowers usually replaced by bulblets.   27. *Poa bulbosa*
      Flowers developing normal grains, not bulblets.
        Lemma with fine, tangled hairs at base.
          Plants of moist ground; panicle 1—3 dm. (4—12 in.) long,
            flexuous.   28. *Poa palustris*

Plants of dry ground; panicle 8—15 cm. (3—6 in.) long, stiff.
                                                    29. *Poa interior*
Lemma without tangled basal hairs.
    Lemma hairy on keel or edge.
        Leaves rough, not over 1 mm. (1/25 in.) wide. 30. *Poa cusickii*
        Leaves not rough, 2—3 mm. wide.   31. *Poa epilis*
    Lemma without hairs on keel or edge.
        Leaf usually folded, stiff; ligule 5—7 mm. long.
                                                    32. *Poa longiligula*
        Leaf not folded, soft; ligule 2 mm. long.   33. *Poa stenantha*
Spikelets little flattened, lemma rounded on the back.
    Lower part of lemma with very short but coarse hairs.
        Stems 3 dm. or less high; basal shoots numerous.   34. *Poa secunda*
        Stems 5 dm. or more high; basal shoots few.   35. *Poa canbyi*
    Lower part of lemma without hairs or with short fine ones.
                                                    36. *Poa ampla*

22. *Poa annua* L. ANNUAL BLUEGRASS. Bright green; stems 1—3 dm. high,
    usually spreading; leaves 1—3 mm. wide, thin and soft; panicle 3—7 cm.
    long, pyramidal. This was quite common in edges of lawns in Fargo
    about 1920 and again in 1943-48. We have it elsewhere only from Bis-
    marck, but it may be expected anywhere in eastern part of State. It
    makes quite a dense growth in wet years in spots where lawns had been
    bare. It survived the winter of 1944-45 freely and began to bloom
    about May 1, 1945.

23. *Poa compressa* L. CANADA BLUEGRASS. Bluegreen; stems 2—4 dm. high,
    distinctly flattened; leaves 1—4 mm. wide, rather firm; panicle oblong,
    3—7 cm. long, branches short. Native of Europe, sometimes planted
    and locally abundant in old fields or along roadsides. This has a dark,
    bluegreen color as compared to the bright green of Kentucky Bluegrass
    and flowers about two weeks later.

24. *Poa pratensis* L. KENTUCKY BLUEGRASS. (Fig. 136). Bright green;
    stems 3—6 dm. (1—2 ft.) high; leaves 2—4 mm. (⅛ in.) wide, blunt at
    tip and of equal width throughout the length; panicle pyramidal, 5—10
    cm. (2—4 in.) long, branches widely spreading or drooping; spikelets
    dull brown at maturity. Very common in eastern third of State, local
    in moist places farther west. Matures in mid-June.
        This is the standard lawn grass for northern U. S. where moisture
    is sufficient. It is regarded as introduced from Europe but has spread
    so rapidly that it appears like a native plant. It invades and practically
    takes possession of moist prairie. Coulees in pastures often have a solid
    turf of this grass, while the higher ground has little. It is shallow
    rooted and dries out easily. Often called "Junegrass" (see No. 71).

25. *Poa arida* Vasey. PLAINS BLUEGRASS. Much like Kentucky Bluegrass;
    panicle often compact like that of Canada Bluegrass, or slender and
    elongated. It seems to inhabit especially flat coulee bottoms which are
    distinctly saline. We have it from nearly all parts of State except Red
    River Valley. Many specimens were referred by Bergman to *P. pseu-
    dopratensis* Rydb., which is regarded as the same species.

26. *Poa glaucifolia* Scribn. and Will. SWALLEN BLUEGRASS. Much like *P.
    arida* in appearance and habitat. It seems rather common but was not
    recognized until recently. We have it from Cavalier, McLean, Dunn and
    Billings Counties. We have taken the liberty of naming it for Mr. J. R.
    Swallen of the U. S. Department of Agriculture, who spent two weeks
    in the State in 1939 trying to clear up the identity of some of our blue-
    grass species.

27. *Poa bulbosa* L. BULBOUS BLUEGRASS. Tufted grass, similar to the next
    species but usually forming bulblets instead of seeds; bulblets purplish,
    2 mm. long, the lemma and palea becoming 5—15 mm. Introduced from
    Europe; escaped and apparently well established at the Dickinson
    branch station since 1912.

28. *Poa palustris* L. FOWL BLUEGRASS. FALSE REDTOP. Forming tufts up to 2
    dm. thick; stems 3—6 dm. high; panicle elongated, 1—3 dm. long, fine
    and loose; mature lemma straw color with a coppery band toward tip.
    Very common in sloughs, pot holes, low prairie and other wet ground.

136

137

136. Kentucky Bluegrass (*Poa pratensis*).
137. Western Wheatgrass (*Agropyron smithii*).

Matures late in July, the latest of our bluegrasses. A good hay grass. Formerly called *P. flava* or *P. serotina.*

29. *Poa interior* Rydb. INLAND BLUEGRASS. Resembles No. 28 but has a smaller panicle and often grows on hillsides or other relatively dry prairie; distinguished by a short ligule, less than 1 mm., compared to 2—3 mm. for *P. palustris.* Specimens of this were referred by Bergman to *P. nemoralis* L. which it closely resembles.

30. *Poa cusickii* Vasey. Tufted grass with a narrow, dense panicle. One specimen from Walhalla was at one time referred to this species but it probably belongs to some other species as *P. cusickii* is known chiefly from the Rocky Mountains.

31. *Poa epilis* Scribn. This is another Rocky Mountain species but we have one specimen from Ward (or Mountrail) County which Mr. Swallen regarded as undoubtedly this species.

32. *Poa longiligula* Scribn. and Will. This western species was reported for the State in Hitchcock's Manual, but we have not found it.

33. *Poa stenantha* Trin. A specimen from Williams County was at one time placed here but Mr. Swallen thought it better regarded as *P. canbyi.* This and the three preceding may not occur in North Dakota, but are included here to aid in further study of this difficult group.

34. *Poa secunda* Presl. SANDBERG BLUEGRASS. "WESTERN BLUEGRASS". Forming dense, erect tufts up to 1 dm. thick; stems 2—4 dm. high; panicle, narrow, dense, 2—10 cm. long. One of the common early summer grasses of western North Dakota. We have it from as far east as McHenry and Bottineau Counties. It was previously called *P. sandbergi* Vasey and *P. buckleyana* Nash. Some authors regard *P. secunda* as a distinct species of Chile.

35. *Poa canbyi* (Scribn.) Piper. CANBY'S BLUEGRASS. This is so similar to the preceding that separation of the two species is doubtful. Four specimens from Burleigh, Dunn and Billings Counties were placed here by Mr. Swallen. *P. lucida* Vasey and *P. laevigata* are other names which have been applied to plants of this sort.

36. *Poa ampla* Merr. One specimen from Golden Valley County was identified as this western species by Mr. Swallen.

37. *Eragrostis cilianensis* (All.) Link. STINKGRASS. Widely spreading annual; stems 3—6 dm. (1—2 ft.) long; panicle 5—15 cm. (2—6 in.) long, densely short branched; spikelets 5—15 mm. (⅕—⅗ in.) long, very flat, 10—40-flowered; grains easily removed from lemma, red, rounded, 0.5 mm. long. Common, strong scented, weedy grass. Formerly called *E. major* Host. TEFF (*E. abyssinica*), a food plant of Africa, has a similar grain.

38. *Eragrostis pectinacea* (Michx.) Nees. More slender than the preceding, spikelets narrow, 5—8 mm. long. Usually in sandy soils. We have it from Cass and McHenry Counties only. Formerly called *E. purshii* Schrad.

39. *Eragrostis hypnoides* (Lam.) BSP. CREEPING LOVEGRASS. Annual, but extensively creeping by prostrate stems which root at nodes; flowering stems 5—20 cm. high; panicles 1—5 cm. long, rather dense; spikelets 5—10 mm. long. Wet soil, pond or river banks. We have it from Richland, Ransom, Cass, Stutsman and Pembina Counties.

40. *Catabrosa aquatica* (L.) Beauv. BROOKGRASS. Perennial; stems 1—4 dm. long, mostly floating in water with tips upright; leaves numerous, soft, 5—10 cm. long. Quite local, usually in cold water from springs. Records from Richland, Pembina, Mercer and Mountrail Counties. The panicles much resemble those of redtop.

41. *Redfieldia flexuosa* (Thurb.) Vasey. BLOWOUT GRASS. Perennial from long rhizomes; leaves 3—6 dm. long, 4—5 mm. wide at base, tapering to a slender point, tough and wiry; stems 3—8 dm. high; panicle 1—3 dm. long, the branches long and slender; spikelets 5—7 mm. long. It is the first grass to invade bare sand and is useful to check blowing, though too tough and wiry to be of much value as forage. The leaves come

from near the ground and arch over, their tips often resting on the ground. Known only from sand dune areas in Richland, Ransom and McHenry Counties (fig. 17).

42. *Distichlis stricta* (Torr.) Rydb. SALTGRASS. Perennial from hard, white rhizomes; leaves 5—10 cm. long, 2—3 mm. wide, thick and stiff; seed-bearing stems 5—10 cm. high, sterile ones 10—20 cm.; spikelets 6—10 mm. long, crowded into a dense, oblong panicle 2—6 cm. long. Common on saline soils, especially flats where little else grows. It forms a dense turf, coarser than buffalograss, but is not eaten readily by livestock. Formerly included in *D. spicata* (L.) Greene, which is now considered as limited to the sea coast.

43. *Dactylis glomerata* L. ORCHARD GRASS. Perennial in clumps 2—3 dm. thick; leaves numerous, 2—6 dm. long, 2—8 mm. wide; stems 6—12 dm. high, rather bare; panicles 5—20 cm. long, branches few, stiff, the lower widely spreading; spikelets 8—12 mm. long, crowded in one or more rounded or oblong, 1-sided clusters on each branch. Commonly planted for hay in some regions, rarely here. Occasionally found escaped. Some thrifty plants were found near summer cottages at Lake Metigoshe in the Turtle Mts. in 1942.

44. *Phragmites communis* Trin. COMMON REED. Perennial from thick rhizomes; stems 1—3 m. (3—10 ft.) high, 5—10 mm. (⅕—⅖ in.) thick, leafy; leaves 2—4 dm. (8—16 in.) long, 1—5 cm. (½—2 in.) wide; panicle 1—4 dm. long, half as wide, dense and fluffy at maturity, spikelets slender, 10—15 mm. long, shorter than the basal tuft of hairs. Our largest grass in ponds and wet places, often where ground is constantly wet from seepage. In old lake beds it frequently is cut for hay in dry years (See N. D. Exp. Sta. Bull. 347).

45. *Schizachne purpurascens* (Torr.) Swallen. FALSE MELIC. Tufted perennial; leaves mostly basal, 1—3 dm. long, 2—5 mm. wide; stems 5—10 dm. high, slender, with few, long, drooping branches with 1 or 2 spikelets each; spikelets 2—2.5 cm. long, purplish; lemma 1 cm. long, strongly 7-nerved, with tuft of hairs at base and awn 1 cm. long arising a little below tip. Quite local in moist, aspen woods; Richland, Pembina, Ramsey, Bottineau and Dunn Counties. A striking grass with its tall bare stems and drooping panicles. These appear late in May but mature slowly, ripening in July.

BARLEY TRIBE Hordeae

Usually recognized rather easily by the unbranched, stiff spikes, the spikelets crowded in two rows as in wheat, rye and barley. In barley, 3 spikelets are side by side at each node, but each spikelet has only one flower. In most plants of this group, the spikelet has 2—5 or more flowers.

Key to Species or Genera

Spikelets 1 at each node (2 sometimes in middle of spike).
  Spikelets with edge against stem, 1 basal glume on outer side.
    Perennial; glume shorter than spikelet.   46. *Lolium perenne*
    Annual; glume longer than spikelet.
      Grain plump, 6—8 mm. (¼—⅓ in.) long, 2—3 mm. wide.
          47. *Lolium temulentum*
      Grain flat, slender, 8—10 mm. long, 1—2 mm. wide (fig. 139).
          48. *Lolium rigidum*
  Spikelets with side against stem; 2 nearly equal glumes.
    Perennial; common native grasses.   49-57. *Agropyron*
    Annual; cultivated plant, frequent in fields or along roads.
          58. *Secale cereale*
Spikelets 2—3 at each node, but lateral ones sometimes infertile.
  Spikelets 3 at each node, often only central one fertile (fig. 140).
         59. *Hordeum jubatum*

Spikelets 2 at each node, each 2—6-flowered.
Glumes absent or very small; spike loose, spikelets widely
    spreading.   61. *Hystrix patula*
Glumes as long as lemma; spikelets not widely spreading.
  Spike breaking up when mature.   60. *Sitanion hystrix*
  Spikelets falling away from stem.   62-67. *Elymus*

46. *Lolium perenne* L. PERENNIAL RYEGRASS. Tufted perennial; leaves numerous, 1—2 dm. (4—8 in.) long, 2—4 mm. (⅛ in.) wide; stems 3—6 dm. high; spike 1—3 dm. long, usually curved; spikelets 6—10-flowered; lemma broadly rounded, narrowly oblong, not evidently nerved. Commonly used in lawngrass mixtures for temporary cover and sometimes persisting for a year or two as stray plants. *L. multiflorum* Lam., ITALIAN RYEGRASS, is a little coarser and the lemmas are often awned. The two forms are not readily distinguished.

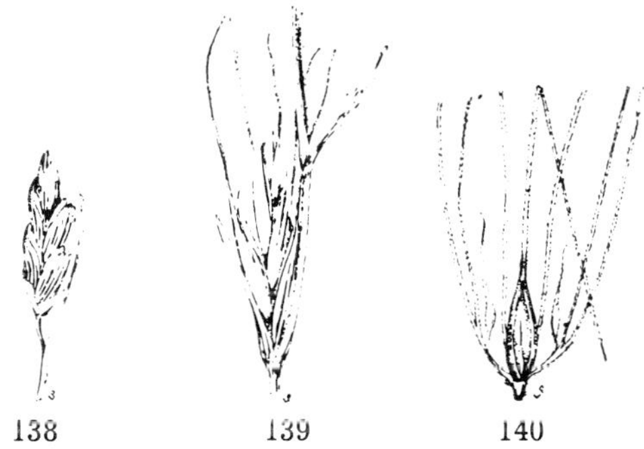

138   139   140

47. *Lolium temulentum* L. DARNEL. Coarse annual, 6—9 dm. high; grain about the size and shape of a small wheat grain, slightly flattened. We have specimens from Walsh, Richland and Stark Counties, but it has not been found for many years. Introduced from Europe and a troublesome weed in some regions.

48. *Lolium rigidum* Gaud. STIFF DARNEL. Annual, more slender than the preceding; stems several, erect, 3—6 dm. high; grains larger than those of *L. perenne,* somewhat more flattened than those of *Agropyron.* This seems to have been introduced with durum wheat in Cavalier County about 1900. We have had reports of its being abundant and troublesome in grain fields and the seeds have occasionally been found in grain from other localities. I found it in Slope County in 1945 and in Mountrail County in 1946. (fig. 139).

Agropyron WHEATGRASS

Perennials, closely related to wheat, differing chiefly in perennial habit and slender grains. Mostly native grasses, the species variable and often hard to identify. Mature specimens are needed because spikelets of flowering specimens shrivel in drying so they do not present characteristic features.

Key to Species

Rhizomes present; forming a sod, but stalks sometimes rather scattered.
  Lemma with a distinct awn 2.5—10 cm. (1—4 in.) long.
    Awn strongly recurved at maturity; lemma hairy.
                      51. *Agropyron albicans*
    Awn straight, not recurved; lemma not hairy. 49. *Agropyron repens*
  Lemma without an awn or with only a short point.
    Glumes rigid, tapering gradually to a slender point.
                      50. *Agropyron smithii*

Glumes not rigid, short pointed or narrowed abruptly.
Leaves thin, soft, sometimes hairy.   49. *Agropyron repens*
Leaves thick, stiff, often rolled, rough above but not hairy.
Lemma covered with fine hairs.   52. *Agropyron dasystachyum*
Lemma not hairy.   53. *Agropyron riparium*
Rhizomes absent; plants tufted.
Spikelets much crowded together in a flattened, twisted spike.
    54. *Agropyron cristatum*

Spikelets not crowded, spike straight, not flattened.
Lemma acute or sharp pointed, not awned.
    55. *Agropyron trachycaulum*

Lemma long awned.
Awn straight, 1—3 cm. long; spike 1-sided.
    56. *Agropyron subsecundum*
Awn curved, 1—2 cm. long; spike not 1-sided.
    57. *Agropyron spicatum*

49. *Agropyron repens* (L.) Beauv. QUACKGRASS. Perennial by vigorous rhizomes; stems leafy, 3—10 dm. (1—3 ft.) high; leaves thin, soft, dark green, 6—10 mm. (⅓ in.) wide; spikes 8—15 cm. (3—6 in.) long; spikelets 10—20 mm. long, glumes nearly as long as spikelet to half as long, "jointed" at base, spikelet readily breaking away at maturity; glumes broad, rough edged and thin at tip or narrowed into a short awn, usually keeled on upper half and strongly 5—7-nerved; lemma boat-shaped, 8—10 mm. long. Introduced, persistent weed, common in eastern half of State, more local westward. Begins to bloom about June 20; seed mature by July 10.

50. *Agropyron smithii* Rydb. WESTERN WHEATGRASS. "GO-BACK GRASS". (Fig. 137). Perennial by rhizomes; leaves usually crowded near ground, thick, stiff, ridged and rough above, 2—4 mm. wide, pale, grayish green; stems 3—10 dm. high with few leaves except at base; spikelets longer and wider than in *A. repens,* glumes not jointed at base, not distinctly nerved.

One of the commonest grasses in a variety of situations. It will grow in saline and hard soils, but more luxuriantly when sod is broken, and if not destroyed by further tillage, occupies the ground almost exclusively. It is an important hay and pasture grass in the drier parts of the State, where few grasses thrive as well.

The specimens referred by Bergman to *A. molle* (S.&S.) Rydb. appear to be *A. dasystachyum* except for one from Grant County (Bell 385) which has rigid glumes, the glumes and lemmas densely long-hairy. In about 50 specimens there is one from Steele County (Hope, Wright 2231a) which has a few short hairs on the lemma.

51. *Agropyron albicans* Scribn. and Smith. Similar to the following except for the awns. Uncommon. Five specimens: lemma strongly hirsute (Renville County); moderately hirsute (Grant, Burke); finely hispidulous (Mercer, McKenzie).

52. *Agropyron dasystachyum* (Hook.) Scribn. and Smith. THICKSPIKE WHEATGRASS. Similar to No. 50, leaves narrower, usually rolled; spikelets densely covered with fine hairs. Occasional in stiff, clay soils as on buttes. Stutsman! (Bergman 41), Mercer, McKenzie, Stark, Billings, Morton and Grant Counties. The lemmas are usually short hirsute. In two specimens from Williams and one from Grant County (Bell 271, 315, 320) the glumes also are hirsute. These sheets have 3 stalks each and in 2 cases the lemmas of 1 spike are nearly glabrous.

53. *Agropyron riparium* Scribn. and Sm. STREAMBANK WHEATGRASS. Similar to No. 50 but rhizomes slender and weak; leaves narrow, often rolled; spikes slender; glumes not rigid, abruptly pointed, thin margined for most of their length. This probably should be considered a variety of *A. dasystachyum.* It is fairly common in western part of State but had been confused with *A. smithii* which it resembles. Where the two species grow together, *A. smithii* shows an obviously more vigorous

growth and wider, flat leaves. The lemmas are nearly smooth or distinctly hispidulous. Specimens from Bottineau, Ward, Bowman, Billings, Morton, and McKenzie Counties.

54. *Agropyron cristatum* (L). Gaertn. CRESTED WHEATGRASS. Tufted, leafy perennial; leaves soft, 1—2 dm. (4—8 in.) long, 5—10 mm. (⅕—⅖ in.) wide; stems slender, numerous, 4—8 dm. high; spikes 5—10 cm. long with 50—100 closely crowded spikelets. Recently introduced as a pasture grass and beginning to appear along roads and other places as an escape. Begins growth very early in spring but stems soon become tough and wiry.

55. *Agropyron trachycaulum* (Link) Malte. SLENDER WHEATGRASS. Erect, tufted perennial; leaves long and soft; spikes slender; spikelets nearly cylindrical, slightly wider at tip; glumes strongly about 7-nerved, nearly covering the spikelet; glumes and lemmas commonly awn-pointed, the awn 1—1.5 mm., sometimes 4—5 mm. long. Common in various locations but not usually making up a large per cent of the hay. It has been grown as a crop plant since 1885, usually under the misleading name "western ryegrass". It has not proved very successful as it does not resist drought. Formerly called *A. tenerum* Vasey and *A. pauciflorum* (Schwein.) Hitchc.

56. *Agropyron subsecundum* (Link) Hitchc. BEARDED WHEATGRASS. Similar to the preceding except for the 1-sided, long-awned spikes which curve or droop. Frequent throughout the State, but less common than *A. trachycaulum*. It is similar to the European *A. caninum* and has also been called *A. richardsonii* Schrad. Dr. M. O. Malte regards it as only a subspecies of *A. trachycaulum*.

57. *Agropyron spicatum* (Pursh) Scribn. and Smith. BLUEBUNCH WHEATGRASS. Leaves fine, stiff, 1—2 dm. long, in large tufts; stems rather leafless, 3—6 dm. high; glumes and lemmas with widely spreading awns. Occasional on dry hillsides, extreme western edge of State. It it well known farther west where it is one of the more productive grasses on dry soils.

58. *Secale cereale* L. RYE. This well known grain is frequently found along roads and railroad tracks but does not persist. *Triticum aestivum* L., WHEAT, often seen in similar places, does not re-seed itself to become established as a wild plant.

59. *Hordeum jubatum* L. WILD BARLEY. "FOXTAIL". Perennial, forming tufts to 2 dm. (8 in.) thick; stems and leaves soft, leaf sheaths soft hairy; stems 3—6 dm. high; spikes 5—10 cm. (2—4 in.) long; spikelets crowded, all three of each node on short stalks, only the central one developed; lemma and all 6 glumes with awns 2.5—6 cm. long; body of fertile lemma 5—7 mm. (¼ in.) long. Var. *caespitosum* (Scribn.) Hitchc. seems common in western part of State; specimens from Mercer, Stark, and Slope Counties. It is more slender with shorter awns 1.5—3 cm. long. A weedy grass, extremely common along roadsides, in waste ground, low prairie and especially sloughs which are occasionally flooded. The seeds usually germinate in late August or September, and if conditions are favorable, the tufts will grow to a diameter of 5—10 cm. the next year. The spikes break up at maturity. The awns are barbed and cause much trouble by penetrating the tissues of animal's mouths. The name "Foxtail" is in general use in North Dakota for this plant but is misleading because it is generally applied to pigeongrass (*Setaria*) in central U. S. and more properly to another grass (*Alopecurus*) in Europe. Common Barley (*H. vulgare*) appears frequently along roadsides or as volunteer plants in fields.

60. *Sitanion hystrix* (Nutt.) J. G. Smith. SQUIRRELTAIL. This is a coarser plant with fewer stems, otherwise much like preceding. It is common farther west and our only records so far are specimens from Medora in 1940, collected by Mrs. Anna Meissner of Mott, N. D. and Bismarck by Stevens in 1946. It can be expected to become frequent in the western part of the State.

61. *Hystrix patula* Moench. BOTTLEBRUSH. Tufted perennial resembling *Elymus canadensis*, but spikes stiff and slender, spikelets widely separated. An eastern woodland species, recorded from Richland, Cass, Barnes, Pembina and Grand Forks Counties.

Elymus   WILD-RYE

Tufted, usually rather coarse perennials with stout spikes; spikelets 2 at each joint; glumes slender, not covering the spikelets; otherwise much like *Agropyron*.

Key to Species

Glumes with an evident thin portion, 5—7-nerved.
  Glumes with well developed awns 1—3 cm. long.
    Awn of lemma 2—5 cm. (1—2 in.) long, widely spreading.
                    62. *Elymus canadensis*
    Awn of lemma 1—2 cm. long, straight.   63. *Elymus virginicus*
  Glumes awnless or nearly so.   *Elymus virginicus*, var. *submuticus*
Glumes thick, narrow, without a thin portion.
  Glumes shorter than lemma.
    Spike slender, erect; awn of lemma 1—4 mm. long.
                    64. *Elymus junceus*
    Spike thick, pendulous; awn 1—3 cm. long.   65. *Elymus interruptus*
  Glumes longer than lemma.
    Spike stiff, erect; leaves stiff, 2—4 mm. wide.   66. *Elymus macounii*
    Spike curved, nodding; leaves lax, 5—10 mm. wide.
                    67. *Elymus villosus*

62. *Elymus canadensis* L. CANADA WILD-RYE. Stems coarse, 6—10 dm. (2—3 ft.) high; leaves thick, rough, 1—2 dm. long, usually 1—2 cm. (½—1 in.) but sometimes only 5—6 mm. (¼ in.) wide; spikes nodding, 1—2 dm. long, 1—3 cm. thick; lemma 2—3 cm. long, flattened, slightly recurved when mature, rough hairy. Common in dry to medium soils. It produces seed abundantly and establishes new plants from seed probably more readily than any of our other grasses. In recent years, it has received considerable attention as a forage grass and selected strains are being grown. *E. robustus* Scribn. and Sm., with stouter, more upright spikes, is considered by some authors as a distinct species, by others as a variety or form of *E. canadensis*. We have been unable to separate it clearly.

63. *Elymus virginicus* L. VIRGINIA WILD-RYE. Not so stout as the preceding; spikes erect or somewhat nodding; glumes hard, thick and curved at base, thin, wide and strongly nerved above. Frequent in woods, apparently throughout the State. Stouter specimens resemble *E. canadensis*, but can be recognized by difference in glumes. One specimen from Wahpeton (Bell 254) has especially stout spikes. The leaves are usually about 10 mm. wide; in one specimen from Fargo they are 15 mm., and in one from Medora, 4—5 mm. Specimens are about equally divided between glabrous and hispidulous lemmas, the latter var. *intermedius* (Vasey) Bush. Var. *submuticus* Hook. Stouter and stiffer, the spikes stiffly upright, more resembling a wheat spike. More common (?) than the typical form, often making quite dense stands in open woods or brushy places. The lemmas are glabrous in all but two of our specimens.

64. *Elymus junceus* Fisch. RUSSIAN WILD-RYE. Leaves numerous, 2—4 dm. long, 3—6 mm. wide, mostly basal, forming clumps 2—3 dm. thick; stems mostly bare, 6—10 dm. high; spikes 1—1.5 dm. long, 6—10 mm. thick, dense; lemma 8 mm. long, its awn 2—6 mm., thinly covered with short hairs. This was introduced from Siberia in 1927 and has been grown experimentally. A few plants were found on the grounds of the Dickinson branch station as early as 1913. We have no further records of its establishment as a wild plant.

65. *Elymus interruptus* Buckl. Similar to *E. canadensis*, taller and coarser, erect, leaves 2—3 cm. wide, spikes hanging down sharply, foliage and stems grayish green. Occasional in woods. We have it from Cass, Barnes, Pembina, Benson and Ramsey Counties. Quite a striking plant, usually 1.5 m. (5 ft.) high. Previously called *E. diversiglumis* Scribn. and Ball. This has recently been pronounced a hybrid between *E. canadensis* and *Hystrix patula* (Rhodora 50:88, 1948).

66. *Elymus macounii* Vasey. MACOUN WILD-RYE. Stems 5—8 dm. high, forming dense, leafy tufts up to 2 dm. wide; leaves mostly erect, 1—2 dm. long, 2—5 mm. wide; spikes 5—15 cm. long, slender; spikelets lying close against stem. Common in low prairie and coulees. It seems not to develop seed and recent work indicates that it is a hybrid between *Agropyron trachycaulum* and *Hordeum jubatum.* The specimen reported as *Elymus glaucus* Buckl. by Bergman appears to be an *Agropyron,* probably an escape or found in planted material.

67. *Elymus villosus* Muhl. SLENDER WILD-RYE. Resembles a slender plant of *E. virginicus,* the spike more slender but densely flowered and awns longer. Rare in woods, eastern part of State. Previously called *E. striatus* Willd.

OAT TRIBE Aveneae

Spikelets usually 2-flowered; lemma awned from below tip. Wild oats might be taken as representing this group, but most species have smaller or much smaller flowers.

Key to Species

Lemma 1—2 cm. (⅖—⅘ in.) long, its awn about the same length.
  Lemma 2 cm. long, much thickened; annual plants.
    Awn on lemma of both flowers stout, twisted, sharply bent.
                       68. *Avena fatua*
    Awn only on lemma of lower flower or on neither.   69. *Avena sativa*
  Lemma 1 cm. long, firm but not thick; prairie perennial.
                       70. *Avena hookeri*
Lemma not over 5 mm. (⅕ in.) long, awn small or absent.
  Lemma compressed, keeled, awn absent or very short from near tip.
                     71. *Koeleria cristata*
  Lemma rounded, not keeled.
    Lemma usually not awned but acute; second glume wider than first.
      Panicle dense; second glume broadly rounded at tip.
               72. *Sphenopholis obtusata*
      Panicle loose; second glume acute or nearly so.
           73. *Sphenopholis intermedia*
    Lemma thin, wide and jagged at tip, with an awn from below middle.   74. *Deschampsia caespitosa*

68. *Avena fatua* L. WILD OAT. Annual; leaves coarse; stems 3—8 dm. high; panicle 1—3 dm. (4—12 in.) long, widely branched, spikelets drooping; glumes thin, strongly nerved, covering the flowers; spikelets 2-flowered, lemma of each flower awned; lemmas dark brown, slate colored or yellow, hairy or nearly smooth, base much enlarged (popularly referred to as a "sucker mouth") and with a tuft of stiff hairs. A very common introduced weed. Grains dropping as soon as mature.

69. *Avena sativa* L. COMMON OAT. Differs from wild oat in that base of lemma is not enlarged nor hairy. The grains are not shed at maturity. The lemma of the lower flower is strongly awned in some varieties, more often with a weak awn or none, the upper flower never awned. Frequent about elevators, roadsides, etc.

70. *Avena hookeri* Scribn. SPIKE OAT. Perennial, erect, tufted, narrow leaved, native grass; stems 2—4 dm. high; panicle narrow, 5—10 cm.

long, branches and spikelets erect. Frequent on prairie, sometimes locally common. Eastern third of State west to Mountrail County. June.

71. *Koeleria cristata* (L.) Pers. PRAIRIE JUNEGRASS. Tufted, erect perennial; lower sheaths usually soft hairy, blades rather gray, 1—3 mm. wide; stems 3—6 dm. (1—2 ft.) high; panicle narrow, dense, spike-like, 5—12 cm. (2—7 in.) long; lemma 3—4 mm. (⅛ in.) long, sometimes short awned. Common on prairie; an important early grass. June. Also widely distributed in Europe. It suffered severely in dry years.

72. *Sphenopholis obtusata* (Michx.) Scribn. PRAIRIE WEDGEGRASS. Similar to *Koeleria*, the tufts less dense and taller, leaves green, flat, panicles heavier; spikelets 2—2.5 mm. long, crowded, widely spreading. Frequent, sometimes abundant on low prairie. In the Big Coulee at Esmond, Pierce County, I noted it as almost the chief grass in some areas. Flowering later than *Koeleria*.

73. *Sphenopholis intermedia* Rydb. SLENDER WEDGEGRASS. Taller and more slender than the preceding, panicle branches loose and spreading; spikelets 3—4 mm. long, erect. Specimens from Richland, Dickey, Walsh, Pembina and Williams Counties. It seems to grow in open places in wooded areas and probably is more common than shown by specimens for it might be mistaken for a *Poa* or *Calamagrostis*. Difficult to distinguish from the preceding. Bergman 708 from Oakes, has slender, spike-like panicles but the spikelets are slender and erect instead of crowded and spreading as in *obtusata*. Bergman 610 from Mandan is an exceptionally robust plant. Bergman referred this species to *S. pennsylvanica* (DC.) Gray, which is *S. nitida* (Spreng.) Scribn., reported by Hitchcock (Man. Grasses of U. S.) for North Dakota but probably by error.

74. *Deschampsia caespitosa* (L.) Beauv. TUFTED HAIRGRASS. Tufts 1—2 dm. wide with many, fine, soft leaves; stems bare, 3—6 dm. high; panicle loose, 1—2 dm. long; spikelets purplish; lemma hairy at base, awn somewhat longer to twice as long. First detected near Rugby, Pierce County, in 1920; later collected in Walsh, Kidder, Benson, Nelson and Williams Counties. Local in wet, hummocky sloughs. The thick tufts, with leaves somewhat like those of Kentucky Bluegrass, are striking. June and early July.

TIMOTHY TRIBE Agrostideae

Panicles varying from spike-like to widely branched; spikelets 1-flowered, often quite small with thin lemmas. A large and difficult group.

Key to Species or Genera

Lemma awnless or awned on back.
  Flower cluster a spike; glumes flattened and fringed with hairs.
    Glumes awned; spike rough; spikelets not falling readily.
                    75. *Phleum pratense*
    Glumes not awned; spike soft; spikelets falling quickly.
      Spikelets 5—6 mm. (¼ in.) long; coarse grass 3—8 dm.
        (1—2½ ft.) high.   76. *Alopecurus ventricosus*
      Spikelets 2—3 mm. long; slender, soft grasses 2—6 dm. high.
        Perennial; awn hardly longer than lemma.
             77. *Alopecurus aequalis*
        Annual; awn twice as long as lemma. 78. *Alopecurus carolinianus*
  Flower cluster a dense or loose panicle; glumes not flattened nor
    fringed.
    Lemma with a basal tuft of hairs.
      Lemma awned on back.   79-82. *Calamagrostis*
      Lemma not awned.   83. *Calamovilfa longifolia*
    Lemma without basal hairs.

Grain readily separating from lemma at maturity.
101-104. *Sporobolus*
Grain not readily separating from lemma.
Lemma with a short awn from just below tip; stamen 1.
84. *Cinna latifolia*

Lemma awnless; stamens 3.
Stems and leaves very fine; panicle loose, flexuous.
85. *Agrostis scabra*
Stems and leaves medium; panicle branches short and stiff.
86. *Agrostis alba*

Lemma awned from tip or at least sharp pointed; grain closely
enclosed by lemma.
Lemma with 3 awns 5—8 cm. (2—3 in.) long; short, harsh, tufted
grass. 87. *Aristida longiseta*
Lemma with 1 awn or only sharp pointed.
Awn stout, twisted, 2—10 cm. long, persistent.
Glumes 6—8 mm. (¼ in.) long; awn 2—3 cm. long, slender.
88. *Stipa viridula*

Glumes 2—3.5 cm. (1—1½ in.) long; awn 10—15 cm. long, stout.
Glumes 1.5—2 cm. long; lemma 8—12 mm. 89. *Stipa comata*
Glumes 2.5—3.5 cm. long; lemma 15—25 mm. 90. *Stipa spartea*
Awn slender, 1 cm. or less long, not persistent.
Lemma thick and hard, nerves not evident. 91-94. *Oryzopsis*
Lemma usually thin, nerves prominent. 95-100. *Muhlenbergia*

75. *Phleum pratense* L. TIMOTHY. Tufted, rather coarse perennial, stem
bases somewhat bulbous; stems 4—10 dm. (16—40 in.) high, stiff; spikes
cylindrical, 5—10 cm. (2—4 in.) long, 6—8 mm. (¼ in.) wide, spikelets
closely crowded; lemma thin, white; grain yellowish, rounded, 1 mm.
long, easily removed from lemma. Formerly planted in eastern part
of State, rarely in recent years; frequently seen in grassland or waste
ground.

76. *Alopecurus ventricosus* Pers. MEADOW FOXTAIL. This plant looks like
timothy but has rhizomes. The glumes have soft hairs instead of the
fringe of rather stiff hairs of timothy, and the lemma has an awn on
its back. Specimens were received in 1935 from Gust Steinhaus of
Max, McLean County, and were identified by J. R. Swallen, who com-
mented that it was the first record for United States. It has since been
recorded from Newfoundland by Ivan J. Green (Rhodora 46:387).
This form is often considered a subspecies of *A. pratensis* L., the common Meadow Foxtail often planted in Europe. We have a single specimen of that collected at Fargo in 1903. It was probably only an escape,
though I recall a small field in the outskirts of Fargo in 1910. This
specimen has the spikelets ascending instead of strongly spreading, as
in timothy and the *ventricosus* specimen. In a previous report (N. D.
Agr. Exp. Sta. Bull. 300:103) the identification of the Steinhaus specimen was erroneously attributed to Dr. A. S. Hitchcock and the authority for the name incorrectly given as "(Gouan) Huds.", which happens
to belong to a similar but different grass (*Gastridium ventricosus*) occurring in the Pacific Coast region of the United States as a weed from
Europe.

77. *Alopecurus aequalis* Sobol. SHORTAWN FOXTAIL. Tufted perennial;
stems 2—4 dm. high; leaves soft, 1—2 dm. long, 2—4 mm. wide; spikes
2—7 cm. long, 4 mm. wide. Frequent in wet places, especially edges of
ponds where ground is covered with water at times. The mature spikelets fall away first at the top of the spike, leaving the stalk bare.

78 *Alopecurus carolinianus* Walt. Similar to preceding, but quite distinct
by thicker spikes with rather prominent awns. A specimen from Wat-
ford City, McKenzie County, collected by Victor Lundeen in 1938, was
identified by J. R. Swallen. We have no other records.

Calamagrostis REEDGRASS

Mostly marsh grasses with rather large panicles of small spikelets. *C. montanensis* is an upland species resembling *Koeleria*. Fine tuft of hairs at base of lemma about as long as lemma.

Key to Species

Panicle open, lower branches 5—10 cm. long, spreading or drooping.
79. *Calamagrostis canadensis*
Panicle dense, lower branches 2—5 cm. long, erect or nearly so.
  Stems 4—10 dm. high.
    Leaves stiff, rough; ligule 4—6 mm. long, panicle stiff.
80. *Calamagrostis inexpansa*
    Leaves soft; ligule 1—3 mm. long; panicle lax.
81. *Calamagrostis neglecta*
  Stems 1—4 dm. high; panicle narrow, spike-like.
82. *Calamagrostis montanensis*

79. *Calamagrostis canadensis* (Michx.) Beauv. BLUEJOINT. Perennial from short rhizomes, forming large, dense, leafy clumps; stems 6—12 dm. (2—4 ft.) high; leaves 3—6 dm. long, 4—8 mm. (¼ in.) wide, curved and drooping; panicle 1—2.5 dm. long. Infrequent in wet ground, usually in or near woods. Specimens from Richland, Cass, LaMoure, Pembina and Bottineau Counties.

80. *Calamagrostis inexpansa* A. Gray. NORTHERN REEDGRASS. Perennial by long rhizomes, forming a dense sod; stems 4—10 dm. high; leaves stiff, 1—3 dm. long, rough above, grayish green; panicle oblong, dense. 5—15 cm. long, 2—4 cm. wide. Common in low ground, often forming a dense stand just above the *Fluminea* area in pot holes. Bergman had referred many of our specimens to *C. hyperborea* and its varieties but Dr. A. S. Hitchcock reviewed them and placed them all in *C. inexpansa*. The leaves are much like those of western wheatgrass, but longer.

81. *Calamagrostis neglecta* (Ehrh.) Gaertn. Much more slender and soft than the preceding. This had been overlooked until recently. We have it from Steele, Benson and McHenry Counties. Quite local in wet, hummocky meadows with *Deschampsia caespitosa*, forming dense tufts 1—2 dm. wide.

82. *Calamagrostis montanensis* Scribn. PLAINS REEDGRASS. Perennial by rhizomes; stems often single or a few together, sometimes crowded; leaves stiff, rough above, grayish green, 1—2 dm. long, 2—4 mm. wide; panicle dense, 5—10 cm. long, 1—2 cm. wide. Common on fairly dry prairie, especially clay slopes. More common westward, east to Sargent, Ransom and Pembina Counties. The panicle is much like that of *Koeleria* but the stems are solitary or few, leaves stiff and rough.

83. *Calamovilfa longifolia* (Hook.) Scribn. BIG SANDGRASS. Coarse, hard leaved perennial from long, stout rhizomes; stems 1—2 m. (3—7 ft.) high; leaves 3—6 dm. (1—2 ft.) long, 4—8 mm. (¼ in.) wide; tapering to a long, slender tip; panicle 1—3 dm. long, branches erect or spreading; spikelets 6—7 mm. long. Common on prairie and roadsides, especially in dry, sandy soils. It is too wiry to be a good forage grass but is useful to check soil blowing. Hitchcock listed *C. gigantea* for North Dakota, but this was found to be an error.

84. *Cinna latifolia* (Trev.) Griseb. DROOPING WOODREED. Tufted, soft perennial; stems 8—15 dm. high; leaves 1—2 dm. long, 1 cm. wide, lax; panicle 1—2 dm. long, the slender branches ascending or drooping; spikelets numerous, 5 mm. long. Very local and rare in wet woods; Pembina, Bottineau and Rolette Counties.

85. *Agrostis scabra* Willd. TICKLEGRASS. Tufted, very slender perennial; leaves 5—15 cm. long, not over 1 mm. wide; stems 3—6 dm. high, half of which is panicle; branches of panicle long and slender with a few spikelets at ends; spikelets 1.5—2 mm. long. Grains smaller than those of redtop; lemma thin and delicate, palea absent. Frequent on prairie, neglected fields or waste ground. Of no importance. Formerly referred

141. Feather Bunchgrass (*Stipa viridula*).
142. Needle-and-thread. (*Stipa comata*).

to *A. hiemalis* (Walt.) BSP., which is now considered distinct and restricted chiefly to southeastern U. S.

86. *Agrostis alba* L. REDTOP. Perennial by short rhizomes; stems 6—10 dm. high; leaves soft, dark green, 1—2 dm. long, 5—10 mm. wide; panicle 5—15 dm. long, branches stiff, 2—4 cm. long, tending to be in whorls; spikelets 2 mm. long; lemma and palea thin, delicate. Frequent in wet meadows, roadside ditches, etc. Planted for pasture and often escaped. A good grass for wet ground.

87. *Aristida longiseta* Steud. RED THREEAWN. Stiff, hard perennial in tight tufts 5—15 cm. wide; stems 2—3 dm. high; leaves 5—15 cm. long, rolled, not over 1 mm. wide; panicle 1—2 dm. long, the branches slender; lemma 10—15 mm. long, slender, rounded, 1 mm. wide, awns 5—8 cm. long. Frequent on dry hillsides west of Missouri River; collected at Lisbon, Ransom County, in 1947 (Stevens 1033). A hard, wiry grass of little value.

    Aristida basiramea Engelm. was reported for North Dakota in Hitchcock's Manual. I was informed that the specimen in the National Museum was labeled "Blue River" and that someone had changed it to read "Red River". It probably referred to the Blue River in Nebraska.

88. *Stipa viridula* Trin. FEATHER BUNCHGRASS. GREEN NEEDLEGRASS (Fig. 141). Perennial in tufts 1—3 dm. (4—12 in.) wide; leaves 1—3 dm. long, 1—3 mm. (1/25—⅛ in.) wide, tapering to a slender tip; stems 3—15 dm. high; panicle 1—2 dm. long, 2—3 cm. (⅘—1⅕ in.) wide; lemma elliptical, 5—6 mm. long, hairy, pointed at base but not sharp; awn 2—3 cm. long, slender, bent in 2 places, easily breaking off. A common and valuable grass.

89. *Stipa comata* Trin. & Rupr. NEEDLE-AND-THREAD. (Fig. 142). Smaller than the preceding; stems 3—6 dm. high, panicle 1—3 dm. long, the slender branches widely spreading or drooping; lemma 8—12 mm. long, cylindrical, very sharply pointed at base and with a tuft of stiff hairs; awn strongly twisted, 1—1.5 dm. long. A common grass westward; frequent on dry ridges or roadsides in eastern N. D. The sharp grains mature in early July. They penetrate tissues of mouths of animals and cause injury. The grass can best be cut or pastured before grains mature or after they drop. They are commonly called "hay needles".

90. *Stipa spartea* Trin. PORCUPINE GRASS. Similar to the last but taller and stouter; stems 5—10 dm. high; panicle branches not widely spreading but usually arching, the heavy grains hanging in a close group. All over the State but more common eastward.

Oryzopsis RICEGRASS

These resemble species of *Stipa* in the plump, hard grains, but awns of lemma are not so stout and they drop off readily. The grains of No. 91 were used by Indians for food.

Key to Species

Lemma covered with dense, silky hairs.   91. *Oryzopsis hymenoides*
Lemma smooth or with very short, stiff hairs.
    Spikelets 2—2.5 mm. long; leaves 1—2 mm. wide.
        92. *Oryzopsis micrantha*
    Spikelets 6—9 mm. long; leaves 3—10 mm. wide.
      Lower stem leaves 1—3 dm. long, upper blades not over 1 cm.
        93. *Oryzopsis asperifolia*
      Lower stem leaves early drying up; upper blades 1—3 dm. long.
        94. *Oryzopsis racemosa*

91. *Oryzopsis hymenoides* (R. & S.) Ricker. INDIAN RICEGRASS. INDIAN MILLET. Wiry perennial in tufts 1—2 dm. (4—8 in.) wide; leaves 2—6 dm. long, 2—4 mm. (⅛ in.) wide, usually rolled and tapering to a slender tip; panicle 1—3 dm. long, loose, the slender branches widely spreading, the final pair at nearly 180° from each other; glumes thin, ovate, 6—7 mm. long; lemma nearly spherical but pointed, 3 mm. wide,

black, hard, densely covered with soft, white hairs which rub off easily; awn 4 mm. long, falling so quickly as to be easily missed. Frequent west of Missouri River. On Bad Land clay slopes or hillsides it may be hardly 3 dm. high, on sandbars 6—8 dm. Formerly called *Eriocoma cuspidata* Nutt.

92. *Oryzopsis micrantha* (Trin. & Rupr.) Thurb. LITTLE RICEGRASS. Slender perennial in dense tufts 1—2 dm. wide; leaves 2—5 dm. long, 1—2 mm. wide; stems 3—6 dm. high; panicle 5—20 cm. long, branches slender and widely separated; lemma smooth, yellowish green, 2—2.5 mm. long, awn slender, 5—10 mm. long. Missouri River and westward. This was reported by Bergman only from Morton (=Grant) County. One specimen from Barnes County (Bergman 340), mis-labeled *Sporobolus heterolepis,* is our only record east of Missouri River. It forms dense mats on wooded coulee banks, chiefly under cedars or bushes.

93. *Oryzopsis asperifolia* Michx. ROUGH-LEAVED RICEGRASS. Tufted perennial; lower leaves 1—2 dm. long, 3—8 mm. wide; stems 2—7 dm. long, widely spreading and bare of leaves; panicle slender, 5—8 cm. long; lemma 6—8 mm. long, yellowish at maturity, with a basal tuft of hairs and an awn 5—10 cm. long. Local in aspen woods; Ramsey, Rolette and Dunn Counties. Early May.

94. *Oryzopsis racemosa* (J. E. Smith) Ricker. BLACKSEED RICEGRASS. Similar to the preceding, but stems leafy and panicle branches spreading; lemma nearly black at maturity, awn 15—25 mm. long. Woods; Cass, Ransom and Barnes Counties. July.

Oryzopsis bloomeri (Boland.) Ricker, was reported in Bulletin 300 from the Bad Lands, but such plants have been demonstrated by Weber to be hybrids. *O. bloomeri* occurs from Washington and California to Montana and Colorado.

Muhlenbergia MUHLY

Wiry perennials of small to medium size; spikelets small, in dense or loose panicles; lemma membranous, closely adherent to grain. The species are of small value and are difficult to identify though our chief forms can be recognized fairly easily.

Key to Species

Panicle branches widely spreading. 95. *Muhlenbergia asperifolia*
Panicle branches short or erect.
  Rhizomes not present, stems densely tufted; usually on hillsides.
        96. *Muhlenbergia cuspidata*
  Rhizomes present, plants forming a sod; mostly on low ground.
    Leaves 1—2 mm. wide, often rolled; panicles 2—3 mm. wide.
        97. *Muhlenbergia richardsonis*
    Leaves often 3—6 mm. (⅛—¼ in.) wide, not rolled; panicles
      5—15 mm. wide.
      Glumes longer than lemma with stiff tips; panicles dense,
        terminal. 98. *Muhlenbergia racemosa*
      Glumes shorter than lemma with soft tips; lateral panicles
        common.
        Stems smooth below nodes; panicle branches somewhat
          spreading. 99. *Muhlenbergia mexicana*
        Stems hairy below nodes; panicle branches erect or nearly.
          100. *Muhlenbergia foliosa*

95. *Muhlenbergia asperifolia* (Nees & Mey.) Parodi. SCRATCHGRASS. Perennial by rhizomes, forming patches; stems erect or spreading and only tips or panicles erect; leaves small, 2—5 cm. (1—2 in.) long, 1—2 mm. (1/25—$\frac{1}{12}$ in.) wide; panicle 5—15 cm. long, with slender branches; spikelets 1.5—2 mm. long. Frequent on saline flats. We have it from eastern third of State and Bad Lands but it probably occurs all over. The panicles break away easily at maturity, thus it is one of the "tickle" grasses. This species has often been placed in *Sporobolus.*

96. *Muhlenbergia cuspidata* (Torr.) Rydb. PLAINS MUHLY. Dense, wiry tufts 1—2 dm. wide; stems 2—4 dm. high with knotted bases; leaves 2—5 cm. long, 1—2 mm. wide; panicle 5—10 cm. long, slender; spikelets 3 mm. long; glumes about equal in length. Common on dry hillsides. The grain is often parasitized by an insect which causes it to develop into a hard, yellow, rounded body 1 mm. thick.

97. *Muhlenbergia richardsonis* (Trin.) Rydb. MAT MUHLY. Perennial by rhizomes, forming a dense mat of fine stems which are erect or spreading at base, 1—2 dm. high; panicle very narrow, 5—10 cm. long. Common in moist or fairly dry places.

98. *Muhlenbergia racemosa* (Michx.) BSP. MARSH MUHLY. WILD TIMOTHY. Perennial from short, hard, scaly rhizomes; stems leafy, branched, 3—6 dm. high; leaves 5—10 cm. long, 2—5 mm. wide; panicle 3—10 cm. long, 1 cm. wide, somewhat separated into rounded sections. Common on low prairie, marshy places or in woods. This is the best of the group for hay but it rarely occurs in quantity. It is quite variable. Plants in wet ground often resemble *M. richardsonis* in having very slender panicles.

99. *Muhlenbergia mexicana* (L.) Trin. WIRESTEM MUHLY. Similar to preceding but with long branches from many nodes; panicle branches slender, loose. It grows in woods and we have it only from Ransom and Cass Counties.

100. *Muhlenbergia foliosa* (Roem. & Schult.) Trin. Resembles slender forms of the two preceding. Three specimens from Benson County (Lunell in 1906, '14, '15) seem to be this species.

Sporobolus   DROPSEED

Medium sized to small grasses, somewhat like *Muhlenbergia* but the grain readily separates from the lemma (as in wheat).

Key to Species

Slender, often prostrate annual; panicles slender, often enclosed by
  leaf sheath. 101. *Sporobolus neglectus*
Upright perennials; panicles spreading or enclosed.
  Spikelets 2—2.5 mm. ($\frac{1}{10}$ in.) long; panicle 1—2 dm. long, widely
    branched, often enclosed by leaf sheath.
        102. *Sporobolus cryptandrus*
  Spikelets 3—6 mm. long; panicle spreading or enclosed.
    Panicle slender, mostly enclosed by leaf sheath. 103. *Sporobolus asper*
    Panicle open, spreading, not enclosed.   104. *Sporobolus heterolepis*

101. *Sporobolus neglectus* Nash. SMALL DROPSEED. Small annual, stems 1—4 dm. (4—16 in.) long, usually spreading; upper leaves 2—5 cm. (1—2 in.) long, 1—2 mm. wide; panicles numerous, 2—5 cm. long, slender, partly or wholly enclosed by leaf sheath; lemma thin, white; grain 1.5 mm. ($\frac{1}{16}$ in.) long, nearly as wide, translucent with pinkish flecks and a prominent black spot at base. Dry soil, common at least in southeast. Often forming a thick, temporary turf in bare places in lawns or developing long stems if not crowded.

102. *Sporobolus cryptandrus* (Torr.) Gray. SAND DROPSEED. Perennial, tufts somewhat spreading at base; leaves 1—3 dm. long, 2—5 mm. wide, a conspicuous tuft of white hairs at base of blade; stems 3—10 dm. high; panicle 1—2 dm. long, sometimes mostly covered by leaf sheath, sometimes with branches spreading; lemma thin, delicate; grain pink, rounded, 1.25 mm. wide. Common in sandy soils. Not a very palatable species; produces seed freely.

103. *Sporobolus asper* (Michx.) Kunth. ROUGH DROPSEED. Tufts up to 1 dm. thick; stems stiff, erect, 5—10 dm. high; leaves as in last but without hair tuft at base; panicles slender, 1—2 dm. long, mostly enclosed by leaf sheath; grain as in *neglectus* but about twice as long as wide. This was not recognized until 1920, apparently because it matures very

late and panicles remain enclosed. It is common locally on lower slopes or sandy soils, but is of little value. Specimens from Richland to Ramsey and Mercer Counties.

104. *Sporobolus heterolepis* A. Gray. PRAIRIE DROPSEED. Very dense tufts 1—2 dm. wide; leaves numerous, soft, 1—3 dm. long, 2 mm. wide; stems few, 3—6 dm. high; panicle 5—15 cm. long, 3—8 cm. wide, dark colored; grains less conspicuous than in other species. Common, especially in slight depressions of the prairie. A good forage grass with a striking, pungent odor when in bloom.

GRAMA TRIBE Chlorideae

Spikelets several flowered, arranged in several, slender, 1-sided spikes, usually closely crowded, but widely separated in No. 112. The terminology of spike and spikelet is complicated in this group.

Key to Species

Flowers dioecious; low plants with runners (fig. 143).
       105. *Buchloe dactyloides*
Flowers perfect; plants without runners.
  Stems prostrate; flower clusters dense, head-like, spiny.
       106. *Munroa squarrosa*
  Stems erect, at least at first; flower clusters slender, not spiny.
    Panicle becoming a long, very loose spiral, 2—3 dm. (8—12 in.)
      long.   107. *Schedonnardus paniculatus*
    Panicle not spiral.
      Spikes numerous, panicle branches spreading, 1—2 dm. long.
        Coarse perennial, 1—2 m. (3—7 ft.) high; panicle branches
          stout, stiff (fig. 146).   110. *Spartina pectinata*
        Small to medium annuals, 2—6 dm. high; panicle branches
          slender.
          Spikelets rounded, crowded and overlapping; branches
            thick and erect.   108. *Beckmannia syzigachne*
          Spikelets narrow, not overlapping; branches slender,
            spreading.   109. *Leptochloa fascicularis*
      Spikes few or many on a single, slender stem.
        Spikes many, loosely hanging from a curved stalk 2—3 dm.
          long (fig. 144).   112. *Bouteloua curtipendula*
        Spikes densely crowded on 1—5 stiff, 1-sided branches.
          Coarse grass 4—10 dm. high from rhizomes; leaves 3—10
           dm. long.   111. *Spartina gracilis*
         Low, tufted grasses; leaves 2—5 cm. (1—2 in.) long (fig. 145).
          Tip of branch bearing flowers projecting as a sharp point.
            114. *Bouteloua hirsuta*
          Tip of branch bearing flowers not projecting.
            113. *Bouteloua gracilis*

105. *Buchloe dactyloides* (Nutt.) Engelm. BUFFALO GRASS. (Fig. 143). Tufted but spreading by prostrate stems (stolons or runners) 3—5 dm. (12—20 in.) long, which root at joints to form solid patches; leaves gray green, 2—10 cm. (1—4 in.) long, 1—2 mm. (1/25—$\frac{1}{12}$ in.) wide; pistillate flowers in head-like clusters on stems 1—3 cm. high; staminate spikes 10—15 mm. long, 2 or 3 on a stem 5—20 cm. high. Quite common in southwest, especially on flats in coulees; uncommon east of Missouri River, rare or absent in northern part of State. We have records from Barnes, LaMoure, Emmons and McLean Counties. A colony was found in native prairie at Fargo in 1920 but was soon destroyed by building operations. A valuable grass but it probably never was as common as often supposed because blue grama has been mistaken for it. It is being used for lawns to some extent, but the gray color makes it rather unattractive. Blooms in June and since the taller stems bear only staminate flowers, these soon fall and no flowers are evident in late summer.

106. *Munroa squarrosa* (Nutt.) Torr. FALSE BUFFALO GRASS. Weedy annual; stems prostrate, wiry, forming circular mats 2—5 dm. wide; flowers in spiny head-like clusters. Rare in dry, sandy, bare soil. Records from Emmons and McLean Counties only (fig. 20).

143. Buffalograss (*Buchloe dactyloides*).

107. *Schedonnardus paniculatus* (Nutt.) Trel. TUMBLEGRASS. Small, densely tufted perennial, leaves 2—5 cm. long, 1 mm. wide; flower stalks finally becoming 2—3 dm. long in a loose spiral, the slender branches far apart; slender spikelets 4 mm. long, lying close to stem. Frequent on dry prairie; a weedy grass, the flower stalks breaking away at maturity.

108. *Beckmannia syzigachne* (Fern.) Steud. SLOUGHGRASS. Stout, leafy annual, 3—10 dm. high; leaves rather thick, 1—3 dm. long, 5—10 mm. wide; panicle dense, narrow, 1—2 dm. long; spikelets nearly circular, 3 mm. long, overlapping in 2 rows on the short branches. Common in low ground; palatable. Formerly included in the European *B. erucaeformis* (L.) Host.

109. *Leptochloa fascicularis* (Lam.) Gray. Tufted, leafy annual; leaves 1—2 dm. long, 3—5 mm. wide; stems 2—8 dm. high; panicle 1—2 dm. long, with many slender branches 2—10 cm. long. Edges of saline ponds or on flats. We have it from Sargent, McIntosh, Burleigh, Morton and Stark Counties. Also called *Diplachne fascicularis* (Lam.) Beauv.

110. *Spartina pectinata* Link. PRAIRIE CORDGRASS. (Fig. 146). Coarse perennial from stout rhizomes; leaves 3—10 dm. (1—3 ft.) long, 5—15 mm. (⅕—⅗ in.) wide, tapering to a slender tip; stems 1—2 m. (3—7 ft.) high; panicle 1—3 dm. long, the 5—15 stiff branches widely spreading at maturity; spikelets 10—15 mm. (½ in.) long, closely crowded together, comb-like. Very common in low ground. Too coarse and harsh for good hay but useful to hold dams and stream banks from washing. Formerly called *S. michauxiana* Hitchc., and still earlier included in *S. cynosuroides* (L.) Roth, which is now considered re-

stricted to the Atlantic Coast. The name "Prairie Cordgrass" merely means it is the cordgrass of the prairie region.

111. *Spartina gracilis* Trin. ALKALI CORDGRASS. Similar to last but smaller, the stems often scattered, not forming a dense stand; flowering branches 2—5, erect, lying close to main stalk. Frequent, especially on saline flats where there is a fair cover of mixed grasses.

144. Side-oats Grama (*Bouteloua curtipendula*).

145. Blue Grama (*Bouteloua gracilis*). 146. Cordgrass (*Spartina pectinata*).

112. *Bouteloua curtipendula* (Michx.) Torr. SIDE-OATS GRAMA. (Fig. 144). Perennial with short rhizomes, forming tufts 1—3 dm. wide; leaves 5—15 cm. long, 2—4 mm. wide; stems erect or curved, 2—5 dm. long, bearing 25—50 hanging spikes 1—2 cm. long; each of these spikes with 5—8 spikelets. Frequent to common, especially on lower hill slopes. A valuable grass, striking in appearance, but not resistant to drought nor close grazing.

113. *Bouteloua gracilis* (HBK.) Lag. BLUE GRAMA. (Fig. 145). Perennial in dense tufts 5—25 cm. wide; leaves 3—10 cm. long, 1—2 mm. wide; stems 2—5 dm. high with 2 or 3 spikes 2—5 cm. long toward top; spikes comb-like, smooth along the back, usually purplish. One of the most common prairie grasses, also in openly wooded areas. Our most valuable native pasture grass for drier soils. It has been commonly called "buffalo grass" which it resembles, but buffalo grass (No. 105) can easily be distinguished by its patchy growth and stolons. Grama blooms in late July, the flowers are perfect and remain on the stems the rest of the summer.

114. *Bouteloua hirsuta* Lag. HAIRY GRAMA. Similar to last; spike puffier, a row of hairs along the back. Rather rare and local, southern part of State, especially in sandy soils. We have specimens from the sand-hills in Richland County and from Morton County. Mr. J. T. Sarvis stated that it occurs in the southwestern part of the State.

CANARYGRASS TRIBE Phalarideae

Spikelets with one perfect flower and remains of 1 or 2 others: grains flattened, lemmas hard and smooth.

Key to Species

Panicle as wide as long, its branches drooping; flowering in May.
115. *Hierochloe odorata*
Panicle dense, stiff, head-like or spike-like; flowering June, July.
  Panicle head-like, nearly as wide as long; annual grass.
116. *Phalaris canariensis*
  Panicle several times as long as wide; perennial grass.
117. *Phalaris arundinacea*

115. *Hierochloe odorata* (L.) Beauv. SWEETGRASS. Perennial by slender, yellow rhizomes; leaves soft, 2—5 mm. wide, lower ones 1—3 dm. long, upper 3—5 cm.; stems 2—5 dm. high; panicle 5—10 cm. long, nearly as wide; spikelets 5 mm. long, nearly as wide, 2 lower flowers staminate, 1 upper perfect. Frequent in wet grassland. Conspicuous in late May when in bloom. The Indians used the pliable, fragrant leaves in baskets, mats and other weaving.

116. *Phalaris canariensis* L. CANARY GRASS. Annual; stems stiff, 3—6 dm. high; leaves rather coarse; panicle dense, ovate or oblong, 2—4 cm. long; glumes with white edges. This produces the seeds used in bird feed. It has been found a few times, probably only where seed happened to be scattered.

117. *Phalaris arundinacea* L. REED-CANARY GRASS. Coarse perennial with stout rhizomes; leaves stiff, 1—2 dm. (4—8 in.) long, 5—15 mm. (⅕—⅗ in.) wide; stems 1—2 m. (3—7 ft.) high; panicle 7—16 cm. (3—7 in.) long, branches spreading to 2—4 cm. wide in flower, erect and compact at maturity; spikelets 4—6 mm. long, narrow, pointed; lemma dark colored, smooth and hard. Frequent in wet ground, sometimes in large clumps in shallow water. This had considerable publicity in recent years as a forage grass. It produces heavily but requires a good supply of water and seeds shatter readily. Var. *picta* L., Ribbon Grass, is frequently grown as an ornamental on account of the white-edged leaves.

RICE TRIBE Oryzeae

Spikelets 1-flowered, strongly flattened laterally (right angle to front-to-back). Rice (*Oryza sativa*) belongs to this group.

Key to Species

Leaves very rough on edges; spikelets 5 mm. long.   118. *Leersia oryzoides*
Leaves not rough on edges; spikelets 3 mm. long.   119. *Leersia virginica*

118. *Leersia oryzoides* (L.) Swartz. RICE CUTGRASS. Perennial by rhizomes; stem bases often rooting at nodes; leaves 5—25 cm. long, 5—15 mm. wide, soft but having sharp teeth pointing backward along edges; stems weak, 5—15 dm. high; lower panicles (Sept.) largely enclosed in leaf sheaths; spikelets flat, oblong, 5 mm. long, soon falling. Locally abundant along stream or pond banks. We have it from Cass, Ransom, Stutsman, Steele, Pembina and Slope Counties. The leaves are sharp enough to cut hands and clothing.

119. *Leersia virginica* Willd. WHITEGRASS. Perennial by very short, scaly rhizomes 1—3 cm. long; stems slender, upright or spreading, 3—10 dm. long; leaves soft, 1—2 dm. long, not rough on edges; panicle branches few; spikelets 3 mm. long. Collected in 1947 (Stevens 1143) in a deep, wooded gully above the Sheyenne River near Kindred in Richland County.

WILD RICE TRIBE Zizanieae

120. *Zizania aquatica* L. WILD RICE. Annual, 1—2 m. (3—7 ft.) high; leaves 3—6 dm. (1—2 ft.) long, 1—5 cm. (½—2 in.) wide; panicle 3—5 dm. long, lower branches staminate, widely spreading, upper branches pistillate, erect; grain cylindrical, 8—12 mm. (⅓—½ in.) long, nearly

black. Occasional along banks of Red River and lower parts of its tributaries. Flowers in August; grains drop quickly and must be kept wet and cold to retain their vitality.

MILLET TRIBE Paniceae

Spikelets 1-flowered, grain plump, enclosed by hard, thick lemma and palea and often by 2 or more thin glumes. An additional glume, the remnant of another flower, is sometimes present.

Key to Species or Genera

Spikelets with bristles or spines below flowers.
　Spikelets in rounded, spiny burs 4—6 mm. wide.
　　　　　　　　　　　　　　　　121. *Cenchrus longispinus*
　Spikelets in a dense, spike-like panicle 5—15 cm. long. 122-125. *Setaria*
Spikelets without bristles or spines, but scales sometimes awned.
　Branches of flower cluster 1-sided; spikelets crowded along one side.
　　Flowering branches thick, glumes with stiff, short hairs; lemma
　　　usually awned.　　　　　　　　126. *Echinochloa crusgalli*
　　Flowering branches slender, hairs on glumes very small, if any;
　　　lemma not awned.
　　　Leaf sheaths smooth; lemma dark brown. 127. *Digitaria ischaemum*
　　　Leaf sheaths velvety hairy; lemma greenish.
　　　　　　　　　　　　　　　128. *Digitaria sanguinalis*
　Branches of flower cluster not 1-sided; spikelets on slender stalks.
　　　　　　　　　　　　　　　129-139. *Panicum*

121. *Cenchrus longispinus* (Hackel) Fernald. SANDBUR. Annual; stems spreading, 2—5 dm. long; leaves 6—12 cm. long, 2—7 mm. wide; flower clusters of 5—10 burs, each having 2 spikelets and stiff spines. Sandy soil; common in a few localities in southern part of State. A great nuisance on account of the spiny burs which fall off when mature. Formerly listed as *C. tribuloides* L. or *C. pauciflorus* Benth.

Setaria BRISTLEGRASS.

FOXTAIL GRASS.   PIGEONGRASS

Annuals; flower clusters spike-like or with short branches; spikelets rounded, 1—several bristles coming from below spikelet.

Key to Species

Flower cluster unbranched, cylindrical; bristles 5 or more below
　spikelet.　　　　　　　　　　122. *Setaria lutescens*
Flower cluster with very short branches, tapering; bristles 1—3.
　Bristles with sharp hooks which catch on clothing.
　　　　　　　　　　　　　　　124. *Setaria verticillata*
　Bristles rough but not hooked.
　　Bristles purplish; grains falling out of glumes when mature.
　　　　　　　　　　　　　　　125. *Setaria italica*
　　Bristles green or yellowish; entire spikelet falling when mature.
　　　　　　　　　　　　　　　123. *Setaria viridis*

122. *Setaria lutescens* (Weigel) Hubb. YELLOW PIGEONGRASS. Leaves 1—2.5 dm. (4—10 in.) long, 1 cm. (⅖ in.) wide, gradually tapered and twisted once around; spikelets ovate, 3 mm. (⅛ in.) long, palea flat, lemma strongly curved and cross ridged; grain ovate, very flat on one side, greenish (fig. 147). Common weed, long known as *S. glauca.*

123. *Setaria viridis* (L.) Beauv. GREEN PIGEONGRASS. Leaves broader, shorter and stiffer than in preceding, not twisted; flower cluster slightly tapered, branching scarcely evident; spikelets oblong, 2—2.5

mm. long, palea convex, lemma more strongly convex, faintly roughened; grain oblong, flattened on one side, greenish or white (fig. 148).
Common weed.

124. *Setaria verticillata* (L.) Beauv. BUR PIGEONGRASS. Very much like
S. viridis except for the hooked bristles. First found at Fargo in 1922,
since at Grand Forks, Valley City, Edgeley and Lisbon as a weed in
gardens or about streets. It usually can be detected by the fact that
the flower clusters catch upon each other and become tangled together.

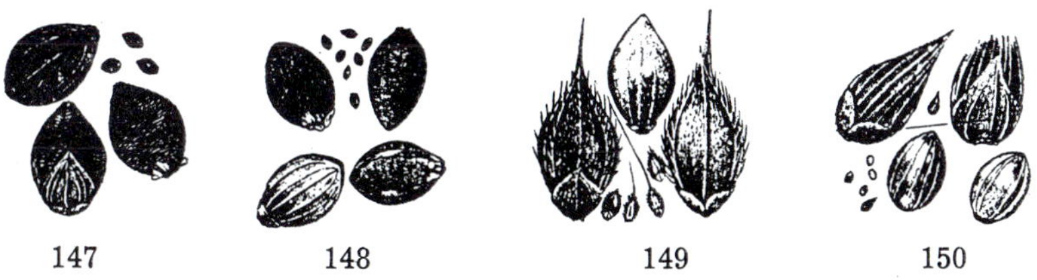

147   148   149   150

125. *Setaria italica* (L.) Beauv. FOXTAIL MILLET. Like *S. viridis* but usually coarser and does not shatter readily. The larger "heads" are
distinctly branched to appear lobed; grain yellowish. It occurs frequently in fields or along roadsides but scarcely is established as a
wild plant. Numerous cultivated strains include Siberian, Kursk,
Hungarian, etc.

126. *Echinochloa crusgalli* (L.) Beauv. BARNYARD GRASS. Coarse annual,
1—2 m. high; leaves 1—6 dm. long, 5—15 mm. wide; panicles 1—2 dm.
long, branches 2—10 cm. long; spikelets ovate, 3 mm. long, closely
crowded; glumes strongly nerved and with scattered, stiff hairs; third
glume (sterile lemma) often long awned; palea flat, lemma convex,
both hard, smooth and shining, greenish yellow; grain lenticular, translucent white (fig. 149). Common weed in low ground, often making
a dense stand where crops have been flooded. Quite variable, spikelets often deep purple. "Billion-dollar grass" or Japanese millet,
sometimes planted, is var. *frumentacea* (Roxb.) Wight. It has more
rounded spikelets, the lemma and palea gray.

127. *Digitaria ischaemum* (Schreb.) Muhl. SMOOTH CRABGRASS. Annual;
stems spreading, 2—5 dm. long; leaves 2—10 cm. long, 2—6 mm. wide,
smooth; 2—6 branches of flower clusters 4—10 cm. long, widely
spreading, purplish; spikelets 2 mm. long. Quite common in lawns,
where it is a troublesome weed. We have it from only Cass and
Richland Counties but it doubtless occurs elsewhere.

128. *Digitaria sanguinalis* (L.) Scop. CRABGRASS. Larger than the last,
extensively rooting at lower joints of stems; leaves finely soft hairy;
branches of flower clusters usually 8—10, not purple. One of the
most troublesome weeds in the central states. I found some plants
in a boulevard at Fargo in 1920, and there is a specimen without date
collected at Fargo by Bolley. A specimen was received in 1944 from
Rugby, Pierce County, and plants were found in a garden at Fargo in
1945 (Stevens 851).

Panicum

Annuals or perennials of various sizes and habits; panicles usually widely branched and symmetrical; spikelets rounded or slender, plump; lemma and palea hard, smooth and shining. Includes
weeds, native and cultivated grasses. The smaller native species
are difficult to identify. There are over 100 species, chiefly in
southern U. S.

Key to Species

Glumes not hairy.
  Leaves without hairs; coarse perennial, 5—15 dm. high.
                              129. *Panicum virgatum*
  Leaves, especially sheaths, hairy; annuals.
    Spikelets 5 mm. (⅕ in.) long; coarse, cultivated plant.
                              130. *Pancium miliaceum*
    Spikelets 2—3.5 mm. long; weeds.
      Spikelets 2—2.5 mm. long, abruptly pointed (fig. 150).
                              131. *Panicum capillare*
      Spikelets 3—3.5 mm. long, gradually narrowed at tip.
                              132. *Panicum flexile*
Glumes with fine hairs, these sometimes few or not prominent.
  Spikelets 3—4 mm. long.
    Leaves 6—12 mm. wide, spreading,
      Leaves hairy above; glumes densely covered with hairs which
        have thickened bases.   133. *Panicum leibergii*
      Leaves smooth above; glumes with few, slender hairs.
                         134. *Panicum scribnerianum*
    Leaves 3—4 mm. wide, erect.
      Upper leaves 7—10 cm. long; panicle branches slender, erect.
                       135. *Panicum perlongum*
      Upper leaves 5—8 cm. long; panicle branches stiff, spreading.
                     136. *Panicum wilcoxianum*
  Spikelets 1.5—2 mm. long.
    Spikelets 2 mm. long; leaves and stems densely hairy.
                     137. *Panicum praecocius*
    Spikelets 1.5 mm. long; leaves and stems sparsely hairy.
      Leaves moderately hairy on upper side.   138. *Panicum huachucae*
      Leaves almost without hairs on upper side.
                   139. *Panicum tennesseense*

129. *Panicum virgatum* L. Switchgrass. (Fig. 151). Stout perennial, forming thick clumps from short rhizomes; leaves 1—6 dm. (4—24 in.)
long, 3—15 mm. wide; stems 5—15 dm. high; panicles 1—3 dm. long,
nearly as wide; spikelets ovate, 3.5—5 mm. (⅙ in.) long. Common,
chiefly in fairly moist soil. Aug. A fair hay grass.

130. *Panicum miliaceum* L. Broomcorn Millet. Proso. Coarse annual;
leaves 2—3 dm. long, 1—2 cm. wide, hairy; stems 5—10 dm. high;
panicle 1—3 dm. long, branches drooping when mature; spikelets 4—5
mm. long, nearly as wide; lemma gray, yellowish, red or dark brown
in various strains. Often grown for the grain and frequently found
in fields and along roadsides.

131. *Panicum capillare* L. Witchgrass. Coarse annual, leaves very hairy,
1—2.5 dm. long, 5—15 mm. wide; stems erect or spreading, 3—8 dm.
long; panicle 1—4 dm. long with slender branches; spikelets 2—2.5 mm.
long; lemma greenish (fig. 150). Common weed in fields, grassland
and waste ground. Panicles break away at maturity and are blown
by wind. Probably most of our plants are var. *occidentale* Rydb., but
I have not been able to separate them satisfactorily.

132. *Panicum flexile* (Gattinger) Scribn. Similar to slender plants of the
last, panicle only half as wide as long. I found this along the shore
of Wood Lake, near Tokio, Benson County in 1940. A specimen collected by Lunell in a similar location in the same county was reported
as *P. barbipulvinatum* (=*capillare*, var. *occidentale*).

133. *Panicum leibergii* (Vasey) Scribn. Tufted perennial with short, unusually broad leaves; stems 2—6 dm. long, usually spreading; leaves
5—10 cm. long, 7—15 mm. wide, hairy; panicle 8—15 cm. long with
few branches; spikelets rounded oblong, 4 mm. long. Frequent in
lower prairie. June.

134. *Panicum scribnerianum* Nash. Similar to last, shorter, more compact;
leaves narrower, more upright; panicle dense, stiff. Occasional, especially in dry, sandy soil. We have it from Cass, Ransom and Grant
Counties.

151. Switchgrass (*Panicum virgatum*).

135. *Panicum perlongum* Nash. Very slender, upright, tufted perennial, 1—3 dm. high; spikelets nearly without hairs. I found this in some abundance in native prairie near Fargo, June 24, 1920, and specimens were verified by Dr. Hitchcock.

136. *Panicum wilcoxianum* Vasey. Low, densely tufted perennial with small leaves and short, dense panicles 2—5 cm. long. Frequent on dry, sandy prairie.

137. *Panicum praecocius* Hitchc. & Chase. Forming dense, spreading tufts; stems and leaves densely hairy, hairs standing at right angles to stem; leaves 5—9 cm. long, 4—6 mm. wide; stems 2—4 dm. long, branching. We have this only from the sandhills in northern Richland County where it was first detected in 1934 and verified by J. R. Swallen.

138. *Panicum huachucae* Ashe. A single collection of this in the same area as the last was identified by Mr. Swallen.

139. *Panicum tennesseense* Ashe. Collected July 18, 1940, on a gravelly lake beach at Mctigoshe State Park, Turtle Mts. and identified by Mr. Swallen. This specimen is 3 dm. high, slender, with purple spikelets; descriptions do not indicate that these features are distinctive as compared with other species.

SORGHUM TRIBE **Andropogoneae**

Spikelets in pairs, one sessile and fertile, the other stalked and sterile; flower clusters usually quite hairy; slender awns usually present; late flowering.

Key to Species

Coarse, introduced grasses; lemma hardened; panicle 2—3 dm.
(8—12 in.) long, not hairy.
Annual, planted for forage.   140. *Sorghum vulgare*
Perennial weed spreading by rhizomes.   141. *Sorghum halepense*
Native perennials; lemma not hardened; flower clusters smaller, hairy.
Flower cluster a brown, feathery panicle 1—2 dm. long.
     142. *Sorghastrum nutans*
Flowers in slender spikes, 2—10 cm. (1—4 in.) long, 2 or 3 together;
hairs gray or yellowish.
Plants 3—8 dm. high, fine, wiry; spikes 3—5 cm. long.
     143. *Andropogon scoparius*
Plants 1.5—2 m. (5—7 ft.) high, coarse; spikes 5—10 cm. long.
Flower spikes purplish; leaves green and purple.
     144. *Andropogon furcatus*
Flower spikes yellowish; leaves gray.   145. *Andropogon hallii*

140. *Sorghum vulgare, var. sudanense* (Piper) Hitchc. SUDANGRASS. Coarse annual, 1—2 m. (3—7 ft.) high; leaves 2—5 dm. (8—20 in.) long, 1—3 cm. (⅖—1⅕ in.) wide: panicle 2—3 dm. long, widely branched; spikelets 5—7 mm. (¼ in.) long; lemma greenish. Often planted and occasionally escaped in fields or along roadsides, not established as a wild plant. The sorgos, grown for hay, grain or sorghum, are larger seeded strains of the same species. They have compact panicles and rounded grains.

141. *Sorghum halepense* L. JOHNSONGRASS. Much like sudangrass but perennial from rhizomes; spikelets smaller, reddish brown. This is a very troublesome weed in the southern states but usually is not winter hardy north of Kansas. A plant grew for several years beside a building in an alley at Williston. A specimen was verified by J. R. Swallen.

142. *Sorghastrum nutans* (L.) Nash. INDIAN GRASS. Perennial from short rhizomes; stems 1—2 m. high; leaves 3—6 dm. long, 5—10 mm. wide: panicles ovate or lanceolate, 1—3 dm. long. Late Aug. Common in some localities, especially southern part of State, in rather low prairie. We have a specimen from Pembina County and one from McHenry County but it seems uncommon northward. A good forage grass. The coppery colored plumes are very decorative.

143. *Andropogon scoparius* Michx. LITTLE BLUESTEM. (Fig. 152). Perennial in tufts 2—3 dm. wide; stems branching, fine, wiry, 3—8 dm. high; leaves 2—6 dm. long, 3—6 mm. wide; spikes 3—6 cm. long, soon becoming curved, the spreading white hairs conspicuous. Common, especially in dry soils and on hillsides. One of the poorer forage grasses. The plants become quite reddish and are conspicuous on the hillsides.

144. *Andropogon furcatus* Muhl. BIG BLUESTEM. (Fig. 153). Stems coarse, 1—2 m. high; leaves often strongly purplish, 3—8 dm. long, 5—10 mm. wide: flower spikes 2 or 3 together, 5—10 cm. long, hairy but not usually conspicuously white as in Little Bluestem nor yellowish as in Hall's Bluestem. Common in coulees and moist prairie, not on dry ground. One of the most palatable of all our grasses, easily destroyed by close grazing.

145. *Andropogon hallii* Hack. HALL'S BLUESTEM. SANDHILL BLUESTEM. Much like Big Bluestem but stems and leaves quite gray; hairs of flower clusters more prominent and yellowish. Found chiefly in sand dune areas where it is locally common. We have numerous specimens from the Richland-Ransom County area, the McHenry County area, from west of Missouri River and one from Wells County.

SEDGE FAMILY Cyperaceae

Grass-like plants, usually with 3-angled, solid stems, 3-ranked leaves and closed sheaths; flowers in spikelets with chaffy bracts, no petals nor sepals; stamens 3, pistil 1, stigmas 2 or 3; fruit a 3-angled, flattened or rounded achene.

152

153

152. Little Bluestem (*Andropogon scoparius*).
153. Big Bluestem (*Andropogon furcatus*).

This is a large group of plants not commonly distinguished from grasses. Most of them grow in low meadows or bogs where they often comprise a large part of the vegetation but are mostly of low value for hay. The seed-like fruits of a few are a considerable item in wild duck food. The descriptive characters are quite technical, require mature material and a good lens for study.

Key to Genera

Flower clusters in fruit with a conspicuous tuft of white hairs about
   3 cm. (1½ in.) long.   15. *Eriophorum viridicarinatum*
Flower clusters without such a tuft of hairs.
  Pistillate and staminate flowers in separate spikes or parts of
    spikes; achene enclosed by a sack-like scale (fig. 154). 36-82. *Carex*
Flowers with both pistil and stamens; achene not so enclosed.
  Base of style persistent as a tubercle on top of achene (fig. 155).
                8-14. *Eleocharis*
    Achenes without such tubercle.
      Scales of spikelets in 2 rows; no bristles present.   1-7. *Cyperus*
      Scales spirally arranged; several bristles in place of flower parts.
        Plants slender, 1—2 dm. (4—8 in.) high.
               16. *Hemicarpha micrantha*
        Plants stout, 3—10 dm. high.   17-25. *Scirpus*

Cyperus

Very small to large plants, perennial or annual; flowers in spikelets which are usually slender or flattened and grouped in a dense cluster at top of stem; stigmas 2 or 3; achenes flat or 3-angled.

Key to Species

All scales falling away at maturity, leaving a bare stalk.
  Stigmas 2; achenes flat; spikelets flat.   1. *Cyperus rivularis*
  Stigmas 3; achenes 3-angled.
    Spikelet stalk narrow, not winged.
      Plants annual, 1—3 dm. (4—12 in.) high; stamens 1 in each
        flower.
        Scales awned or awn tipped.   2. *Cyperus inflexus*
        Scales acute but not sharp pointed.   3. *Cyperus acuminatus*
      Plants perennial, 3—6 dm. high; stamens 2 or 3.
              4. *Cyperus schweinitzii*
    Spikelet stalk flat with wing-like edges.
      Plant annual; wing edges falling off.   6. *Cyperus erythrorhizos*
      Plant perennial; wing edges persistent.   5. *Cyperus esculentus*
Only flower scales falling, leaving 2 basal scales.   7. *Cyperus speciosus*

1. *Cyperus rivularis* Kunth. Annual, stems tufted, 1—3 dm. (4—12 in.) high; spikelets usually purplish, very flat, 1—2 cm. (⅖—⅘ in.) long, 4—5 mm. (⅙ in.) wide; achenes 1 mm. long. Wet, sandy soil, Richland and Ransom Counties. It is quite a common plant farther south and this is probably its northern limit. Previously reported incorrectly as *C. diandrus* Torr., which has a much elongated style.

2. *Cyperus inflexus* Muhl. Annual, stems tufted, 5—15 cm. high; spikelets flat, 4—6 mm. long, scales with sharp, recurved points. On wet soil; fragrant when dry. Specimens are from Richland, Ransom, Cass, Stutsman and McHenry Counties, but it probably occurs all over the State.

3. *Cyperus acuminatus* Torr. & Hook. Annual, stems sometimes tufted, 1—3 dm. high; spikelets like the preceding but scales not sharp pointed. Wet places. Fargo is our only record.

4. *Cyperus schweinitzii* Torr. Perennial, stems with enlarged bases, usually tufted, 3—5 dm. high; spikelets slender, 10—15 mm. long; achene 3-angled, 2 mm. long. Frequent in very sandy soil throughout the State. Named for L. D. Schweinitz, a Philadelphia botanist of the early 19th century and an authority on sedges.

5. *Cyperus esculentus* L. NUTGRASS. CHUFA. Perennial by small, brown tubers 1—2 cm. long; stems 2—5 dm. high; leaves yellowish green; spikelets straw-colored: achenes 1.5 mm. long. We have this only from Fargo and Eddy County. It is quite common at Fargo along the edge of the river where it was first collected in 1920. The plants at Sheyenne, Eddy County, were believed to have been introduced with gladiolus "bulbs" (corms) from the west coast. The man told us 3 years later that the plants were all gone. It is a common plant in warmer regions, forms dense patches and is hard to eradicate. The tubers have been used for food in various countries.

6. *Cyperus erythrorhizos* Muhl. REDROOTED CYPERUS. Annual, but rather stout, 3—6 dm. high with dense clusters of slender, reddish brown spikelets; achenes 1 mm. long. We have this only from Cass, Sargent and Stutsman Counties. The great mass of reddish brown roots is quite characteristic.

7. *Cyperus speciosus* Vahl. Annual, otherwise more like No. 5 than No. 6. A specimen from Jamestown, collected by C. C. Schmidt in 1898, has been referred to this species. It is an immature plant and may be incorrectly determined, a chance introduction or mislabeled as to source.

Eleocharis SPIKERUSH

Leafless plants with tufted, slender, green stems; flowers in a short spike[1] at tip of stem; each flower with one chaffy scale and usually several slender bristles; achenes flattened or bluntly 3-angled, base of style persisting as a tubercle. Rather insignificant plants of wet ground. Variations in form of tubercle and markings on achenes make interesting objects for study with lens or microscope.

Key to Species

Style branches 2; achenes flattened, smooth.
  Annual; tubercle of achene much wider than long.
                                           8. *Eleocharis engelmannii*
  Perennials with rhizomes; tubercle longer than wide.
    Basal scales of spike 2 or 3; stems 2—5 mm. thick.
                                           9. *Eleocharis palustris*
  Basal scale 1, enclosing base of spike; stems about 1 mm. thick.
                                          10. *Eleocharis calva*
Style branches 3; achenes 3-angled or rounded.
  Achene with fine ribs in both directions; slender plants 0.5—2 dm.
    high.
    Achene with 3 main ribs and 6 9 smaller ones.
                                          11. *Eleocharis acicularis*
    Achene with 9 about equal ribs   12. *Eleocharis wolfii*
  Achene smooth or rough, not ribbed.
    Achene roughened with low projections; tubercle distinct, shorter
      than achene.   13. *Eleocharis compressa*
    Achene finely netted; tubercle as long as achene and seeming
      part of it.   14. *Eleocharis pauciflora*

8. *Eleocharis engelmannii* Steud. Annual; stems 1—2 dm. (4—8 in.) high, densely tufted, soft; spikes brown, 1—2 cm. (2/5—4/5 in.) long; achenes 1 mm. long, dark gray, tubercle short and wide, nearly covering the

---

[1] These are usually called spikelets. Spike seems better but spikelet is merely a "little spike."

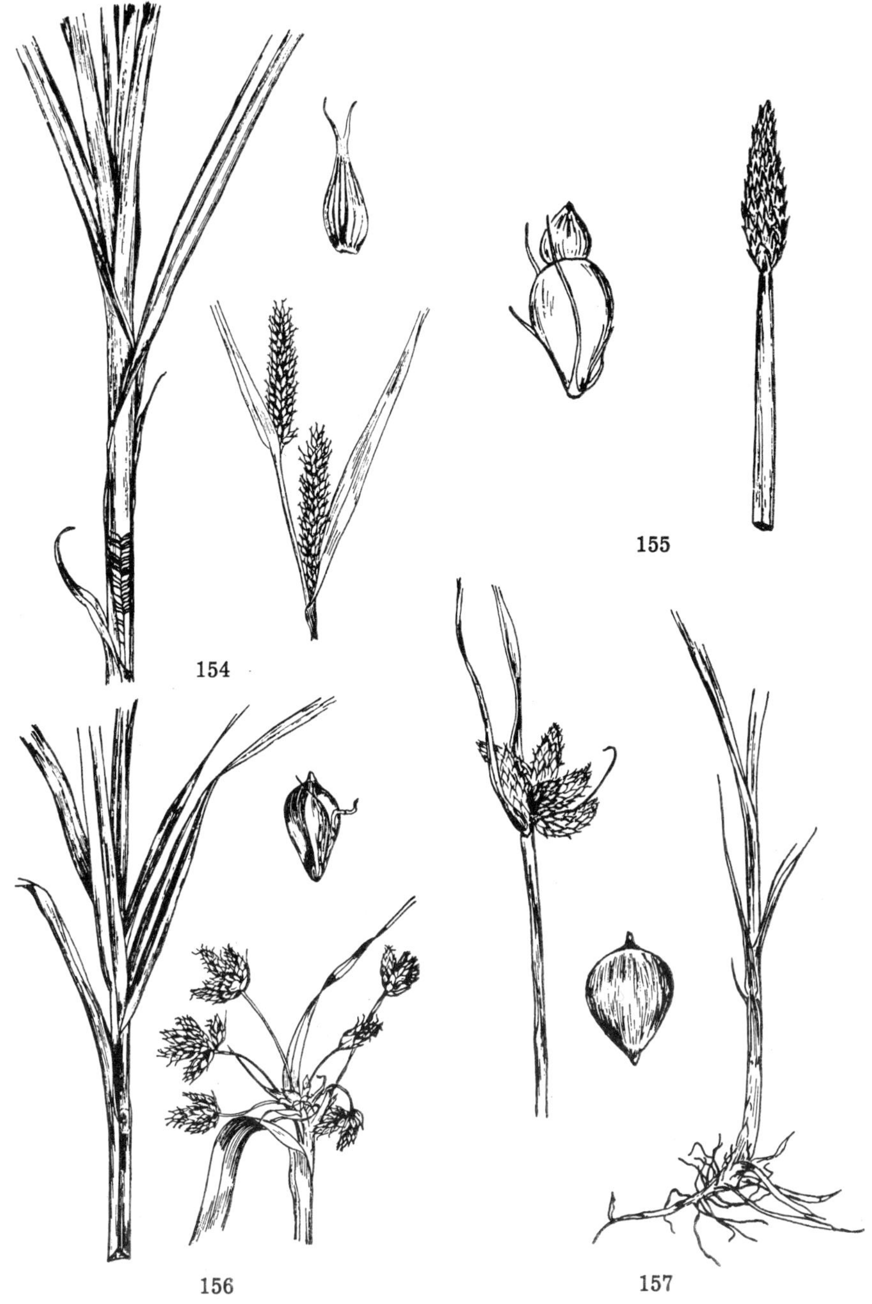

154. Slough Sedge (*Carex atherodes*).
155. Spikerush (*Eleocharis palustris*).
156. River Bulrush (*Scirpus fluviatilis*).
157. Prairie Bulrush (*Scirpus paludosus*).

broad top. Low, bare soil where water has stood for some time. We have it only from Richland, Cass and Benson Counties but it may be much more general. Usually abundant where it occurs.

9. *Eleocharis palustris* (L.) R. & S. (Fig. 155). Perennial from rhizomes; stems soft, 2—5 mm. thick, 1—6 dm. high; spike 1—3 cm. long; achene plump, brown, 1—2 mm. long. The largest and one of the commonest species, often forming a dense stand in low ground or shallow water.

10. *Eleocharis calva* Torr. More slender than the preceding, stems hardly over 1 mm. thick; spike 5—10 mm. long. It apparently grows more in boggy places where other vegetation is dense, in contrast to hard soils where *E. palustris* may be the predominant plant.

11. *Eleocharis acicularis* (L.) R. & S. Stems thread-like, 0.5—2 dm. high; spikes 3—6 mm. long; achenes oblong, narrowed at lower end, 0.75 mm. long, beautifully marked with longitudinal ridges and finer cross lines. A very common plant, forming a moss-like carpet on drying pools and wet ground.

12. *Eleocharis wolfii* A. Gray. Similar to preceding, a little coarser, spikes flattened; achenes with coarser, equal ridges. This was found at Fargo in 1901 and 1910 but I have been unable to find it since.

13. *Eleocharis compressa* Sull. Perennial from rhizomes; stems slender, wiry, 1—2 dm. high; spikes brown, 1—2 cm. long; achenes reddish brown, 1 mm. long, obtusely 3-sided. Low grassland or ditches, perhaps widely distributed though we have specimens only from Richland, Cass, Ransom and Grant Counties; abundant at Fargo. Has usually been called *E. acuminata* (Muhl.) Nees.

14. *Eleocharis pauciflora* (Lightf.) Link. Somewhat like No 11, differing as indicated in the key. One collection from McHenry County by Wm. A. Kluender in 1936. This species was formerly placed in *Scirpus* because the tubercle is not distinct from the body of the achene.

15. *Eriophorum viridicarinatum* (Engelm.) Fernald. COTTONGRASS. Perennial from rhizomes; leaves grass-like; stem 3—6 dm. high, bearing at the top several spikelets; flower parts divided into hair-like filaments about 2 cm. long. Local in wet meadows or boggy places, eastern half of State. The drooping spikelets with white, cottony tufts are conspicuous. Formerly referred to *E. polystachyon,* an European species.

16. *Hemicarpha drummondii* Nees. Tufted annual plant resembling a *Cyperus,* 1—2 dm. high; spikelets rounded, 3—5 mm. long, 2 or 3 together near the top of a leaf-like stem. One record from Leonard, Cass County, in 1923.

Scirpus BULRUSH

Perennials with grass-like or much reduced leaves; spikelets in umbels, heads or solitary; each flower with a scale and usually several bristles; achenes flat or 3-angled.

Key to Species

Involucre of only one bract (a short leaf just below flower cluster,
    often seeming a continuation of stem).
  Spikelets 1—10 in a head-like cluster.
    Stems round.   17. *Scirpus nevadensis*
    Stems sharply 3-angled.   18. *Scirpus americanus*
  Spikelets numerous in compound umbels.
    Style 2-cleft; achene flat.
      Spikelets rounded, 5—10 mm. (⅕—⅖ in.) long, reddish brown.
                                                     19. *Scirpus validus*
      Spikelets cylindric, 7—15 mm. long; dark gray. 20. *Scirpus acutus*
    Style 3-cleft; achenes 3-angled.   21. *Scirpus heterochaetus*
Involucre of 2 or more leaves with flat blades.
  Spikelets 15—25 mm. long in clusters of 3—15.

Achene flat; plants 3—8 dm. (1—2½ ft.) high. 　22. *Scirpus paludosus*
Achene 3-angled; plants 10—15 dm. high. 　23. *Scirpus fluviatilis*
　Spikelets 3—10 mm. long, numerous in compound umbels.
　　Style branches 3; achenes 3-angled. 　24. *Scirpus atrovirens*
　　Style branches 2; achenes flat. 　25. *Scirpus microcarpus*

17. *Scirpus nevadensis* S. Wats. Perennial, 2—4 dm. (8—16 in.) high; leaves several from near base of stem, hard and narrow, their edges mostly rolled inward (to a rounded form); spikelets 1—5, ovoid-oblong, 5—15 mm. (⅕—⅗ in.) long; achenes 2.5 mm. long, plano-convex. This species was reported by F. P. Metcalf (Journ. Wash. Acad. Sci. 10:188-198, 1920) as common in North Dakota. I have seen it at only three places, but in some abundance, on saline, sandy shores of Strawberry Lake, near Esmond, Benson County, Girard Lake, Pierce County and Powers Lake, Burke County. Another specimen is from the Lostwood Lakes, Burke County (F. M. Uhler and W. A. Warren, No. 1024).

18. *Scirpus americanus* Pers. CHAIRMAKERS RUSH. THREE SQUARE. Perennial, forming large but often not dense patches; stems sharply 3-angled, the terminal bract appearing a continuation of stem, the small cluster of spikelets 1—2 dm. below the tip; spikelets 7—20 mm. long; achenes thick, plano-convex, 2.5—3 mm. long. One of the commonest species, often forming beds along pond margins.

19. *Scirpus validus* Vahl. COMMON BULRUSH. SOFTSTEM BULRUSH. (Fig. 129D). Stems cylindrical, tapering only slightly, spongy, 1—2 m. (3—7 ft.) high, with a tassel-like flower cluster at tip; achenes 2 mm. long, obovoid, gray. In shallow water, on marshy ground or on solid ground after drying up of water.

20. *Scirpus acutus* Muhl. HARDSTEM BULRUSH. Similar to preceding but stems taller and harder; branches of flower clusters fewer and stiffer, spikelets not breaking up so readily at maturity; achenes larger, 3 mm. long. Previously known as *S. occidentalis* Chase. Both of these species are common and frequently grow together or in distinct areas in the same pond. They form dark green islands near the shore or may be scattered all over shallow ponds. The fruits are an important source of wild duck food. The stems may be grazed by cattle in times of feed scarcity but usually are not eaten.

21. *Scirpus heterochaetus* Chase. Similar to last but even more slender; flower clusters with few spikelets. We have it only from Fargo where it is rather frequent in ponds and drainage ditches.

22. *Scirpus paludosus* A. Nels. PRAIRIE BULRUSH. (Fig. 157). Stems stout, sharply triangular, 3—6 dm. high from tuber-like roots; leaves 1—3 dm. long, 1 cm. wide, their bases enclosing the stems; spikelets 1—2.5 cm. long, ovate to cylindric, either sessile or short stalked; achenes light brown, flat, 3 mm. long. A common plant around prairie ponds, especially very shallow, saline ones which usually dry out early.

23. *Scirpus fluviatilis* (Torr.) A. Gray. RIVER BULRUSH. (Fig. 156). Much taller than the preceding, 1—1.5 m. high, leafy; spikelets more numerous and some of the clusters on stalks 3—10 cm. long; achenes 3-angled, 4 mm. long. Pond margins and ditches, probably all over the State.

24. *Scirpus atrovirens* Willd. DARKGREEN BULRUSH. Stems slender, 1—2 m. high, leafy, darkgreen; spikelets numerous in a dense, compound umbel; achenes nearly white, 1 mm. long. A common plant in wet meadows, along streams or ponds; striking in appearance because of its tall, slender clusters of stems.

25. *Scirpus microcarpus* Presl. Similar to the preceding but weaker; leaves pale green, often red spotted; branches of flower clusters longer and looser. This is more distinctly a swamp plant. We have it from a number of widely separated localities but it can hardly be called common.

Carex SEDGE

Grass-like perennials with 3-angled stems and 3-ranked leaves; flowers in several spikes at upper part of stem; pistillate and staminate flowers separate, both in one spike or in separate spikes; pistil completely enclosed in one scale (perigynium), another flat scale just below it; stigmas 2 or 3, fruit a flat or 3-angled achene (fig. 154).

One of the largest and most difficult groups of flowering plants with over 1000 known species, most abundant in cool, northern regions where they compose much of the grassland vegetation, especially in wet soils. Some are useful for forage but on the whole they are less useful than grasses. Distinctions are based largely on details of fruiting parts and specimens in flowering condition are not suitable for identification. Vegetative parts are important also, especially underground stems. Some species develop large tufts, others have long rhizomes which send up stalks at intervals forming a loose or compact sod.

Key to Species

A. Spikes only 1 on each stem.
    Low plants about 1 dm. high; achenes 3-angled or rounded.
      Perigynia green or gray, densely hairy; tufted plants with-
        out rhizomes.   59. *Carex filifolia*
      Perigynia nearly black, smooth; plants with rhizomes.
                              60. *Carex obtusata*
    Taller plants 2—4 dm. high; achenes flattened.
      Very slender, tufted plants; spikes 4—15 mm. long, staminate
        above.   57. *Carex leptalea*
      Fairly stout plants from rhizomes; spikes 2—4 cm. long,
        usually either all staminate or all pistillate.
                         62. *Carex scirpiformis*
AA. Spikes 2 to many, sometimes crowded in a dense cluster.
  B. Spikes alike, usually not stalked, staminate at either tip or
      base; achenes flattened.
    C. Long rhizomes present, stalks appearing singly or form-
        ing patches.
      Flowering stems 1 dm. high; leaves 0.5—2 mm. wide;
        dry prairie plants.
      Spikes usually dioecious; flower scales ovate, sharp
        pointed.   36. *Carex douglasii*
      Upper flowers of each spike staminate; scales
        obovate, not sharp.   37. *Carex eleocharis*
      Flowering stems 3—6 dm. high; leaves flat, 3—5 mm.
        wide; low land plants.
      Perigynia strongly ribbed on inner side.
                   39. *Carex sartwellii*
      Perigynia not so ribbed.   38. *Carex praegracilis*
  CC. Rhizomes not present or only short; plants forming tufts.
    D. Staminate flowers at top of spike.
      Perigynia abruptly narrowed into beak.
        Spikes 10 or less; perigynia green.
          Leaves 3.5—5 mm. wide.   42. *Carex gravida*
          Leaves 1—3 mm. wide.
            Perigynia widely spreading at maturity,
             much thickened at base.   40. *Carex rosea*
          Perigynia erect, not thickened at base.
                   41. *Carex hookerana*
      Spikes many; perigynia yellowish or brown.

Leaf sheath with cross ridges; perigynia
yellowish.   43. *Carex vulpinoidea*
Leaf sheath without ridges; perigynia brownish.
                              44. *Carex diandra*
Perigynia gradually narrowed into a flat beak.
                              45. *Carex stipata*

DD. Staminate flowers at base of spike.
Edge of perigynia rounded or with a sharp edge.
Perigynia widely spreading at maturity.
                              46. *Carex interior*
Perigynia erect, lying close together at maturity.
                              47. *Carex deweyana*
Edge of perigynia thin, wing-like.
Bract (leaf just below head) much longer than head.
Bract leaf-like, several times as long as head.
                              56. *Carex synchnocephala*
Bract very narrow, 1.5—6 cm. long.
                              55. *Carex athrostachya*
Bract rarely longer than head.
Beak of perigynia flattened and minutely
toothed along sides.
Perigynia lanceolate, 2½ or more times as
long as wide.
Leaves not over 3 mm. wide; spikes
oblong; perigynia erect. 48. *Carex scoparia*
Leaves 3—7 mm. wide; spikes rounded;
perigynia widely spreading.
                              53. *Carex cristatella*
Perigynia ovate, not over twice as long as
wide.
Perigynia 3—4 mm. long.
Spikes crowded together, rounded at
base.   49. *Carex bebbii*
Spikes loose, narrowed at base.
                              50. *Carex tenera*

Perigynia 4—7 mm. long.
Perigynia 4—5.5 mm. long, faintly nerved
on inner side.   51. *Carex brevior*
Perigynia 5.5—7.5 mm. long, strongly
10-15-nerved on inner side.
                              52. *Carex bicknellii*
Beak of perigynia scarcely flattened or
toothed, either at sides or tip.
                              54. *Carex xerantica*

BB. Spikes of 2 kinds, often long stalked, the staminate ones
usually at tip of stem.
Achenes more or less flattened; stigmas 2.
Low, tufted plants; perigynia rounded, soft, orange
yellow at maturity, becoming thin and papery when dry.
                              64. *Carex aurea*
Tall plants forming thick beds; perigynia dry, flattened.
Flowering stems coming from center of old leaf tufts.
                              75. *Carex aquatilis*
Flowering stems coming up beside the old tufts.
                              76. *Carex emoryi*

Achenes 3-angled or rounded; stigmas 3.
Pistillate spikes drooping on slender stalks.
Beak as long as, or longer than body of perigynium
which is thin, not ribbed.   71. *Carex sprengelii*
Beak shorter than body of perigynium which is
strongly ribbed.
Perigynia crowded or reflexed.
                              78. *Carex pseudo-cyperus*
Perigynia not crowded nor reflexed.

Pistillate spike 10—15 mm. wide; beak 2 mm.
    long.   77. *Carex hystricina*
Pistillate spike 2—3 mm. wide; beak none.
                              69. *Carex gracillima*

Pistillate spikes erect or spreading at maturity.
   Achene plump, rounded at tip; spikes leafy, shorter
      than lower leaves.   58. *Carex saximontana*
   Achene usually 3-angled and narrowed at tip.
     Beak of perigynium less than ¼ as long as body,
       rounded or with 2 very short teeth at tip.
      Perigynia covered with short hairs.
                         61. *Carex pennsylvanica*
     Perigynia not hairy.
       Perigynia 3-angled, narrowed at base.
         Achenes nearly black; leaves 1—2 mm. wide.
                         63. *Carex eburnea*
         Achenes light brown; leaves 3—15 mm. wide.
           Leaves 3—5 mm. wide; prairie plants.
                         65. *Carex tetanica*
           Leaves 5—15 mm. wide; woodland plants.
                         66. *Carex blanda*

       Perigynia nearly circular in cross section,
        rounded at base.
        Leaves not hairy, usually 3—9 mm. wide.
         Plants tufted, no rhizomes.
                      67. *Carex granularis*
         Plants from slender rhizomes.
                      68. *Carex crawei*
       Leaves hairy, 1—3 mm. wide.   73. *Carex torreyi*

Beak of perigynium from ¼ as long to as long
   as body, usually with 2 distinct teeth at tip.
   Beak not over ¼ as long as body of perigynium.
     Perigynia densely hairy; spikes 1—3 cm.
       long, widely separated. 74. *Carex lanuginosa*
     Perigynia not hairy, spikes 4—15 mm. long,
       close together.   72. *Carex viridula*
   Beak ½ as long to equalling body of
     perigynium.
     Plants forming stolons (leafy stems rooting
       at tips).   70. *Carex assiniboiensis*
     Plants without stolons.
      Plants with long rhizomes; base of style
       straight.
       Perigynia ovoid, teeth not over 2 mm. long.
                      79. *Carex laeviconica*
       Perigynia lanceolate, teeth 2—4 mm. long.
                      80. *Carex atherodes*
      Plants tufted; base of style S-curved.
       Pistillate spikes well separated, erect.
                      81. *Carex rostrata*
       Pistillate spikes crowded and spreading;
        perigynia reflexed.   82. *Carex retrorsa*

36. *Carex douglasii* Boott. Stems about 1 dm. high from long rhizomes;
leaves hard, 1—2 mm. wide; 5—15 spikes, 5—15 mm. long, in dense
cluster 2—3 cm. long and fully as wide; staminate spikes separate from
pistillate. Dry prairie. Specimens from Stutsman, Grant, Emmons and
Stark Counties.

37. *Carex eleocharis* Bailey. NEEDLELEAF SEDGE. (Fig. 159). Similar to preceding but cluster of spikes 1—2 cm. long and about half as wide. One
of the commonest sedges on high prairie, forming distinct patches.
Early May. Of considerable forage value. Previously known as *C.
stenophylla* Wahl.

38. *Carex praegracilis* W. Boott. Tall, long leaved sedge from extensive dark brown rhizomes; stems 3—6 dm. high; leaves 1—3 mm. wide; spikes many, in a dense, oblong or pointed cluster 2—5 cm. long; single spikes rounded or oblong, 4—8 mm. long with 6—10 perigynia and a few inconspicuous staminate flowers at tip. One of our commonest sedges in low meadows, usually mixed with grasses. It is the chief one in such places which has stout, brown rhizomes. The stems may come up singly at intervals or form dense clumps. Sometimes it grows on fairly dry banks. The heads are conspicuous in early May when the mass of stamens matures. Our plants have passed under various other names, *C. marcida* Boott., *C. camporum* Mack., and *C. siccata* Dewey, the last now regarded as a closely related but distinct species.

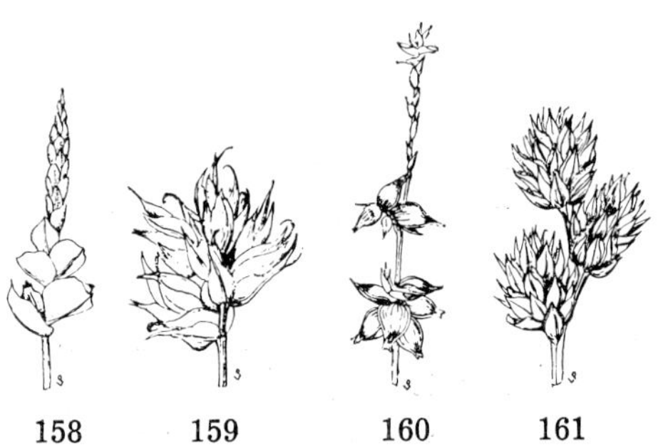

158   159   160   161

39. *Carex sartwellii* Dewey. Quite similar to preceding. Two specimens from Benson County, collected by Lunell. Mackenzie (N. Am. Fl.) records it for North Dakota and perhaps it is more common but has not been distinguished from *C. praegracilis*.

40. *Carex rosea* Schkuhr. (Fig. 160). Tufted; leaves fine and soft, 1—2 dm. long, 1—2 mm. wide; stems slender, 2—5 dm. high; spikes several, rounded, 5—8 mm. wide, usually about 1 cm. apart, perigynia widely spreading at maturity; lower bracts (below spikes) bristle-like. A rather striking little plant in moist, rich woods. Specimens from Richland, Cass, Barnes and Bottineau Counties.

41. *Carex hookerana* Dewey. Slender plant, somewhat resembling *C. praegracilis*, but spikes forming a narrow cluster. We have two specimens collected by Lunell in Benson County in boggy or swampy locations. I found it in Burke County in 1946 (Stevens 874), a large tuft in bottom of a small prairie coulee.

42. *Carex gravida* Bailey. Stout, short to medium height, growing in thick tufts; spikes 10—20, rounded, crowded into an oblong head 1—2.5 cm. long; lower bracts very slender, shorter than the head. Quite common in woods or low prairie.

43. *Carex vulpinoidea* Michx. Fox Sedge. Large tufts, 1—3 dm. thick; stems 3—8 dm. high; spikes many, rounded or oblong, 5—10 mm. long, crowded into dense clusters 2—10 cm. long, 1—2 cm. thick when mature; lower bracts very slender. One of the commonest and most striking sedges in low meadows and ditches.

44. *Carex diandra* Schrank. Dense clumps much as in *C. vulpinoidea* and not easily distinguished from it. Apparently more of a bog species. One specimen from Rolette County grew in several inches of water. Another specimen from a slough in Billings County seems to be correctly labeled as this species. Mackenzie reported the closely related *C. prairea* Dewey from North Dakota.

45. *Carex stipata* Muhl. A coarse but soft plant, similar to *C. vulpinoidea*. Our only specimens are from swampy places in Richland and Cavalier Counties and from low meadow at Fargo.

46. *Carex interior* Bailey. Tufted, slender plant; stems 2—4 dm. high; spikelets rounded, 5 mm. wide, about 3 on a stalk, slightly separated. Specimens from low meadows in Stutsman County.

47. *Carex deweyana* Schw. Soft, tufted plant; stems slender, 3—5 dm. high; spikes 3—5 rather closely clustered, and 1 or 2 further down the stem; lower bract slender. Specimens from Cass, Pembina, Cavalier and Bottineau Counties. It is one of the rare species here because it is restricted to the aspen woods type of vegetation.

48. *Carex scoparia* Schkuhr. Resembles the preceding but lower bract little developed. Bergman had reported one specimen from Stutsman County and Sarvis (N. D. Agr. Exp. Sta. Bull. 308:30, 1941) recorded it as occasional along coulees in Morton County.

49. *Carex bebbii* Olney. Resembles *C. brevior* but is smaller and grows in wet soil. Specimens from Ransom, Emmons, Bottineau and Nelson Counties, the last identified by Dr. J. F. Hermann.

50. *Carex tenera* Dewey. Another member of the *brevior* group, quite slender. I collected it in dry woods at Fargo (Stevens 165), on a ridge through a bog at Walhalla and we have one specimen from McHenry County collected by Lunell.

51. *Carex brevior* (Dewey) Mack. FESCUE SEDGE. (Fig. 161). Tufted, yellowish green, prairie or meadow plant; stems 3—8 dm. high; spikes 3—10 in a dense cluster, spikes rounded or oblong, 7—10 mm. long; perigynia flat, nearly circular, beak less than half as long as body. This is one of the commonest sedges of middle to low prairie. Our plants have often been called *C. festucacea* or var. *brevior*. Mackenzie regards it as distinct from *festucacea*.

52. *Carex bicknellii* Britton. Very similar to *C. brevior*. I do not feel that I can distinguish the two species. Bergman had referred specimens from Richland, Cass and Stutsman County to *C. bicknellii* and Mackenzie records it for North Dakota.

53. *Carex cristatella* Britton. Resembles *C. brevior* but it is a taller plant growing in shaded places and is green, not yellowish. Specimens from Wahpeton, Fargo and Walhalla.

54. *Carex xerantica* Bailey. Another member of the puzzling *brevior* group. Bergman had referred specimens from Valley City and Walhalla to it and Mackenzie lists it for this State.

55. *Carex athrostachya* Olney. A specimen from Lunell, collected in Benson County in 1914, is labeled, "First record east of the Rocky Mountains". It looks something like *C. brevior*, but is taller and more slender, the perigynia longer and narrower.

56. *Carex synchnocephala* Carey. Densely tufted, usually low, leafy; lower bracts leaf-like; perigynia unusually long and narrow, 5—6 mm. long, 1 mm. wide. It seems to grow especially in old lake or river beds and along shores. Richland, Cass, Cavalier, Benson, Wells and Bottineau Counties.

57. *Carex leptalea* Wahl. Very slender plant with long leaves and very few flowers. Collected only in wooded area at Walhalla.

58. *Carex saximontana* Mack. Densely tufted, leafy plant; stems often shorter than lower leaves, spikes few flowered, lower bracts leafy, perigynia obovoid, 4 mm. long. Woods and open coulees. The specimen reported by Bergman from Morton County as *C. durifolia* probably was this species. Since then collected in Richland, Grand Forks, Logan, Emmons, Burleigh, Dunn and Mountrail Counties. Apparently common but easily overlooked.

59. *Carex filifolia* Nutt. THREADLEAVED SEDGE. "NIGGERWOOL." (Fig. 158). Stiffly upright in dense, hard tufts up to 3 dm. wide, made up of smaller tufts 1—2 cm. wide; leaves wiry, edges in-rolled; stems 1—2 dm. high, each with 1 narrow spike 1—3 cm. long, upper part staminate, lower pistillate with unusually large, rounded scales having papery edges; perigynia rounded, 3—3.5 mm. long, 2-ribbed and somewhat hairy. One of the commonest species of dry prairie and hillsides, western part of

State, less common eastward but occurring there on dry hills. Quite an important early forage plant where abundant. Flowers in early May, the cluster of stamens conspicuous. The name "niggerwool" refers to black, kinky roots which are conspicuous when sod is turned over and are resistant to decay.

60. *Carex obtusata* Lilj. One of the small prairie species with long, brown rhizomes; closely resembles *C. pennsylvanica* but distinguished by the dark brown, smooth perigynia. First recognized at Fargo in 1920 but a specimen from Kensal, Stutsman County (Bergman 1744) referred to *C. stenophylla* (*C. eleocharis*), belongs to it. Later collected in other localities in Cass and Stutsman Counties; also in Barnes, Burleigh and Morton Counties.

61. *Carex pennsylvanica* Lam. PENNSYLVANIA SEDGE. YELLOW SEDGE. (Fig. 163). A common, small, prairie species, the leaves flat, 1—3 mm. wide, yellowish green as compared with *C. filifolia* or *C. eleocharis*. It has long rhizomes with purple scales and often on the prairie, single stalks come up from these at intervals. Large tufts also develop, especially in sandy soils or wooded areas. Stems 1—3 dm. high, a staminate spike 1 cm. long at tip and 2 or 3 rounded pistillate spikes below; perigynia rounded or somewhat triangular, 2.5—3 mm. long, covered with short hairs; lower bracts small. Early May. Most of our plants should be *C. heliophila* Mack., but I am unable to separate the two forms satisfactorily. Plants on the prairie are rather stiff and upright, about 1 dm. high, while those in woods develop soft, spreading leaves up to 3 dm. long.

62. *Carex scirpiformis* Mack. Plant with stout rhizomes and dark brown scales on stem bases; leaves erect, 5—20 cm. long, 2—3 mm. wide; stems 2—4 dm. high with a single, narrow spike 2—4 cm. long, 4—5 mm. wide; basal bracts very small; perigynia 2—5 mm. long, about half as wide, somewhat flattened, quite hairy. We have a single specimen from sand dunes in McHenry County, collected by Wm. A. Kluender in 1936. It is quite a distinctive species from the single, slender spike which is almost like an *Eleocharis*. Mackenzie (N. Am. Fl.) records it from North Dakota.

63. *Carex eburnea* Boott. Very slender plant, usually 1—2 dm. high, making an almost moss-like growth on wooded banks of streams; perigynia 1—6 in each spike and spikes 2—4 in a loose, slender cluster; achenes gray to nearly black, 2 mm. long, perigynia thin, closely covering the achenes; scales rounded, thin, translucent. Collected only in Richland County near Kindred, at Kathryn, Barnes County and Park River, Walsh County. Incorrectly reported as *C. tenella* Schk., and one lot (Stevens 332) distributed as *C. disperma* Dewey (= *C. tenella* Schk.).

64. *Carex aurea* Nutt. GOLDEN SEDGE. Low, leafy plant, forming distinct tufts but growing from long rhizomes; leaves light green; stems 0.5—5 dm. high; pistillate spikes several, 1—2 cm. long, in a moderately dense cluster; perigynia rounded, 2—3 mm. long, soft and golden or orange yellow at maturity but losing this color on drying; achene nearly black. Low prairie or other moist soil. The largest amount which I have seen was on a sandy beach at Lake Metigoshe, Turtle Mts.

65. *Carex tetanica* Schkuhr. A prairie sedge, somewhat like *aurea* but taller, usually not tufted, spikes well separated, perigynia green; flowering bracts often brownish purple. Specimens from Cass, Dickey, Kidder, Stutsman, McLean and Eddy Counties.

66. *Carex blanda* Dewey. Broad-leaved, tufted plant in woods; leaves soft, 1—3.5 dm. long, 3—15 mm. wide; stems 1—3 dm. long, often shorter than leaves; spikes usually closely clustered, lower bract leaf-like. We have it from Richland, Cass, Ransom and Barnes Counties, where it is common in woods and a striking plant because of the large tufts of wide leaves. One specimen (Stevens 363) from Leonard is 6 dm. high with narrower leaves and spikes well separated. This species was previously included in *C. laxiflora* Lam.

67. *Carex granularis* Muhl. Tufted, wide-leaved plant of wet prairie, pond margins, etc.; lower bract leaf-like; spikes closely clustered, 1—2 cm. long. Ransom, Barnes, Benson and Pembina Counties.

68. *Carex crawei* Dewey. Small plants from rhizomes; stems 1—3 dm. high; leaves 1—2 dm. long, 1—3 mm. wide; pistillate spikes narrowly oblong, 1—3 cm. long, often well separated. Wishek, Logan County (Stevens in 1946), is our only record. It may have been overlooked because of similarity to *C. granularis* and *C. tetanica*.

69. *Carex gracillima* Schw. Rhizome short, thickly covered with brown leaf bases; stem slender, 2—9 dm. high; spikes very slender, drooping, 1—5 cm. long, the terminal one with staminate flowers on lower part. Woods along Sheyenne River, Richland County in 1945 (Stevens 831, 832).

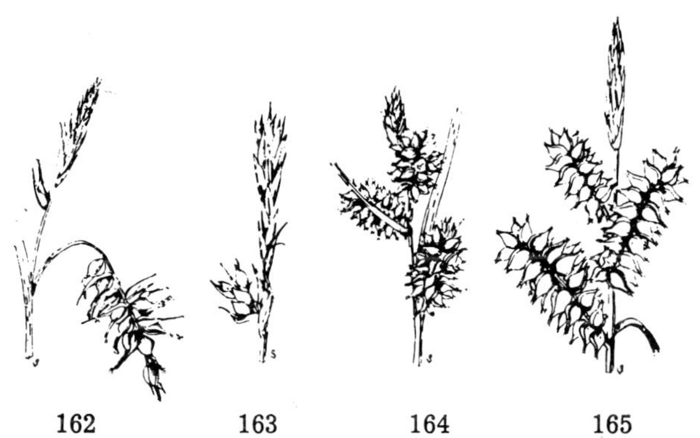

162   163   164   165

70. *Carex assiniboiensis* W. Boott. Slender, tufted plant in woods; fertile stems 3—6 dm. high; sterile stems 6—8 dm. long, curving downward and striking root at tip; spikes well separated, slender, few flowered; perigynia 6 mm. long, beak very slender, longer than body. This has been found in Cass, Richland, Pembina, McHenry and Bottineau Counties, abundant in some places. The leaves on stolons are short, those on outer half directed back toward middle portion.

71. *Carex sprengelii* Dewey. LONG-BEAKED SEDGE. (Fig. 162). Tufted, erect plant, but coming from stout rhizomes; stems slender, 3—8 dm. high, the drooping spikes with conspicuous, slender perigynia beaks. Quite common in wooded areas, especially on sandy banks and ridges. Previously listed as *C. longirostris* Torr.

72. *Carex viridula* Michx. (Fig. 164). Low, tufted plant; stems 1—3 dm. high; spikes oblong, 5—10 mm. long, closely clustered; lower bract well developed, leaf-like. Low grassland and pond edges. Kidder, Eddy and Bottineau Counties only are represented but it is probably well distributed through eastern and central part of State.

73. *Carex torreyi* Tuckerm. I am not familiar with this species. One specimen from Mercer (?) County (L. R. Waldron 2242) had been referred to it as *C. abbreviata* Prescott.

74. *Carex lanuginosa* Michx. Slender, long leaved plant in wet grassland; leaves 2—6 dm. long, 2—5 mm. wide; stems commonly 4—6 dm. high; spikes 2—5 cm. long, 5—8 mm. wide. Quite a common plant, producing a dense growth of long, narrow leaves. The rounded, fuzzy perigynia will distinguish it quite easily.

75. *Carex aquatilis* Wahl. Coarse, tufted plant along edges of rivers and marshes; leaves numerous, 2—4 dm. long, 2—5 mm. wide; stems 6—10 dm. high; spikes cylindric, 1—4 cm. long. A common, often abundant plant in wet places, producing a heavy growth of foliage, blooming in early May. Formerly included in *C. stricta* Lam.

76. *Carex emoryi* Dewey. Two specimens from Towner, McHenry County, collected by Lunell, are labeled as this species and Mackenzie reported it for North Dakota. The plant resembles *C. aquatilis*.

77. *Carex hystricina* Muhl. Medium sized plant, 3—6 dm. high; spikes 1—6 cm. long, 10—15 mm. wide, dropping on slender stalks; perigynia 5—7 mm. long. In wet places, especially boggy or seepage areas. A widely distributed species but rather uncommon in North Dakota. We have it from Richland, Ransom, Barnes, Stutsman, Benson, McHenry and Pembina Counties.

78. *Carex pseudo-cyperus* L. Tall, stout plant, growing in large tufts; stems 3—10 dm. high; spikes 2.5—7.5 cm. long, drooping on slender stalks; perigynia numerous, very closely crowded and usually more or less reflexed, 3.5—5 mm. long. A striking swamp species which has been found only as a rare plant near Walhalla, Pembina County and St. John, Rolette County.

79. *Carex laeviconica* Dewey. A rather wide leaved, yellowish green sedge, frequently forming extensive stands in low grassland; leaves 1—5 dm. long, 2—8 mm. wide, often longer than stems, the sheaths without hairs; spikes 2.5—7 cm. long, 8—10 mm. wide; perigynium with a prominent 2-toothed beak. Formerly included in *C. trichocarpa* Muhl.

80. *Carex atherodes* Spreng. SLOUGH SEDGE. (Fig. 154). Similar to preceding but larger; leaf sheaths covered with fine hairs. One of the commonest and most abundant of all slough sedges, frequently forming dense, pure stands 1—1.5 m. high in wet ground or shallow water. It is often used for hay where such areas become dry enough to mow late in summer.

81. *Carex rostrata* Stokes. Stout, bog plant, resembling *C. atherodes* or *C. laeviconica,* but perigynia smaller and more crowded in the spikes. We have it from Richland, Benson and McHenry Counties only.

82. *Carex retrorsa* Schw. (Fig. 165.) Stout plant forming large tufts; stems 2—8 dm. high; spikes 1.5—8 cm. long, 1.5—2.0 cm. wide, often crowded into a dense group and spreading out at a wide angle; perigynia 7—10 mm. long, crowded, somewhat reflexed at least at lower part of spike. River banks, pond edges, especially old channels. Quite a common and striking plant because of the large tufts, 2—4 dm. in diameter and the crowded, stiff, spreading spikes.

ARUM FAMILY Araceae

Leaves usually broad, from a rootstock or corm; flowers very small but in dense spikes often enclosed by a large, colored leaf; petals 0, sepals 0 or 4—6; fruit a berry with one or more seeds. This family contains many tropical species. Calla "lily", grown as an ornamental, is one member. Elephant ears and taro are others.

Key to Species

Leaves with 3 broad leaflets; fruit a group of red berries.
1. Arisaema atrorubens
Leaves long and narrow; berries little fleshy in an oblong spike.
2. Acorus calamus

1. *Arisaema atrorubens* (Ait.) Blume. JACK-IN-THE-PULPIT. (Fig. 81). Perennial from a rounded corm; leaves 2 or 3 on an upright stalk 2—6 dm. (4—12 in.) high; leaflets ovate, 5—10 cm. (2—4 in.) long; flower stalk 2—4 dm. high; flowers enclosed by a cylindrical, green or purplish spathe which spreads out at top into a flaring, pointed part; berries bright red, 5—10 mm. (⅕—⅖ in.) wide. Frequent in woods eastern part of State. May. Sometimes called Indian Turnip but should not be confused with Tipsin. Corms of Jack-in-the-Pulpit contain fine crystals which produce sharp irritation if eaten raw. Formerly called *A. triphyllum* (L.) Schott.

2. *Acorus calamus* L. Sweet Flag. (Fig. 80). Perennial from a rhizome; leaves erect, 3—10 dm. long, somewhat folded, 1—2 cm. wide; flower spike near top of stalk which resembles the leaf; fruiting spike 3—8 cm. long, 1 cm. thick. Our only record is from a small boggy spot in Turtle Mts. near Bottineau, shown to us by Henry Klebe in 1943. It may have been introduced from the east by Indians with whom it was a favorite plant. Mr. Klebe said it had been in the area for at least 35 years. Dr. M. L. Gilmore reported a colony along the Mouse River near Towner, Pierce County.

DUCKWEED FAMILY Lemnaceae

Floating plants consisting of only a tiny leaf and short root (fig. 62). Flowers very minute, rarely seen, of 1 pistil and 1 or 2 stamens. The plants usually multiply by budding. They are eaten by ducks but not extensively.

Key to Species

Leaves long stalked, remaining together to appear 3-lobed.
1. *Lemna trisulca*
Leaves rounded, not stalked.
Root single without a central core. 2. *Lemna minor*
Roots several, with a central core. 3. *Spirodela polyrhiza*

1. *Lemna trisulca* L. Leaf blade oblong, 5—10 mm. (⅕—⅖ in.) long, the stalk nearly as long. Quite common. Usually in tangled masses below the surface.

2. *Lemna minor* L. Leaf nearly round, rather thick, 3—5 mm. long. Common, forming dense floating masses on shallow water. This name is used for the common small species. *L. perpusilla* Torr. has been reported.

3. *Spirodela polyrhiza* (L.) Schleiden. Resembles the last but is distinctly larger, 3—8 mm. long. It is not common but occasionally occurs with *L. minor*.

SPIDERWORT FAMILY Commelinaceae

Perennials with stems bearing 2—4 long, narrow leaves and a terminal cluster of showy flowers. Stems 2—6 dm. (8—24 in.) high; leaves 1—3 dm. long, 1—2 cm. (2/5—4/5 in.) wide at base; sepals 3, petals 3, stamens 6, pistil 1; fruit a several-seeded, oblong capsule 4—8 mm. (1/4 in.) long.

Key to Species

Bracts (2 upper leaves) enlarged at base, wider than other leaves.
1. *Tradescantia bracteata*
Bracts not enlarged at base. 2. *Tradescantia occidentalis*

1. *Tradescantia bracteata* Small. Spiderwort. A common species in meadows, along roadsides or even in fields but not on drier prairie. Flowers 2—3 cm. wide, petals varying from pink to dark lavender. June-Aug. The petals wither by noon on a warm day into a soft, inky mass. The filaments of the stamens bear long hairs consisting of a single row of very large cells which are good objects to show streaming of cytoplasm under the microscope. The plants are easily grown in a garden and tend to develop large bunches.

2. *Tradescantia occidentalis* (Britton) Smyth. Very similar to No. 1 but leaves narrower throughout and flowers always blue. Common in sandy soils.

PICKEREL-WEED FAMILY Pontederiaceae

Water or mud plants with thick, soft stems. Sepals 3, petals 3, similar in color; stamens 3—6, pistil 1; fruit a capsule.

1. *Heteranthera dubia* (Jacq.) MacM. WATER STARGRASS. (Fig. 71). Floating or creeping; leaves grass-like, 3—15 cm. long, 3—5 mm. wide; flowers yellowish, 15—20 mm. wide, solitary from leaf bases on slender stalks which rise to surface of water. Records from only Wahpeton and Valley City.

RUSH FAMILY Juncaceae

Wiry, grass-like plants in wet places. Flowers small; sepals and petals 3 each, similar, brown, chaffy; stamens 3 or 6, pistil 1; fruit a small, rounded capsule with many very minute seeds. Common along stream or pond banks, ditches, or in wet meadows. Of little value for forage. (Fig. 166).

Key to Species

Leaves apparently none except one which seems a continuation of
    the stem, the flower cluster arising at its base.
  Stamens 3; flower green or becoming yellowish; plants tufted.
                                                1. *Juncus effusus*
  Stamens 6; flower parts purplish, green only medially; stems from
    long rhizomes.   2. *Juncus balticus*
Leaves present, not appearing as a continuation of the stem.
  Leaves flattened, at least the lower part.
    Stem widely branched from base; annual.   3. *Juncus bufonius*
    Stem erect, 1—4 dm. (4—16 in.) high, branched only or chiefly
        in the flower cluster.
      Flowers 5—6 mm. (¼ in.) long, in several, 3—8- flowered
        heads.   4. *Juncus longistylis*
      Flowers 3—4 mm. long in larger, loose clusters.
        Auricles of leaf sheaths thin and whitish.   5. *Juncus interior*
        Auricles of leaf sheaths thick and brownish. 6. *Juncus dudleyi*
  Leaves round, hollow with occasional cross walls.
    Upper leaf extending beyond the dense, rounded flower head.
      Flower heads 7—10 mm. wide; leaves erect.   7. *Juncus nodosus*
      Flower heads 10—15 mm. wide; leaves spreading. 8. *Juncus torreyi*
    Upper leaf shorter than the elongated, loose flower cluster.
      Stamens 6; sepals 2—2.5 mm. long.   9. *Juncus brachycephalus*
      Stamens 3; sepals 3—4 mm. long.   10. *Juncus alpinus*

1. *Juncus effusus* L. BOG RUSH. Resembles No. 2 except for more tufted habit of growth. A specimen by Brannon in 1895 was recorded by Bergman.

2. *Juncus balticus* Willd. BALTIC RUSH. Stems 3—6 dm. (1—2 ft.) high, very wiry, closely placed on a thick rhizome which is covered with light brown scales; flowers usually appearing blackish. A widely distributed and very common species along pond shores and in other wet places. Often abundant in low meadows but not regarded as of much forage value.

3. *Juncus bufonius* L. TOAD RUSH. Annual, widely branched from base, 5—20 cm. (2—8 in.) high; leaves 2—5 cm. long, 1 mm. (1/25 in.) wide, flat or rolled; flowers borne singly on branches but usually appearing somewhat clustered at tips. Frequent along stream or pond margins.

4. *Juncus longistylis* Torr. Perennial, stems few in a tuft, stiff and erect, 3—6 dm. high; flowers in 3—10 heads of 3—8 flowers each, near top of stem. We have this only from Ransom, Kidder, Nelson and Williams Counties. It is rather distinctive in the few, rather large flowers in definite heads.

5. *Juncus interior* Wiegand. Perennial, densely tufted, 2—4 dm. high; flowers placed quite closely along branches of cluster at top of stem. A common species in wet or only moist soil; frequent in meadows. This and the next were formerly included in *J. tenuis* Willd.

6. *Juncus dudleyi* Wiegand. Very similar to last in appearance and apparently less common though I have not been able to compare them carefully.

166   167

166. Rush (*Juncus*).
167. Leopard Lily (*Fritillaria atropurpurea*).

7. *Juncus nodosus* L. KNOTTED RUSH. Perennial; stems slender, tufted, 1.5—6 dm. high; flowers in 1—30 distinct, rounded heads 6—10 mm. wide. Frequent.

8. *Juncus torreyi* Coville. TORREY'S RUSH. Similar to last but larger, heads denser, 10—15 mm. wide; leaves 2—4 dm. long.

9. *Juncus brachycephalus* (Engelm.) Buch. SMALL-HEADED RUSH. Perennial from a rhizome but stems tufted, 3—6 dm. high, widely branched in flower cluster; heads 2—5-flowered, flower parts not sharp tipped as in most species. Records from Ransom, Benson and McHenry Counties only.

10. *Juncus alpinus* Vill. Stems densely tufted or matted, 1—4 dm. high; flower cluster with 5—12 heads, each 3—10-flowered. We have this

only from shore of Lake Metigoshe near Bottineau. Previously identified as *J. canadensis,* Dr. F. J. Hermann has determined it as *J. alpinus,* var. *fuscescens* Fernald.

LILY FAMILY Liliaceae

Plants with erect, leafy stems or with few leaves from bulbs or rhizomes. Flowers often showy, most often white; sepals 3, petals 3, usually similar; stamens 6, pistil 1, ovary 3-celled; fruit a capsule or berry.

Key to Species

Flowers dioecious; leaves rounded; stem partly climbing by tendrils.
21.　*Smilax herbacea*
Flowers perfect; leaves longer or much longer than wide; not climbing.
　Flowers quite small, 3—5 mm. (⅛ in.) wide, not longer than wide; fruit a berry.
　Leaves absent, stems divided into slender branches 1—2 cm. long.
14.　*Asparagus officinalis*
　Leaves ordinary, 1—3 cm. (⅖—1⅕ in.) wide; flowers in a terminal cluster.
　　Leaves single or 2 on base of flower stalk 1 dm. (4 in.) high (fig. 169).
18.　*Maianthemum canadense*
　　Leaves several on a stem 2—6 dm. high.
　　　Flowers 3—10 in a simple, terminal raceme (fig. 170).
15.　*Smilacina stellata*
　　　Flowers many in a thick, branched cluster.
16.　*Smilacina racemosa*
　Flowers larger, 0.6—8 cm. (¼—3 in.) wide, often longer than wide.
　Flowers 0.6—1 cm. wide.
　　Flowers bell shaped, longer than wide; leaves ovate.
17.　*Disporum trachycarpum*
　　Flowers not longer than wide; leaves long and narrow.
　　　Flowers in a long, branching cluster.
　　　　Flowers greenish white, 10 mm. wide.　　1.　*Zigadenus elegans*
　　　　Flowers yellowish, 8 mm. wide.　　2.　*Zigadenus gramineus*
　　　Flowers in an umbel on a long stalk.
　　　　Leaves flat, 2—3 cm. wide; flowers in July after leaves have dried.　　4.　*Allium tricoccum*
　　　　Leaves narrow and rounded, 5—7 mm. thick; flowers and leaves present at same time.
　　　　　Flowers white, appearing in late May.　　5.　*Allium textile*
　　　　　Flowers lavender, appearing in August.　　6.　*Allium stellatum*
　Flowers 1—8 cm. (½—3 in.) wide, often tubular and longer than wide.
　　Leaves stiff, 2—4 dm. (8—16 in.) long, crowded on a woody base.
11.　*Yucca glauca*
　　Leaves not esepcially stiff, sometimes long and slender; no woody base.
　　　Leaves narrow, 1 cm. or less wide, 1—3 dm. long.
　　　　Sepals narrow; petals 2—5 cm. wide, creamy white with purple base.　　10.　*Calochortus nuttallii*
　　　　Sepals and petals alike.
　　　　　Plants stemless; flowers white.　　3.　*Leucocrinum montanum*
　　　　　Plants leafy stemmed; flowers yellow or greenish.
　　　　　　Flowers yellowish green with purple spots, solitary at upper leaf bases (fig. 167).　　8.　*Fritillaria atropurpurea*
　　　　　　Flowers yellow, 1 or 2 on a central stalk.
9.　*Fritillaria pudica*
　　　Leaves 2—5 cm. wide, 5—15 cm. long.
　　　　Leaves 5 cm. wide, 3 at top of a stalk 3—5 dm. high; sepals green.
20.　*Trillium cernuum*

Leaves 1—3 cm. wide, not so placed; sepals and petals alike.
Flowers 5—8 cm. wide, red, funnel shaped.
7. *Lilium philadelphicum*
Flowers 1—2 cm. wide, yellow or yellowish, bell shaped, drooping.
Flower parts yellow, twisted (fig. 171).
12. *Uvularia grandiflora*
Flower parts nearly white, not twisted (fig. 172).
Stem 5—10 dm. high, leafy, usually arched.
19. *Polygonatum commutatum*
Stem 2—3 dm. high, erect, forked above.
13. *Uvularia sessilifolia*

1. *Zigadenus elegans* Pursh. CAMAS. Perennial from an elongated bulb; leaves 2—4 dm. (8—16 in.) long, 1—2 cm. (⅖—⅘ in.) wide, waxy, somewhat folded; flower stalk 3—6 dm. high, flowers in a simple or branching cluster; flower parts oblong, nearly white with a yellow spot at base; fruit an oblong, 3-pointed capsule, 2 cm. long. Frequent in coulees or other moist grassland. Early July.

2. *Zigadenus gramineus* Rydb. A smaller plant than the preceding, with smaller, more yellowish flowers, the leaves green, not waxy. We have this only from Williams and Dunn Counties. In 1945, we found it common in Williams and did not find it farther south. It seems to bloom in mid-June, earlier than Z. *elegans*. It is similar to and as poisonous as Z. *venosus*, known as Death or Poison Camas, a livestock poisoning plant from western Montana to Eastern Washington. Z. *elegans* is only slightly poisonous and not considered dangerous. The camas eaten by the Indians of Idaho was a related plant with blue flowers, *Camassia quamash*.

3. *Leucocrinum montanum* Nutt. SAND LILY. Perennial from a short crown; leaves many, 1—2 dm. long, 2—8 mm. wide; flowers pure white, 3—4 cm. wide, on a slender tube 5—10 cm. high in May. Our first record of this western species came from Mrs. W. H. Atkinson at Vim, Slope County in 1935. It was also found near Golva, Golden Valley County, by E. C. Moran and J. Clayton Russell. It is a good plant for cultivation, and can be increased by either division of old plants or from the black seeds which are borne in capsules just under ground.

4. *Allium tricoccum* Ait. WOOD LEEK. Perennial from an oblong, thin-coated bulb; leaves large, dark green, 1—3 dm. long, 2—4 cm. wide, appearing in early spring; flowers white, inconspicuous, on a stalk 1—2 dm. high, in early July. Rather rare in woods. We have it from Cass and Ransom Counties only, but it should be present in the other Red River Valley counties.

5. *Allium textile* Nels. & Macbr. WHITE WILD ONION. Perennial from a fibrous-coated bulb; leaves 1—1.5 dm. long, 2—4 mm. wide, rounded, not flattened; flower stem 1—2 dm. high. Common but sometimes local, especially on stony hills. May. This was formerly called *A. reticulatum* Fraser, but that name had been used for a different species.

6. *Allium stellatum* Ker. PINK WILD ONION. Similar to the last but taller and more slender; bulb membranous-coated. Frequent to common in better soil than A. *textile*. Aug. We have no specimens from west of Missouri River.

Allium canadense L. MEADOW GARLIC, has been found at Muskoda, Minn., about 15 miles east of Fargo. It resembles No. 6 but the flowers are followed by bulblets instead of seed capsules and blooms in June.

7. *Lilium philadelphicum* L. WILD LILY. (Fig. 168). Perennial from a deep, white, scaly bulb; stem leafy, 3—6 dm. (1—2 ft.) high; leaves alternate, 2—5 cm. (1—2 in.) long, 3—7 mm. (⅛—¼ in.) wide; flowers 1—3 at top of stem. Frequent in coulees and moist prairie, blooming about July 1. This is popularly called "tiger lily" but that is a garden plant. Most authors regard our plant as a form of the eastern wood lily, var. *andinum* (Nutt.) Ker. It is often called *Lilium umbellatum* Pursh, but the "umbellatum lily" of the gardeners is a different plant.

163. Wild Lily (*Lilium philadelphicum*).

8. *Fritillaria atropurpurea* Nutt. LEOPARD LILY. (Fig. 167). Perennial from a bulb of thick, fleshy scales; stems 1—4 dm. high with a few leaves 3—8 cm. long, 1—3 mm. wide; flowers 1—4 at upper leaf bases, saucer or bell shaped, 1.5—3 cm. wide; capsule 1.5 cm. long, nearly as wide, 6-sided. Hillsides from Missouri River west. May. One record from Jamestown in 1891 may be an error. In some localities it is called "rice-root", the bulb scales resembling grains of rice.

9. *Fritillaria pudica* (Pursh) Spreng. YELLOW BELL. Stem 1—2 dm. high, with several leaves below and 1 or 2 yellow flowers above; leaves 5—10 cm. long, 5—10 mm. wide. Our first knowledge of this came from the late Fannie M. Heath of Grand Forks, who had it in her garden in 1931. Russell Reid confirms her statement that it grew near Huff, south of Mandan at the foot of the bluffs, but we still have no bona fide specimen.

10. *Calochortus nuttallii* T. & G. MARIPOSA LILY. MARIPOSA TULIP. Perennial from an onion-like bulb; stem slender, 2—5 dm. high; leaves few, very narrow, 1—3 dm. long, somewhat folded or rolled; flowers 1 or 2, terminal, 5—8 cm. wide; petals 3, broadly wedge shaped, white with a dark purple base; sepals ovate or lanceolate 1 cm. long, green with whitish edges; capsule 3-angled, 2—3 cm. long, 5 mm. wide, somewhat narrowed toward the top. Coulees and lower slopes of hills; quite local but sometimes abundant. June. Extreme western edge of State, from Marmarth into McKenzie and Slope Counties (fig. 11).

11. *Yucca glauca* Nutt. YUCCA. Stems short, very stout and woody; leaves 2—5 dm. (8—20 in.) long, 1—2 cm. (¾ in.) wide, very stiff and sharp pointed; flowers on a simple or branched central stalk 5—10 dm. high; flowers greenish white, rounded, 3—5 cm. wide; capsules stout, woody, 6—10 cm. long, persisting for a year or more; seeds black, thin, 1 cm. wide. Frequent on hillsides, Missouri River westward (fig. 11). June, July. The names beargrass, soapweed and Spanish bayonet are also applied to this plant, but the last really belongs to a related species, *Y. baccata*, of the Southwest, the capsules of which are somewhat fleshy and are roasted and used for food.

12. *Uvularia grandiflora* J. E. Smith. LARGE BELLWORT. (Fig. 171). Perennial from a rhizome; stems upright, leafy, 3—8 dm. high; leaves oblong, 3—5 cm. long, the base clasping the stem; flowers yellow, drooping, 2—3 cm. long, on ends of short upper branches; capsule 3-lobed, 8—10 mm. wide, hardly as long. Woods. Early May. We have it only from Cass County and a specimen from Bismarck in 1892. We do not know the history of the Bismarck specimen and incline to regard it as cultivated.

13. *Uvularia sessilifolia* L. SMALL BELLWORT. Stem 1—3 dm. high, simple or branched above; flowers cream colored. We have this only from Fargo where it is now rare or extinct.

14. *Asparagus officinalis* L. ASPARAGUS. Commonly escaped from cultivation. The red berries are eaten by birds which scatter the seeds and plants spring up in woods, hedges and windbreaks.

15. *Smilacina stellata* (L.) Desf. FALSE SOLOMON'S SEAL. (Fig. 170). Perennial from a rhizome; stems 3—6 dm. high; leaves oblong, 5—8 cm. long; several, small, white flowers at top of stem in May; fruit a sort of fleshy capsule, rounded, 6—8 mm. wide, usually striped with red. This is by far the commonest plant of this sort, occurring in woods in eastern part of State and in brushy or open coulees all through the State.

16. *Smilacina racemosa* (L.) Desf. FALSE SPIKENARD. A larger, stouter plant than the last, leaves usually somewhat curled; berries red, 4—6 mm. wide, more fleshy than those of No. 15. Rare in woods. Records from Cass and Sargent Counties only. July.

17. *Disporum trachycarpum* S. Wats. Similar to No. 12 or 15, but leaves pointed at tip, broadest at base; flowers solitary or 2 or 3 together at top of plant, greenish or yellowish, 6—8 mm. wide; fruit a bright red berry 8—10 mm. wide. Aspen woods, quite local. We have it from Pembina, Rolette, Bottineau, Stark and Dunn Counties (fig. 11).

18. *Maianthemum canadense* Desf. FALSE LILY-OF-THE-VALLEY. (Fig. 169).
Perennial from a shallow rhizome which bears broadly oblong or
rounded leaves, 3—6 cm. long; flower stem 1—1.5 dm. high, bearing 2,
ovate, pointed leaves and a flower spike 3—5 cm. long; fruit a red
berry, 4 mm. wide. Aspen woods, recorded from Richland, Pembina
and Bottineau Counties (fig. 11). It often forms extensive beds. Late
May.

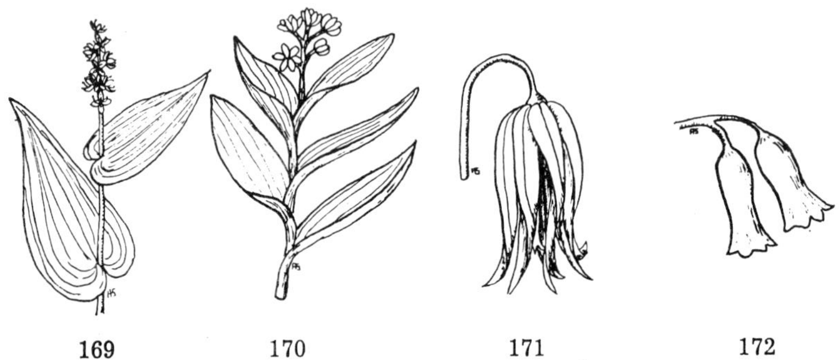

| 169 | 170 | 171 | 172 |

19. *Polygonatum commutatum* (Schult.) A. Dietr. SOLOMON'S SEAL. (Fig.
172). Perennial from a thick rhizome; stems arched, 6—15 dm. (2—5 ft.)
high; leaves oblong, 7—15 cm. (3—6 in.) long; flowers yellowish white
or greenish, 1—1.5 cm. long, hanging from leaf bases in clusters of 2—6;
fruit a bluish black berry, 1 cm. wide. Woods and thickets, widely
distributed and quite common; our largest plant of this sort. June.

20. *Trillium cernuum* L. NODDING WAKE ROBIN. Perennial from a short
rootstock; stem stout, 2—4 dm. high, bearing 3 broad leaves close together at top, the single flower coming out above but curving between
and underneath them; petals white, 2—3 cm. long; fruit a rounded,
ridged, pointed, pink berry, 1—1.5 cm. wide. Woods. Recorded for
Cass and Pembina Counties only, but probably occurs in other Red River
Valley counties. Early June.

21. *Smilax herbacea* L. CARRION FLOWER. Perennial from a rhizome; stem
arching or drooping if not supported, 1—2 m. (3—7 ft.) high, with
tendrils from leaf bases; leaves rounded ovate, 5—15 cm. long with
several prominent ribs; flowers in umbels at leaf bases on a stalk 5—10
cm. long; flower parts small and green; fruit a bluish black berry,
6—8 mm. thick, containing several, hard, rounded, coral red seeds.
Occasional to common in woods throughout the State. We even have
a specimen from a slight depression on short-grass prairie in Bottineau
County, though quite near the Turtle Mts. Flowers late in May, fruit
ripe in fall and rather conspicuous after leaves have fallen.

AMARYLLIS FAMILY Amaryllidaceae

Plants closely resembling lilies but distinguished by petals attached to top of ovary (ovary inferior). A number of house plants
belong to this group but we have only the following native species.

1. *Hypoxis hirsuta* (L.) Coville. YELLOW STARGRASS. Perennial from an
ovoid corm, 5—8 mm. (¼ in.) long; leaves 8—15 cm. (3—6 in.) long,
2—3 mm. wide, with a few, long hairs; flower stalk about as long as
leaves, 1—6 golden yellow flowers 1.5—2 cm. wide at top; fruit a capsule with black seeds. Moist prairie or coulees, usually abundant where
present, but not easily seen except when in bloom. June, July. We
have it west to Logan and Pierce Counties.

IRIS FAMILY **Iridaceae**

Lily or grass-like plants with variously shaped flowers, the flower cluster enclosed at first by a pair of bracts (small leaves or scales); stamens 3; ovary 3-celled, inferior; fruit a capsule. The garden iris belongs here and two native species come close to us: *Iris versicolor*, Blue Flag, in Minnesota and *Iris missouriensis* in the Black Hills of South Dakota. Neither is known to occur in North Dakota. Our only species are grass-like plants with star shaped, 6-parted flowers, common on prairie.

Key to Species

Flower stalk forked toward the top, bearing 2 flower clusters.
2. *Sisyrinchium graminoides*

Flower stalk not forked, bearing only 1 cluster.
1. *Sisyrinchium angustifolium*

1. *Sisyrinchium angustifolium* Mill. BLUE-EYED GRASS. (Fig. 173). Perennial from a crown; leaves numerous, 1—2 dm. (4—8 in.) long, 2—4 mm. (⅛ in.) wide; flowers several in a cluster but opening 1 or 2 at a time, about 1 cm. (⅖ in.) wide, deep blue to white; capsule rounded, 3—5 mm. wide, seeds black. Common on prairie. May, June. There may be more than one species as the distinctions between them are technical.

173

174

173. Blue-eyed Grass (*Sisyrinchium angustifolium*).
174. Coral Root (*Corallorrhiza striata*).

2. *Sisyrinchium graminoides* Bicknell. This we found in North Roosevelt
Park, McKenzie County in 1943. It formed distinct colonies, 1—2 m.
wide and seemed a taller, stouter plant than the common species.

ORCHID FAMILY Orchidaceae

Perennials, usually from short, fleshy roots; flowers usually
showy but structure complicated; flower parts 5 or 6, petals usually
different from sepals and not all alike; stamens united with pistil,
anthers usually 1, pollen cohering in 2—8 masses; ovary inferior,
3-angled but 1-celled; fruit a capsule with many, very small seeds.
Of this large tropical family we have only a few species, mostly
local and rare.

Key to Species

Flowers 1—3 at top of leafy stem; middle petal (lip) large, pouch or
    slipper shaped.
  Lip white, marked with pink or purple.
    Lip 2.5—5 cm. (1—2 in.) long; stout, rough hairy, swamp plants.
        1. *Cypripedium reginae*
    Lip 2 cm. long; slender plants on wet prairie.
        2. *Cypripedium candidum*
  Lip yellow, 2—4 cm. long; usually in woods. 3. *Cypripedium parviflorum*
Flowers several to many in elongated clusters; middle petal not un-
    usually large, often spurred at base.
  Plants green, leafy; a bract (small leaf) below each flower.
    Bracts as long as or longer than flowers.
      Spur half as long as lip which is toothed.   4. *Habenaria bracteata*
      Spur as long as lip which is entire.   5. *Habenaria hyperborea*
    Bracts much shorter than flowers.
      Lip broad, prominently fringed.   6. *Habenaria leucophaea*
      Lip not fringed.
        Leaves alternate along stem.   7. *Spiranthes cernua*
        Leaves 2 at base of flower stalk.   8. *Liparis loeselii*
  Plants not green; leaves represented by purplish scales.
    Lip 3-lobed.   9. *Corallorrhiza maculata*
    Lip not lobed, but edge slightly wavy.   10. *Corallorrhiza striata*

1. *Cypripedium reginae* Walt. SHOWY LADYSLIPPER. Plant stout, 3—6 dm.
(1—2 ft.) high, covered with coarse hairs; leaves broadly elliptic,
5—15 cm. (2—6 in.) long; flower white, marked with pink or purple,
lip 2.5—5 cm. long. Known only from swampy areas near Walhalla,
Pembina County; Fort Totten, Benson County; Leonard, Richland
County. Blooms about July 1. The names *C. hirsutum* Mill. and *C.
spectabile* Salisb. have also been applied to this plant. The plants
cause severe skin poisoning to some people.

2. *Cypripedium candidum* Willd. SMALL WHITE LADYSLIPPER. Stems 1—3
dm. high, usually in clumps; stems and leaves smooth; leaves elliptic or
lanceolate, 6—15 cm. long; lip of flower 2 cm. long, white with some
purple inside. Moist prairie, blooming June 1. There are old records
for Fargo and for Rutland, Sargent County. It is quite common about
15 miles east of Fargo in Minnesota. In 1935, I found it near Park
River, Walsh County and later in Richland County near Leonard.

3. *Cypripedium parviflorum* Salisb. YELLOW LADYSLIPPER. (Fig. 175).
Stems single or in clumps, 2—5 dm. high; lip of flower 2—4 cm. long,
deep yellow; sepals green or yellowish, striped with purple. Woods or
wet prairie. June. Recorded from Ransom, Walsh, Pembina, Benson
and McHenry Counties. Large flowered plants are often called *C.
pubescens* Willd. Recent authors consider both as varieties of the European *C. calceolus* L.

175. Yellow Ladyslipper (*Cypripedium pubescens*).

4. *Habenaria bracteata* (Willd.) R. Br. LONG-BRACTED ORCHIS. Stem leafy, 1—3 dm. high; leaves 5—10 cm. long, 1—2 cm. wide at base, usually tapering to a point; flowers greenish, 8—10 mm. long in a terminal spike; bracts 2—4 cm. long, lanceolate. Frequent in woods. June, July. Records from Morton, Stark, McLean and eastern counties. Recent authors regard this a variety of *H. viridis* R. Br.

5. *Habenaria hyperborea* (L.) R. Br. TALL GREEN ORCHIS. Similar to No. 4 but larger. Records from Ransom, LaMoure, Barnes and Bottineau Counties.

6. *Habenaria leucophaea* (Nutt.) Gray. PRAIRIE FRINGED ORCHIS. Stems 3—10 dm. (1—3 ft.) high; leaves narrow, 1—2 dm. long; spike 5—15 cm. (2—6 in.) long, densely flowered; lip white, 10—15 mm. (½ in.) long, 3-lobed and long fringed; spur slender, 2—3 cm. long. We have this only from low, sandy prairie in northern Richland County.

7. *Spiranthes cernua* (L.) Rich. LADIES TRESSES. Stem 1—2 dm. high; leaves narrow, 5—15 cm. long; flowers white, 1 cm. long, in a dense spike 5—10 cm. long. Moist prairie. Late Aug. We have it only from Richland, Ransom and Stutsman Counties. It is a striking little plant, the flower spike slightly curving and twisting, the flowers in three rows.

8. *Liparis loeslii* (L.) Rich. LOESEL'S TWAYBLADE. Leaves 2, elliptic-lanceolate, 5—15 cm. long; flower stalk 5—20 cm. high with several greenish flowers 4—6 mm. long. Wet, wooded places. Walhalla, Pembina County is our only record.

9. *Corallorrhiza maculata* Raf. CORAL ROOT. Stem 2—4 dm. high, purplish, with several scales in place of leaves; flowers purplish, 1 cm. long, in a terminal spike. The plants grow on roots of trees. Some specimens collected by Mrs. Anna Meissner in an aspen thicket along the Heart River south of Richardton, Stark County, seemed to represent both this and the next species.

10. *Corallorrhiza striata* Lindl. (Fig. 174). Very similar to last. One old record from Neche, Pembina County. The plants are easily overlooked and may occur in other localities in aspen woods where leaf mold, rather than grass, is the ground cover.

WILLOW FAMILY Salicaceae

Trees or shrubs with soft wood, alternate leaves and dioecious flowers in catkins (drooping spikes); sepals and petals 0, stamens few or many, pistil 1; fruit a small capsule with many, small, hair-tufted seeds.

Key to Genera

Buds covered by several scales; bracts of catkins fringed; stamens
  many.　　　　　　　　　　　　　　　　　　　　　1-4. *Populus*
Buds covered by 1 scale; bracts entire; stamens 2—10.　　　5-18. *Salix*

Populus POPLARS and COTTONWOODS

Large trees with broad leaves. In addition to the following, many selections or hybrids are planted and they are difficult to identify.

Key to Species

Petioles flattened laterally; leaves rounded or sharp pointed.
  Leaves triangular, straight across base, long pointed.
　　　　　　　　　　　　　　　　　　　1. *Populus deltoides*
Leaves rounded, nearly circular.　　　　2. *Populus tremuloides*
Petioles rounded, not flattened; leaves ovate to lanceolate.
  Capsules not hairy.　　　　　　　　　3. *Populus balsamifera*
  Capsules covered with fine hairs.　　　4. *Populus trichocarpa*

1. *Populus deltoides* Marsh. COTTONWOOD. Large tree, trunk up to 1 m. (3 ft.) or more thick; old bark deeply furrowed, young bark smooth, greenish; leaves thick and smooth, 7—12 cm. (3—5 in.) long, with few, rounded or angular teeth; staminate catkins bright red, 4—8 cm. long, becoming loose and twice as long before falling; fruiting catkins 1—2 dm. (4—8 in.) long, loose, capsules 10 mm. wide. Chiefly along streams; abundant along Missouri and Little Missouri Rivers, rare in eastern part of State but everywhere planted. Early May.

    The correct name for the cottonwood is a puzzle. Some authors have adopted *P. balsamifera* L. The WESTERN COTTONWOOD, *P. sargentii* Dode, distinguished by yellowish branchlets and finely hairy buds, should occur in the western part of the State but we have been unable to form a definite opinion on it.

2. *Populus tremuloides* Michx. ASPEN. Small to medium sized tree, trunk 1—2 dm. in diameter where growing in dense stands, reaching 4 dm. in large trees; young bark almost as white as birch; leaves 2—5 cm. long with very fine, rounded teeth; staminate catkins very hairy. Mid-April. Most abundant in Turtle and Pembina Mts.; local elsewhere along streams, especially in deep, sheltered valleys. Along northern edge of State colonies of small trees in the open are frequent.

3. *Populus balsamifera* L. BALSAM POPLAR. Similar to aspen; young bark green; leaves ovate to lanceolate, 5—10 cm. long, pale or rusty below, edges finely toothed. In same places as aspen, much less common. Both species spread by roots. Leaves of balsam poplar on young sprouts may be rounded, nearly as in basswood. The name *P. tacamahacca* Mill. is used for this tree by those who apply *P. balsamifera* to the cottonwood.

4. *Populus trichocarpa* Hook. BLACK COTTONWOOD. On May 15, 1941, I collected specimens from a colony in a gully above the Little Missouri River in southern Billings County (Twp. 138, R. 102), a place locally known as "Popple Springs". A specimen from these was identified by C. H. Mueller, U. S. Dept. Agr., as this species. It resembles cottonwood but has leaves more like those of balsam poplar. The leaves on this specimen are rounded ovate, abruptly acuminate. This species occurs in the Rocky Mountains, but mostly still farther west. We know of no other record in this area.

Salix WILLOWS

Mostly shrubs with narrow leaves. A difficult group because of numerous species and small differences between them. Dr. C. R. Ball, who had given special attention to these plants, collected in the State in 1928, and generously presented us with many named specimens. Aside from these and a few specimens which Dr. Ball had identified for Bergman, we have given little attention to the group. Collections need to be made from the same plants at different dates to secure flowers, leaves and fruit. Most collections are made at a single date and give very incomplete information. A local student could contribute much to our knowledge of these plants by studying carefully the plants in his locality, collecting good specimens at different dates and determining dates of flowering and maturing of fruit.

Length and shape of leaf varies considerably on any plant. Lower leaves on a twig are usually shorter, broader and more rounded than upper ones. Leaves on a sprout or other vigorous shoot are usually larger and wider than those on ordinary terminal branches. Still, the prevailing form of leaf and teeth of margin are among the most useful characters in identifying species. Except in the short leaved *S. tristis*, there is not much difference in length of leaf between species, and this is omitted in descriptions given here. Because several of our forms are rare and local, the following key to common ones should be useful.

Key to Common Willows

Leaves 2—5 mm. ($\frac{1}{12}$-⅕ in.) wide, linear, nearly same width through-
  out.
  Teeth of edge rather separated and prominent; plant spreading
    by roots.   9. *Sandbar Willow*
  Teeth fine and close together; plants not spreading by roots.
                                 10. *Slender Willow*
Leaves 6—10 mm. (¼—⅖ in.) wide; oblong, obovate or lanceolate.
  Leaves widest toward base, gradually tapering to sharp tip.
    Small to large tree; no stipules.   5. *Peach-leaved Willow*
    Large shrub; rounded stipules present.   12. *Heart-leaved Willow*
  Leaves widest, or equally wide, toward tip, suddenly narrowed or
    rounded.
    Leaves green; rounded stipules present.   13. *Pussy Willow*
    Leaves white, often hairy below; no stipules.   15. *Beaked Willow*

Complete Key to Willows

Leaves not hairy, or only slightly when young.
  Catkins on new leafy branches of the year.
    Pistillate catkins slender and loose; large tree.   5. *Salix amygdaloides*
    Pistillate catkins thick and dense; shrubs.
      Leaves lanceolate or ovate-lanceolate.
        Capsules maturing in early summer.
          Leaves long acuminate, shining on both sides. 6. *Salix lucida*
          Leaves acute or short acuminate, paler below.
              7. *Salix pentandra*
        Capsules maturing in autumn.   8. *Salix serissima*
      Leaves linear or linear-lanceolate.
        Spreading by roots, forming dense thickets.   9. *Salix interior*
        Not spreading by roots; stems single or in a dense clump.
          Stems many in a clump; leaves narrowed at both ends.
             10. *Salix petiolaris*
          Stems single; leaves wider at base.   11. *Salix lutea*
  Catkins on very short branches which are usually without leaves.
    Flowers appearing with the leaves; capsules not hairy.
             12. *Salix cordata*
    Flowers appearing before the leaves.
      Twigs of the year finely hairy; leaves whitish below.
             14. *Salix missouriensis*
      Twigs not hairy; leaves not whitish below.   13. *Salix discolor*
Leaves densely hairy or white at least below.
  Leaves oblong, ovate or ovate-lanceolate.
    Leaves with dense hairs, lying flat when young, nearly smooth
      but white when old.   15. *Salix bebbiana*
    Leaves permanently white with fine, tangled hairs. 16. *Salix humilis*
  Leaves linear, narrowly lanceolate or ovate-lanceolate.
    Leaves 4—12 cm. (2—5 in.) long, very gray with fine hairs.
             17. *Salix candida*
    Leaves 2—5 cm. (1—2 in.) long, gray below but not above.
             18. *Salix tristis*

5. *Salix amygdaloides* Anders. Peach-leaved Willow. Large tree, trunk to 5 dm. (20 in.) thick; branches slender, yellowish; leaves lanceolate or ovate-lanceolate, 5—10 cm. (2—4 in.) long, finely toothed. Frequent along stream banks. Blooming about May 10. Golden Willow, *S. vitellina* L., commonly planted, is similar, but has golden yellow or coppery twigs, leaves whitish below. Only staminate trees are seen because they are propagated by cuttings and thus blowing seeds are avoided.

6. *Salix lucida* Muhl. Shining Willow. Shrub 2—5 m. (7—16 ft.) high; leaves thick, green and glossy, ovate-lanceolate, 5—10 cm. long, finely toothed. In boggy places. Three specimens from Dunseith, Rolette County (2480-2), collected by Dr. Ball. I have placed here also one from Walhalla (Stevens 367).

7. *Salix pentandra* L. Laurel-leaved Willow. Shrub or small tree; twigs yellow or reddish; leaves glossy above, elliptic-ovate, 5—12 cm. long, finely toothed. A specimen collected at Valley City in 1899 by Miss Laura L. Perrine was identified as this species by Dr. Ball. It was no doubt a planted specimen as the species is introduced from Europe.

8. *Salix serissima* (Bailey) Fern. Autumn Willow. Shrub, 2—4 m. high with dark brown branches, otherwise similar to last, differing in longer period required to mature seed. A northern plant found in cold bogs; collected only at Walhalla and named by Dr. Ball.

9. *Salix interior* Rowlee. Sandbar Willow. Small shrub, 1—4 m. high, stems slender, not much branched below; leaves linear or linear-lanceolate, 5—10 cm. long, 4—8 mm. wide, sharply toothed, the teeth larger, farther part than on most willows. Late May and June. Common, forming dense thickets along river banks, ditches and other low places.

10. *Salix petiolaris* J. E. Smith. SLENDER WILLOW. Small shrub, 1—3 m. high, often in clumps of fine stems which are purple or dark brown; leaves a little wider and less coarsely toothed than those of Sandbar Willow. Frequent in low places, especially in flat, poorly drained areas.

11. *Salix lutea* Nutt. YELLOW WILLOW. Similar to next but twigs gray, leaves less narrowed at base. We had not recognized this species but Dr. Ball sent us specimens collected in Stutsman, Morton, Rolette, Renville and Ward Counties. It is a western species very similar to *S. cordata* and probably some of our specimens labeled *cordata* are *lutea*. Dr. Ball described (Bot. Gaz. 71:426) var. *famelica* as having small, strongly nerved leaves. He cited one specimen from Marmarth, North Dakota, and stated that *lutea* was abundant on the flood plain of the Yellowstone River at Forsythe, Montana, where he found only one clump of var. *famelica* in 15 or 20 examined.

12. *Salix cordata* Muhl. HEART-LEAVED WILLOW. Large, widely branched shrub, 2—5 m. (7—16 ft.) high, with fine, gray or light brown twigs; leaves oblong-lanceolate, the wider ones often rounded at base. Blooms about May 10. One of the commonest willows. The name "heart-leaved" refers to the rounded stipule about 1 cm. wide, found at leaf base (similar stipules occur on Nos. 13 and 14). Likely to be called "pussy willow" because the catkins are fuzzy, but smaller and less showy than in No. 13. Stem buds are often infested by a small fly, causing the "pine cone gall", an ovoid, scaly growth, 2—4 cm. long, which resembles a pine cone.

13. *Salix discolor* Muhl. PUSSY WILLOW. Large shrub with rather coarse, reddish brown or purplish twigs; leaves oblong or oblong-lanceolate; bud scales nearly black, catkins very hairy. The leaves tend to be as wide or wider toward the tip and to be narrowed abruptly, instead of tapering gradually as in *S. cordata*. Blooms about May 1, the earliest willow. The "pussys" are the catkins just after growth has begun before flowering. Frequent but common only locally. Specimens from Richland, Ransom, Cass, Rolette, and Bottineau Counties.

14. *Salix missouriensis* Bebb. MISSOURI WILLOW. DIAMOND WILLOW. Very similar to Pussy Willow and not easily distinguished. Common along Missouri River, frequent elsewhere. We have two specimens, one collected near Fargo and identified by Dr. Ball, the other collected at Mandan by Dr. Ball. Used for fence posts, also carved for canes and ornaments, leaving the diamond-shaped areas projecting. The diamonds are deformations of the stem, resulting from a branch dying.

15. *Salix bebbiana* Sarg. BEAKED WILLOW. Large shrub, 2—4 m. (7—13 ft.) high, usually much branched above; twigs brown to reddish brown; leaves obovate to elliptic, broad as compared to most other willows, strongly netted veined and white below; capsules 7 mm. long, the beak unusually long. Frequent, locally abundant in marshy areas, blooming about May 1. All specimens collected by Dr. Ball are labeled var. *perrostrata* (Rydb.) Schn.

16. *Salix humilis* Marsh. PRAIRIE WILLOW. Small shrub, 0.5—3 m. high; young branches gray hairy; leaves oblong to lanceolate, edges wavy with few or no teeth; catkins 1—2 cm. long or longer in fruit, appearing before the leaves. Three specimens, all without flowers or fruit, may belong to this species. One from Ransom County (Bell 124) has been labeled "bebbiana", and one from Rolette County (C. B. Waldron 675) was considered by Bergman a hybrid between *bebbiana* and *discolor*.

17. *Salix candida* Fluegge. HOARY WILLOW. Small shrub, 0.5—2 m. high; young twigs gray hairy; leaves narrow, edges wavy, not toothed, very white hairy below. In swampy spots. Specimens from Barnes, Pembina, Benson and Burke Counties. One from Benson County shows flowers and young leaves May 20, mature fruit June 27.

18. *Salix tristis* Ait. DWARF GRAY WILLOW. Small shrub 1 m. or less high; leaves small, usually narrower toward base, quite gray below. We have only one specimen from Hankinson, Richland County.

BIRCH FAMILY Betulaceae

Shrubs or trees with alternate leaves; flowers monoecious, staminate in long, slender catkins, pistillate in catkins, upright spikes or in smaller groups; stamens 2—several; pistil 1, with 2 stigmas; fruit a nut, nutlet or achene (thin with winged edge in birch and alder).

Key to Species

Fruit a nut, 1 cm. (⅖ in.) thick; pistillate flowers few in cluster, close
    to stem.
  Cup around nut open, 1—2 cm. long.   1. *Corylus americana*
  Cup forming a tube 3—4 cm. long.   2. *Corylus cornuta*
Fruit a nutlet or achene 2—5 mm. ($\frac{1}{12}$—⅕ in.) long; pistillate flowers
    in spikes or catkins 1—3 cm. long.
  Nutlet rounded, enclosed in a loose sack.   3. *Ostrya virginiana*
  Nutlet flattened, not enclosed; a scale below each nutlet.
    Fruiting spike stiff, the scales woody, persistent.   4.   *Alnus rugosa*
    Fruiting spike drooping, the scales falling with the fruits.
      Older bark white, peeling easily in horizontal strips.
          5.   *Betula papyrifera*
      Older bark brown or brownish, peeling little or irregularly.
        Wing of achene wider than body; small trees, western.
          6.   *Betula fontinalis*
        Wing of achene narrower than body; clustered shrubs,
        eastern.   7.   *Betula pumila*

1. *Corylus americana* Walt. HAZELNUT. (Fig. 176). Shrub, 1—2 m. (3—7 ft.) high, stems slender, growing singly but forming thickets; leaves 5—10 cm. (2—4 in.) long, broadly ovate or cordate, hairy below, edges with a few large, and many small teeth; staminate catkins 2 or 3 together, 3—5 cm. long; pistillate flowers sessile at leaf base, inconspicuous except bright red stigmas when in bloom about April 15. Frequent but local in wooded places, chiefly eastern part of State. A sterile specimen from East Rainy Butte, Hettinger County, is probably this species.

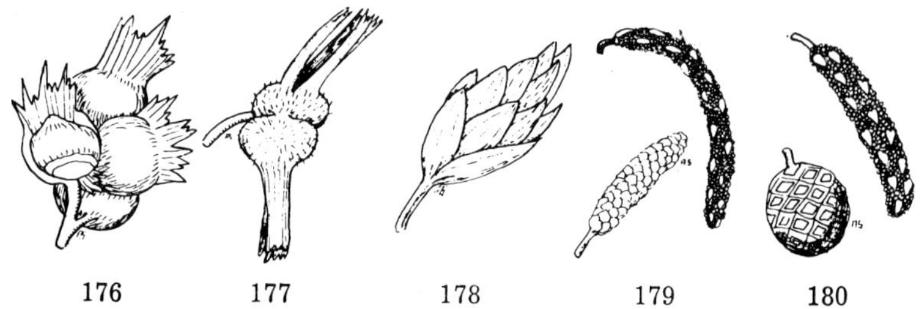

176   177   178   179   180

2. *Corylus cornuta* Raf. BEAKED HAZEL. (Fig. 177). Like the last except for slender, cylindrical beak which encloses nut. Still more local. It is the prevailing species in Pembina and Turtle Mts. We have it from Sheyenne River in Barnes County. Formerly called *C. rostrata* Ait.

3. *Ostrya virginiana* (Mill.) Koch. IRONWOOD. HOP-HORNBEAN. (Fig. 178). Small tree, trunk up to 2 dm. (8 in.) thick and 10 m. (34 ft.) high, usually slender with small, horizontal branches and finely shredded, dark bark; leaves thin and soft, oblong or ovate, 5—10 cm. (2—4 in.) long, finely toothed; staminate catkins 2—3 cm. long, 1—3 together at ends of branches; pistillate catkin similar, single, scale larger and developing in fruit into sack-like body 1 cm. long, enclosing the rounded, hard, smooth nutlet. Local along streams; Cass, Richland, Barnes and Pembina Counties.

4. *Alnus rugosa* (DuRoi) Spreng. Speckled Alder. (Fig. 180). Widely spreading shrub, 2—4 m. high; leaves oblong or ovate, 5—9 cm. long, with a few, coarse, rounded teeth and many fine ones; staminate catkins much like those of hazel, blooming about same time; fruiting spikes oblong, 1—1.5 cm. long, persisting after achenes are shed. In swamps or edges of streams. We have it from Cass, Ransom, Walsh and Pembina Counties. Often considered the same as the European *A. incana* (L.) Willd.

5. *Betula papyrifera* Marsh. Paper Birch. (Fig. 179). Small to large tree; bark at first brown, becoming white after about 10 years; leaves ovate, 4—8 cm. long, smooth, finely toothed; staminate catkins 3—10 cm. long, 5 mm. wide, the pistillate more slender when in bloom but becoming 8 mm. wide in fruit; achene 2 mm. long, its 2 thin wings about that wide; scale below each achene 3-lobed, 4 mm. long. Local on steep, protected slopes of hills, Richland, Barnes and Dunn Counties; quite common in Pembina and Turtle Mts. Blooms about May 10. Many botanists regard this as a variety of the European *B. alba*. Some recognize *B. cordifolia* Regel, or *B. alba*, var. *cordifolia*, which according to Rydberg (Fl. Pl. Pr.) should be the form found in North Dakota, though Rosendahl (Minn. Trees and Shrubs) restricts it in Minnesota to the extreme northeast. The European *B. pendula*, or *B. alba*, var.' *pendula*, and its cut-leaved form, is often planted as an ornamental. The native birch propagates very poorly and seedlings are almost never seen. The character of the bark peeling into thin sheets is too well known. Living trees, when peeled, never recover their beautiful bark but the strips remain as unsightly, black rings. Some trees which show puzzling characters have been regarded as hybrids (*B. sundbergi* Britton) between Paper and Dwarf Birch. Our specimens of this tend to have elliptical leaves and the bark seems darker than in *papyrifera*, not peeling as readily.

6. *Betula fontinalis* Sarg. River Birch. Shrub or small tree, 2—5 m. high; bark brown or becoming buffy on old trunks; leaves similar to those of paper birch or more rounded at tip. This is a puzzling form. The name "River Birch" is not appropriate in North Dakota. The plants grow in shrubby clumps on wet hillsides along small tributaries of the Missouri River and on the Killdeer Mts. More tree-like specimens are occasional along the Sheyenne River in Richland and Ransom Counties. It is known locally as "Black Birch".

7. *Betula pumila*, var. *glandulifera* Regel. Dwarf Birch. Small shrub, 1—2 m. high in thick clumps; leaves rounded to elliptic, 1—3 cm. long, more or less toothed, both leaves and twigs dotted with large, rounded glands when young. Swampy places and edges of streams, forming thickets under favorable conditions. We have it from Richland, Ransom, Eddy and Pembina Counties. It does not occur in Turtle Mts. so far as we have found.

OAK FAMILY Fagaceae

Trees with alternate leaves; staminate flowers in slender, interrupted catkins; pistillate flowers single or clustered at leaf base on new growth; stamens 4—12, pistil 1 with 3 stigmas; fruit a rounded nut, partly enclosed in a cup.

1. *Quercus macrocarpa* Michx. Bur Oak. (Fig. 51.) Small to large tree; leaves 1—2 dm. (4—8 in.) long, obovate, deeply and coarsely pinnately lobed, the lobes rounded; nut in our plants oblong, 1—2 cm. (2/5— 4/5 in.) long; cup about half covering well developed nuts, its edge fringed with many slender scales. Frequent to common on higher wooded ground. The oak of our region has been called *Q. mandanensis* Rydb., but most botanists consider it only a small form of *Q. macrocarpa.*

ELM FAMILY Ulmaceae

Trees with alternate leaves; flowers perfect or monoecious; sepals 3—9, petals 0, stamens 3—9, pistil 1, stigmas 2; fruit 1-seeded, a nut, drupe or achene.

Key to Species

Flowers on new growth; fruit a drupe, 5—8 mm. (¼ in.) wide; leaves
   ovate-lanceolate.   1. *Celtis occidentalis*
Flowers on last year's twigs; fruit a winged achene; leaves oblong,
   elliptic or obovate.
     Leaves flat, edge of fruit wing fringed with hairs.   2. *Ulmus americana*
     Leaves somewhat folded; wing not fringed.   3. *Ulmus rubra*

1. *Celtis occidentalis* L. HACKBERRY. Large tree, bark light gray with narrow, sharp ridges; leaves 6—12 cm. (2—5 in.) long, 4—7 cm. wide, usually widest and quite unsymmetrical at base, finely hairy below, edges finely toothed; staminate flowers in clusters 1—2 cm. or less long; upper flowers perfect, solitary at leaf bases; fruit a reddish brown drupe with thin, dry flesh, remaining on tree all winter. Quite common along Red River at Fargo, rare but occasional elsewhere. We have specimens from Logan, Pierce and Oliver Counties. It has been planted in shelterbelts recently and probably will become more commonly distributed because birds carry the seeds. Blooms about May 10; sensitive to frost, hence often is unable to produce fruit.

2. *Ulmus americana* L. AMERICAN ELM. WHITE ELM. Large tree with rough, dark gray bark, branches usually spreading; primary branch leaf veins prominent and parallel, both surfaces of leaf rough hairy; fruit flat, ovate, 1 cm. long. Blooms about April 25; fruit matures about May 20. A common tree along larger streams, rare in Turtle Mts. Siberian or "Chinese" Elm (*Ulmus pumila* L.) has been widely planted in recent years. The branches and leaves are finer, buds rounded, fruit not hairy.

3. *Ulmus rubra* Muhl. RED ELM. SLIPPERY ELM. Less finely branched, leaves rougher, fruit not fringed. Rare along Red River, no farther west; quite common 40 miles east of Fargo in Minnesota. The Fargo specimen cited by Bergman probably came from a planted tree though I have seen a few trees on the Minnesota side of the river. Has usually been called *U. fulva* Michx.

MULBERRY FAMILY Moraceae

Herbs, vines or trees, often with milky juice; leaves alternate, sometimes deeply lobed or divided; flowers small, green, monoecious or dioecious; sepals and stamens usually 4; petals 0; pistil 1; fruit an achene, drupe or nut.

In mulberry the sepals form the fleshy part of the fruit. In fig, many small flowers are enclosed in a hollow stem tip which becomes fleshy. In osage orange of central U. S. and breadfruit of the tropics, fruits develop much as in mulberry.

Key to Species

Plant a vine; leaves deeply lobed; nutlet enclosed in a sack-like scale.
                   1. *Humulus americana*
Erect, stout herb; leaves palmately compound; nutlet not so enclosed.
                   2. *Cannabis sativa*

1. *Humulus americana* Nutt. AMERICAN HOP. Herbaceous vine, perennial by rhizomes; stem twining, 2—3 m. (7—10 ft.) high; leaves alternate, rounded, 1—2 dm. (4—8 in.) long, palmately deeply 3—5-lobed; staminate flowers in branched axillary clusters, pistillate cone-like,

5—10 cm. (2—4 in.) long, composed of oblong, sack-like scales, each enclosing 2 rounded nutlets. Frequent in brushy places. Leaves, and especially lower stems, are covered with short, stiff, silicified, 2-pronged hairs. Fruiting clusters contain yellow, resinous particles which bear the characteristic flavor. This is now regarded as distinct from the European hop, *H. lupulus* L.

2. *Cannabis sativa* L. HEMP. (Fig. 181). Dioecious annual; stems tough and straight, 2—3 m. high; leaves alternate, palmately divided into 5—7 narrow, sharply toothed leaflets, 5—15 cm. long; flowers in clusters at upper leaf bases; sepals 5, stamens 5, pistil 1; fruit a hard, smooth nutlet, 5 mm. long. Pistillate plants develop rather conspicuous green-bracted seed clusters. Staminate plants drop their leaves and dry up earlier than seed bearing ones. Plants are found occasionally where seed has been dropped. So far as we know it has not become established except to a slight extent in Richland County. It needs a longer growing season than we have in North Dakota.

NETTLE FAMILY Urticaceae

Herbs (in our species) with small, green, perfect or monoecious flowers, some species with stinging hairs; sepals 2—5, sometimes partly united; petals 0; stamens 2—5; pistil 1; fruit a small, flattened achene.

181. Hemp (*Cannabis sativa*).   182. Tall Nettle (*Urtica procera*).

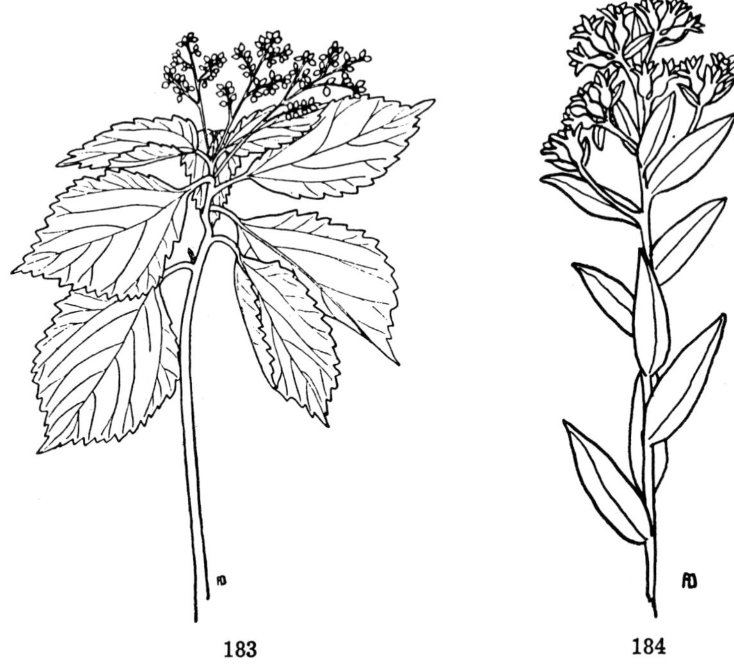

183   184

183. Wood Nettle (*Laportea canadensis*).
184. Bastard Toad-flax (*Commandra*).

Key to Species

Coarse plants, 5—15 dm. (20-60 in.) high, with stinging hairs on
 leaves and stems.
  Leaves opposite, lanceolate, short petioled.   1. *Urtica procera*
  Leaves alternate, rounded, long petioled.   2. *Laportea canadensis*
Weak plants, 1—4 dm. high, without stinging hairs.
  Leaves opposite; stems soft and watery; flowers without involucre.
                                              3. *Pilea pumila*
  Leaves alternate; stems weak but not watery; flower cluster with an
   involucre.   4. *Parietaria pennsylvanica*

1. *Urtica procera* Muhl. TALL NETTLE. (Fig. 182). Perennial by short
   rhizomes, growing in dense clumps or patches; leaves numerous, 7—15
   cm. (3—6 in.) long, finely toothed; sepals 4, staminate and pistillate
   flowers mixed together in branching clusters 5—10 cm. long at upper
   leaf bases; achene 1 mm. long. A very common weed in low ground,
   woods, neglected yards and dump grounds. Formerly called *U.
   gracilis* Ait. Tall nettle is certainly a more suitable name than
   "slender nettle" which had been used for it.

2. *Laportea canadensis* (L.) Gaud. WOOD NETTLE. (Fig. 183). Annual,
   5—10 dm. high; leaves few, near top of stem, 5—15 cm. long, rather
   abruptly narrowed at tip; flowers much as in *U. procera* but larger;
   stamens and sepals 5. Local in dense, moist woods. We have it from
   Richland, Cass, Barnes and Pembina Counties.

3. *Pilea pumila* (L.) A. Gray. CLEARWEED. Annual with watery, almost
   transparent stem; leaves ovate to elliptic, 3—10 cm. long, with rounded
   teeth; flower clusters similar to those of nettles. Wet, shady places.
   Richland, Ransom, Cass and Pembina Counties. Lunell collected plants

in Benson County which he described as *P. opaca* and *P. fontana.* The latter is stated by Deam (Fl. Ind.) as readily distinguished by the white edged fruit and purple inside color of pericarp.

4. *Parietaria pennsylvanica* Muhl. PELLITORY. (Fig. 90). Weak stemmed, slender annual; leaves ovate to lanceolate, 2—5 cm. long, very thin, edges entire; flowers in small clusters at leaf bases; calyx enclosed by 2—6 sepal-like bracts, which may be partly united. Frequent or locally common in partly shaded places. I collected plants in "scoria" on an open butte near Glen Ullin, Morton County (Stevens 144), which were about 1 dm. high and very dense. They were identified by Dr. S. F. Blake as this species.

SANDALWOOD FAMILY Santalaceae

Herbs, shrubs or trees, mostly or partly parasitic; flowers small, white or greenish; calyx 3—6-lobed, on top of ovary; corolla 0; fruit a small nut or drupe. Our species are small herbs. Sandalwood is *Santalum album,* a small, evergreen shrub from India. Some botanists place this family and the next one near the madder family.

Key to Species

Flowers in a terminal, dense, flat cluster; native prairie plant.
1. Commandra pallida

Flowers on slender, solitary stalks from leaf base; introduced.
2. Thesium linophyllon

1. *Commandra pallida* A.DC. BASTARD TOAD-FLAX. (Fig. 184). Perennial, stems stiff, erect, 1—2 dm. (4—8 in.) high; leaves numerous, 1—3 cm. (⅖—1⅕ in.) long, alternate, lanceolate to linear, entire, very smooth, thick, pale green; flowers white, 5 mm. wide; fruit rounded, green, 6—8 mm. (¼ in.) long. Very common on prairie. May, June.

2. *Thesium linophyllon* L. FLAXLEAF. Perennial with many slender stems 2—4 dm. high; leaves narrow, 2—3 cm. long, thin; each flower on a slender stalk from a leaf base; flower 3—4 mm. wide, lobes narrow; fruit narrowly oblong, 4—5 mm. long. Found near Cando, Towner County in 1943 by Wm. J. Leary (Rhodora 46:490). This may have been introduced in grass or other crop seed. It is a common plant in Europe where there are other species of the genus but they are not mentioned in most weed books.

BIRTHWORT FAMILY Aristolochiaceae

Low herbs or vines; flowers perfect or petals absent; calyx usually brownish purple inside; sepals usually 3, partly united; stamens 6, pistil 1, ovary 6-celled; fruit a berry or capsule. The family is perhaps better known for the Pipe Vine (*Aristolochia* spp.), a large vine with cordate leaves, often used as a porch vine.

1. *Asarum canadense* L. WILD GINGER. (Fig. 97). Perennial from a short rhizome which bears 2 leaves and 1 flower between them; leaves kidney shaped, 1—1.5 dm (4—6 in.) wide, soft hairy, their petioles 1—2 dm. long; flower on the ground, cup shaped, 1 cm. (⅖ in.) wide, sepals triangular. Quite local in woods. We have it from Richland, Ransom and Pembina Counties only.

BUCKWHEAT FAMILY Polygonaceae

Herbs with alternate, simple leaves, usually with a membranous stipule surrounding base of petiole; flowers small and green, sometimes colored and showy; sepals 2—6, partly united; petals 0; stamens 2—9; pistil 1, with 2 or 3 styles; fruit a 3 angled, flattened or rounded achene.

Key to Genera

Leaves without stipules; flowers in dense, flat-topped clusters with
an involucre at base of cluster; stamens usually 9.   1-4. *Eriogonum*
Leaves with stipules; flowers in elongated clusters, involucre none;
stamens usually 5—6.
  Sepals 6 in 2 sets, 3 remaining small, 3 enlarged in fruit.   5-11. *Rumex*
  Sepals 5 and similar, not greatly enlarged in fruit.   12-28. *Polygonum*

Eriogonum

Annuals or perennials, often with short, somewhat woody stems;
leaves often densely covered with hairs; flowers usually whitish
or yellowish; fruit a 3-angled achene. This group of plants is without a good common name. There are many species on dry soils in
the Rocky Mountain region. Only one is common in North Dakota.

Key to Species

Main stem slender, 3—6 dm. (1—2 ft.) high, rather bare; flowers
  white.   1. *Eriogonum annuum*
Main stem hardly over 1 dm. high; leaves clustered.
  Leaves green, rounded; flowering branches 1—3 dm. long, slender.
    2. *Eriogonum trichopes*
  Leaves white hairy, oblong; flowering branches short, flowers
    densely clustered.
    Flowers yellow, in a large, flat topped cluster.   3. *Eriogonum flavum*
    Flowers white or pinkish in small, rounded clusters.
      4. *Eriogonum multiceps*

1. *Eriogonum annuum* Nutt.   Annual, densely covered with white hairs;
   stem slender, 3—8 dm. (1—2½ ft.) high, often not branched except at
   flower cluster which spreads out 1—1.5 dm. wide; leaves oblong to
   obovate, 2—5 cm. (1—2 in.) long.   Sandy soil, quite local; Sioux, Grant,
   Bowman and Slope Counties.   It often grows in patches and these are
   conspicuous at some distance because of the very white appearance of
   the plants.

2. *Eriogonum trichopes* Torr.   Annual (or perennial?); leaves mostly
   basal, rounded, 2—5 cm. wide, with slender petioles 2—5 cm. long;
   flower stems widely forking, 1—3 dm. high, with only a few small
   leaves; flowers scattered, calyx yellowish green. One specimen from
   Wade, Grant County (Bell 233).

3. *Eriogonum flavum* Nutt.   Perennial from a thick, woody root, the
   short stem thickly covered with old, brown leaf bases; leaves oblong,
   1—5 cm. long, thick and white felted below, greenish above; main
   flower stem 1—2 dm. high, branched at top and with several small
   leaves; entire flower cluster 2—5 cm. wide.   Common, especially on
   rocky hills, western part of State.   Ward and Rolette Counties are our
   only records east of the Missouri River region but it probably occurs
   locally as far east as Stutsman County. The bright flowers in June
   are quite showy.

4. *Eriogonum multiceps* Nees.   Perennial from a woody root and short,
   prostrate, woody stem, forming a spreading, branched crown as compared with the often single, thick stem of *E. flavum;* entire plant gray
   hairy; leaves linear to oblong, 2—5 cm. long; flower stems 5—15 cm.
   high, clusters 5—10 mm. wide.   Less common than *E. flavum* and apparently closely restricted to heavy clay soil as on bare buttes.   No
   specimens from east of Missouri River.

Rumex DOCK

Mostly weedy perennials; fruiting calyx usually green, not
much enlarged, but in one species large and bright colored, 1 or
more of larger sepals often bearing a seed-like "grain" or "tubercle"
on the back (fig. 185); achene reddish brown, 3-angled, usually
pointed at tip.

Key to Species

Flowers dioecious, all sepals small; slender plants with acid foliage.
5. *Rumex acetosella*
Flowers usually perfect, inner sepals enlarged; leaves not acid.
  Fruiting calyx 1—3 cm. (⅖—1⅕ in.) wide, pink, showy.
6. *Rumex venosus*
  Fruiting calyx 4—8 mm. (⅙—⅓ in.) wide, green, becoming brown.
    Inner sepals with slender teeth along edges.   7. *Rumex persicarioides*
    Inner sepals not toothed or with very small teeth.
      Leaves flat and smooth, pale green.
        Lowest leaves narrowed at base; plant branched from below.
8. *Rumex mexicanus*
        Lowest leaves rounded at base; plant branched only above.
11. *Rumex orbicularis*
      Leaves crisped, dark green.
        Tubercles usually 3; introduced weed 6—10 dm. (2—3 ft.)
          high.   9. *Rumex crispus*
        Tubercles usually 0; native plant 1—2 m. (3—7 ft.) high.
10. *Rumex occidentalis*

5. *Rumex acetosella* L. SORREL DOCK. SHEEP SORREL. Perennial by slender, running roots; leaves mostly near ground, 2—10 cm. (1—4 in.) long, hastate; stems slender, 3—6 dm. (1—2 ft.) high, with a few small leaves; achenes 1.5 mm. (1/16 in.) long and nearly as wide, not pointed, closely covered by calyx. A rare introduced weed. Records from Cass, Stutsman, Stark, Pembina and Bottineau Counties. It is a well known weed in acid, sandy soils in eastern U. S. The leaves are rather fleshy and have a pronounced taste due to the large amount of oxalic acid. The same is true of *Oxalis violacea*, a very different plant, which is also called "sheep sorrel".

6. *Rumex venosus* Pursh. LARGE-FLOWERED DOCK. Perennial by rhizomes; stem stout, 1.5—4 dm. high, usually bushy branched; leaves ovate, oblong or lanceolate, flat and thick, 3—10 cm. long; flowers in a dense, terminal cluster which becomes 1—2 dm. long in fruit; inner calyx lobes rounded, 1—3 cm. wide, pale green or bright pink; achene pointed, 7 mm. long. In very sandy soil, Missouri River westward. A showy plant, blooming in mid-May, fruit mature in early June. It spreads freely but usually grows in soil too sandy for farming.

7. *Rumex persicarioides* L. GOLDEN DOCK. (Fig. 185C). Biennial (?); stems 3—10 dm. high, widely branched; leaves dark green, lanceolate, the lower ovate or oblong, 3—20 cm. long; upper branches with dense, rounded clusters of flowers; inner sepals 1—2 mm. long, with 1—3 slender teeth along each edge; achene yellowish, narrowly oblong, 1.5 mm. long. Common in low ground, frequently forming dense stands on dried out ponds or along pond margins. Blooms in July, fruit ripe Aug., Sept. Rechinger (Field Mus., Bot. Ser. 17:136, 1937) considers our plant *R. fueginus* Philippi.

8. *Rumex mexicanus* Meissn. WILLOW-LEAVED DOCK. (Fig. 185B). Perennial from a stout crown and roots; stems 3—6 dm. high, widely branched; leaves pale green, very flat and smooth, lanceolate or linear-lanceolate, 5—15 cm. long; flowers densely crowded along ends of branches, pale green until mature; inner sepals 5—6 mm. long in fruit, their margins wavy; achene dark reddish brown, pointed, 2—3 mm. long. A common, native weed in wet ground. Usually produces few seeds. Blooms summer and fall. Formerly called *R. salicifolius* and Rechinger regards it as *R. triangulivalvis* (Danser) Reich. f.

9. *Rumex crispus* L. CURLED DOCK. Much like the last but has dark green, curled leaves, the larger ones 3 dm. long; seeds produced very freely. An introduced weed, usually in roadside ditches, not common except in some localities.

10. *Rumex occidentalis* S. Wats. WESTERN DOCK. (Fig. 185A) Large native perennial; stems 1—2 m. high; leaves oblong, curled, the larger

ones 3—6 dm. long; flowers in a dense cluster at top of stem; inner sepals 6—8 mm. long, veins prominent, tubercles often absent. Low meadows or ditches, west to Stutsman and Eddy Counties.

11. *Rumex orbicularis* Gray. GREAT WATER DOCK. Similar to last but leaves flat, the lowest quite rounded, branch veins coming from midrib at nearly a right angle; tubercles usually 3. We have no specimens but F. P. Metcalf reported (Jour. Wash. Acad. Sci. 10:196, 1920) a specimen from McLean County and its identity was confirmed by Rechinger. This seems more of a swamp plant than other species. Formerly regarded as *R. brittanica* L.

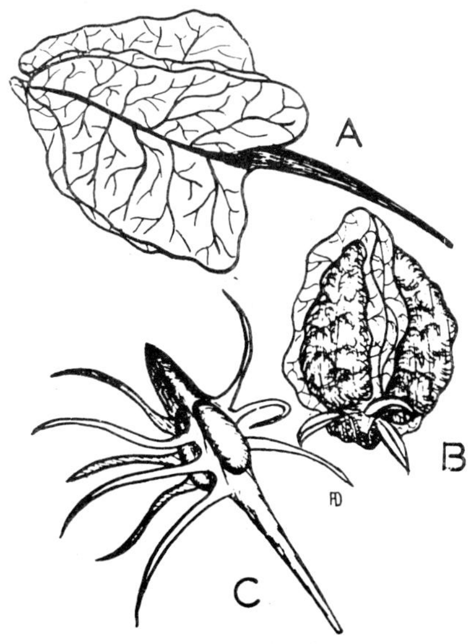

185. Fruits of Dock: A. *Rumex occidentalis;* B. *R. mexicanus;* C. *R. persicarioides.*

Polygonum KNOTWEED. SMARTWEED

Annuals or perennials, mostly weedy; flowers in short, terminal clusters or 1—few in leaf axils; sepals 4—5, green, whitish or pink; petals 0; stamens usually 5; pistil 1 with 2 or 3 stigmas; fruit a flattened or 3-angled achene, dark brown to black. The name knotweed is from the swollen nodes of the stem; smartweed from the acrid juice of some species. Flowers mostly July-Sept.

Key to Species

Twining vines; outer sepals keeled or winged.
  Common annual in fields; outer sepals slightly keeled.
                              12. *Polygonum convolvulus*
  Native perennial in brush; outer sepals strongly keeled.
                              13. *Polygonum scandens*
Stems erect or spreading, not twining; sepals not keeled.
  Stems with sharp, recurved prickles; rare in woods.
                              14. *Polygonum sagittatum*
  Stems without recurved prickles.
    Flowers conspicuous, in terminal clusters.
      Clusters 1—2, always terminal; plants often growing in water;

flowers 4—5 mm. (⅕ in.) wide.
Leaves elliptic to oblong, usually floating; penduncles not
  hairy.   15. *Polygonum natans*
Leaves ovate to lanceolate, usually not floating; peduncles
  hairy.
    Leaf sheath with a green spreading top 1 cm. (⅖ in.) wide;
      flower cluster short and thick, 1—2 cm. long.
                  16. *Polygonum natans*, forma *hartwrightii*
    Leaf sheath cylindrical, membranous at tip; flower cluster
      slender, 3—5 cm. long.   17. *Polygonum coccineum*
Clusters many, both terminal and axillary.
  Achenes flattened, the sides concave or convex.
    Flower clusters erect, short oblong; calyx 3—5 mm. (⅙
      in.) wide.
        Peduncles glandular; stipules not bristly.
                          18. *Polygonum pennsylvanicum*
        Peduncles not glandular; stipules ending in bristles.
                              19. *Polygonum persicaria*
    Flower clusters drooping, slender; calyx 3 mm. wide.
                          20. *Polygonum lapathifolium*
  Achenes 3-angled.
    Flower clusters drooping; achenes dull brown, not shining.
                          21. *Polygonum hydropiper*
    Flower clusters erect; achenes shining black.
      Sepals pink, not glandular.   19. *Polygonum persicaria*
      Sepals green, dotted with yellowish glands.
                          22. *Polygonum punctatum*
Flowers small and few at leaf bases; achenes 3-angled.
  Stems prostrate or ascending.
    Stems slender; leaves narrowed at ends.
                          23. *Polygonum aviculare*
    Stems stout; leaves rounded at ends.   24. *Polygonum buxiforme*
  Stems erect.
    Plants very stout, bushy branched, 1—3 dm. (4—12 in.) high.
                          25. *Polygonum achoreum*
    Plants 3—10 dm. (1—3 ft.) high or very slender if 1—3 dm.
      Stipules of upper leaves 5 mm. (⅕ in.) long, extending
        much beyond flowers.   28. *Polygonum autumnale*
      Stipules of upper leaves 3 mm. long, shorter than flowers.
        Stems 1—3 dm. high; achenes 1—2 mm. long.
                          23. *Polygonum aviculare*
        Stems 4—10 dm. high; achenes 3—5 mm. long.
          Achenes numerous, 1.5—2 mm. wide.
                          26. *Polygonum ramosissimum*
          Achenes few, 1—1.5 mm. wide, often long pointed.
                          27. *Polygonum prolificum*

12. *Polygonum convolvulus* L. WILD BUCKWHEAT. Annual; stem twining,
sometimes 2 m. (7 ft.) high; leaves 2—6 cm. (⅘—2½ in.) long, heart
shaped, narrowed suddenly to a slender tip, basal lobes usually sharp
pointed; flowers few in leaf axils, developing unevenly all summer;
achenes 3-angled, dull black, 3—4 mm. long, nearly as wide, closely
covered by calyx which finally becomes brown. A very common
weed.

13. *Polygonum scandens* L. CLIMBING FALSE BUCKWHEAT. Similar to last
but coarser; leaves pale green; sepals distinctly winged; achene black,
smooth and shining. In brushy places. We have it from Barnes, Pembina, Rolette, Bottineau, Grant and Golden Valley Counties.

14. *Polygonum sagittatum* L. TEAR-THUMB. Annual; stem slender, spreading, with spines pointing backward; leaves oblong or lanceolate, sagittate at base, 1—10 cm. long; flowers greenish, 4 mm. long, in terminal
or axillary clusters; achenes 3-angled or flattened, black, shining.
Rare in cold, swampy places. Recorded only from Turtle Mts. and I
have been unable to find it there in recent years.

15. *Polygonum natans* A. Eaton. FLOATING LADY'S THUMB. Perennial from stout rhizomes; stems with roots at joints; leaves floating, flower clusters short and thick, 1—2 cm. long, projecting above water; leaves oblong or elliptic, smooth and shining, 5—10 cm. long; calyx bright pink, 5—7 mm. long. Local in ponds. We have it only from Ransom, Hettinger, Grant and Bottineau Counties. Without flowers the plants might be confused with *Potamogeton natans,* but the Lady's Thumb grows in shallow water and has horizontal stout stems.

16. *Polygonum natans,* forma *hartwrightii* (Gray) Stanford. Perennial by stout rhizomes, usually in marsh or wet grassland; stems and leaves rough hairy; leaves narrowly oblong, without petioles, base with a sheathing stipule which has a rounded, toothed blade 1—2 cm. wide. This is a very different looking plant from No. 15, and has been considered a distinct species by many botanists. We have specimens from 8 counties widely scattered. There has been much confusion regarding the relationships of these plants and they offer a good field for careful study to anyone who has opportunity to watch closely individual plants for several years under varying conditions. These grassland plants are usually without flowers. It would be an interesting experiment to transplant some rhizomes from such a place to the edge of a pond.

17. *Polygonum coccineum* Muhl. LONG-ROOTED SMARTWEED. (Fig. 186). Perennial by stout rhizomes; stems erect, 2—10 dm. (8—40 in.) high, upper part usually hairy; leaves ovate to lanceolate, 1—2 dm. long, 3—6 cm. (1—2½ in.) wide, surface usually with short, stiff hairs; flower clusters slender, 5—10 cm. (2—4 in.) long; flowers bright pink; achenes slightly flattened, nearly circular, plump, black, 2—3 mm. long. Very common in shallow water, ditches, low meadows, potholes, etc. Especially abundant in lakes and ponds in the Devils Lake region where it is showy in July. Flowers freely but often produces few seeds. Sometimes develops horizontal stems with swollen internodes in water, the leafy branches arising from these stems and projecting above water. This has been called *P. emersum* (Michx.) Britt. and *P. muhlenbergii* S. Wats.

18. *Polygonum pennsylvanicum* L. PENNSYLVANIA SMARTWEED. Annual, coarse, widely branched, 4—10 dm. high; leaves lanceolate, 5—20 cm. long, tapering to a slender point; flower clusters oblong, 3—5 cm. long, 1—1.5 cm. wide; flowers pink or nearly white; achenes black, shining, nearly circular, 3—4 mm. long, one side flat with a slight ridge, the other side slightly concave. Sometimes locally common along streams or pond banks, but rare for the most part. We have it from Richland, Ransom, Sargent, and Cass Counties only. It is a very common plant farther south.

19. *Polygonum persicaria* L. LADY'S THUMB. Similar to last but much smaller, usually 3—6 dm. high; leaves usually with a dark, central spot; achenes either ovoid, flattened with one side convex, or 3-angled, very black and shining, 2.5—3 mm. long. Distribution about as for No. 18. We have it from southeastern counties, also from Benson, Mountrail and Golden Valley Counties.

20. *Polygonum lapathifolium* L. PALE OR WILLOW-LEAVED SMARTWEED. Annual, 1—2 m. (3—7 ft.) high; leaves lanceolate, 1—2 dm. (4—8 in.) long; flower clusters 2.5—5 cm. (1—2 in.) long, slender and drooping; flowers usually greenish white, sometimes bright pink; achenes circular, concave on both sides, light brown, 2 mm. ($\frac{1}{12}$ in.) long. Our common species, especially in low fields and wet places. Along pond and stream banks it sometimes forms dense growths up to 2 m. high. In fields it is more often 6—10 dm. high. Several specimens have the leaves finely white wooly below and have been called *P. incanum* Schmidt, *P. tomentosum* (Schrank) Bickn. or *P. lapathifolium,* var. *salicifolium* Sibth.

21. *Polygonum hydropiper* L. WATER PEPPER. Annual, widely branched, 3—6 dm. high; leaves lanceolate, 3—8 cm. long; flowers scattered along drooping branch tips; calyx green, dotted with yellow glands; achenes

3-angled, 2—2.5 mm. long, pointed, dull black. Rather rare, locally
common in wet places. We have it from Richland, Ransom, Cass, Walsh
and Stark Counties. Usually a few flowers are partly enclosed in lower
leaf sheaths.

186

187

186. Long-rooted Smartweed (*Polygonum coccineum*).
187. Knotweed (*Polygonum aviculare*).

22. *Polygonum punctatum* Ell. Annual, erect, rather slender; leaves lanceolate, 5—10 cm. long, glandular dotted below; flower clusters erect, very
slender, 1—6 cm. long; sepals green, glandular; achenes 3-angled, 3 mm.
long, black, smooth, shining. Local in swampy places; Richland County.

23. *Polygonum aviculare* L. KNOTWEED. DOORWEED. GOOSEGRASS. (Fig.
187). Annual, erect to prostrate; stems slender, 1—6 dm. (4—24 in.)
long; leaves elliptic, 1—3 cm. (²⁄₅—1⅕ in.) long; flowers 1 or few at leaf
bases, whitish or pink, 2 mm. ($\frac{1}{12}$ in.) long; achene dark reddish brown,
finely striate, 3-angled, pointed, 2—3 mm. long. Very common weed
along walks and roads, not usually a field weed, though I have seen
it common in grain fields in Pembina and Cavalier Counties. Quite
variable in size and habit of growth. Plants growing in hard trodden
ground, are fine leaved and lie close to the ground. When growing in
grass or in dense stands, the stems are partly supported or stand erect.
The fine leaved, prostrate forms have been called var. *angustissimum*
Meissn. or *Polygonum neglectum* Besser.

24. *Polygonum buxiforme* Small. Annual, prostrate, similar to last but
stems coarse, leaves oblong, pale green. Frequent to common, not
always distinct from *P. aviculare*. It has been sometimes referred to
the European *P. littorale* Link.

25. *Polygonum achoreum* Blake. ERECT KNOTWEED. -Annual, erect or
spreading, stout and densely branched, 1—3 dm. high; leaves oblong or
broadly elliptic, 1—3 cm. long, pale green; calyx green; achenes 3-
angled, somewhat flattened, 3 mm. long and nearly as wide, light brown,
dull. Common along paths, in yards, etc. The leaves are often covered
with mildew. It had formerly been referred to the European *P. erectum*
L.

26. *Polygonum ramosissimum* Michx. BUSHY KNOTWEED. Annual, erect,
widely branched, 3—10 dm. (1—3 ft.) high; leaves oblong to linear,
1—4 cm. (²⁄₅—1³⁄₅ in.) long; flowers clustered at leaf bases of the long

branches; calyx yellowish green; achene sharply 3-angled, reddish brown, smooth, 3 mm. (⅛ in.) long, 2 mm. wide at base. Common, especially along roadsides.

27. *Polygonum prolificum* (Small) Robins. Similar to last but more slender, branches erect, flowers much fewer, sepals white margined (not yellowish); achenes often twice as long as sepals. Frequent in sloughs, mixed with grass or on dried pond and ditch banks. Specimens of this were included by Bergman in *P. ramosissimum* and *P. aviculare* (Bell 640, 1376, 1435, L. R. Waldron 1654, 1687, Wright at Towner and Bergman 2301). It seems quite a distinct species, but the proper name for it is still uncertain. Late fall specimens usually have a large number of "exserted" achenes which would place it in *P. exsertum* Small, but this is probably only an abnormal condition. By that date the lower leaves have fallen so that it is difficult to compare leaves of various forms. Plants growing among grasses are tall and slender, those in open ground, short and widely branched from base.

28. *Polygonum autumnale* Brenckle. Annual, erect, 2—5 dm. high, slender, branching from base; leaves linear or lanceolate, 2—4 cm. long, flowers 1—3 at upper leaf bases, at first covered by stipule; achenes reddish brown, nearly smooth, 3-angled but flattened, 1.2—2 mm. long, 0.8 mm. wide. This recently described species (Bull. Torr. Bot. Cl. 68:495, 1941) may be distinct. It seems to resemble *aviculare* more than *prolificum*, but grows upright. The achenes are more stubby, paler and smoother than those of *aviculare*. The young leaves at ends of branches are nearly covered by shining, white stipules. The only specimens at hand grew on mud of a roadside ditch near McKenzie, Burleigh County.

GOOSEFOOT FAMILY Chenopodiaceae

Chiefly weedy annuals with alternate, often fleshy leaves, some species perennial or shrubby; flowers small, green, single at leaf bases or sometimes clustered; sepals usually 5, petals 0, stamens usually 5, pistil 1; fruit 1-seeded, enclosed by a thin covering, also by calyx and sometimes by bracts; seed horizontal, attached at center of one side, or vertical and attached at edge.

Key to Species or Genera

Shrubby perennials, woody at least at base.
  Tall, spiny shrub, 1—2 m. (3—7 ft.) high; leaves cylindrical, 1—4
    cm. (⅖—1½ in.) long.   35. *Sarcobatus vermiculatus*
  Low shrubs, 3—8 dm. (1—2½ ft.) high, or woody only at very base.
    Leaves 3—5 mm. (⅛—⅕ in.) wide, either fleshy or hairy, soft,
      not rough.
    Leaves very fleshy, not hairy; stems spreading. 33. *Suaeda fruticosa*
    Leaves not fleshy, very hairy; stems erect.   30. *Eurotia lanata*
    Leaves 5—20 mm. wide, thick, gray and rough with short hairs.
      Plant quite woody, old stem tips becoming spiny.
            21. *Atriplex confertifolia*
      Plant woody only at base, not spiny.   20. *Atriplex nuttallii*
Herbaceous annuals but stems sometimes quite hard.
  Upper leaves and bracts of old plants very spiny.   34. *Salsola kali*
  Plants not especially spiny.
    Leaves opposite; flowers all in spikes 2—4 cm. long; plant fleshy,
      brittle.   31. *Salicornia rubra*
    Leaves alternate; flowers not all in spikes.
      Flowers monoecious or dioecious; fruit enclosed by a pair of
        bracts.
      Bracts soft, not united; fruit oblong, short winged at tip.
            29. *Axyris amaranthoides*
      Bracts united at base, often quite hard.
        Stems creeping; bracts flattened at right angles to fruit

which is 2-toothed at tip.   22. *Suckleya suckleyana*
Stems erect or spreading; bracts flattened parallel to fruit
        which is not toothed. (Fig. 188).   14-21. *Atriplex*
Flowers perfect; fruit not enclosed by a pair of bracts.
    Leaves cylindrical, not flattened.   32. *Suaeda depressa*
    Leaves flattened, not cylindrical, but sometimes narrow.
        Fruit laterally flattened, winged, projecting beyond calyx.
                                              25-27. *Corispermum*
        Fruit depressed circular, enclosed by calyx.
            Calyx transversely winged, at least in part.
                Leaves lanceolate, entire; calyx wing narrow.
                                              28. *Kochia scoparia*
                Leaves oblong, toothed; calyx wing prominent.
                                    24. *Cycloloma atriplicifolium*
            Calyx not winged, sepals often keeled lengthwise.
                Sepal 1; stamen 1; low, early maturing plant.
                                        23. *Monolepis nuttalliana*
                Sepals 5; stamens usually 5, rarely 1—2.
                                              1-13. *Chenopodium*

Chenopodium GOOSEFOOT

More or less succulent annuals; leaves often gray with a
"mealy" covering, especially on lower side; flowers small, green,
usually in small clusters along upper branches; sepals 5, rarely 1—2;
stamens 5, rarely 1—2; pistil 1, stigmas 2; fruit 1-seeded, the seed
black, circular, flattened.

Key to Species

Calyx becoming fleshy and bright red; flowers in dense, axillary
    clusters.   13. *Chenopodium foliosum*
Calyx green; flowers in small, usually scattered clusters.
    Plants strong scented, sticky; rare.
        Leaves gray, not finely divided; odor offensive.
            3. *Chenopodium berlandieri*, var. *farinosum*
        Leaves green, finely divided; odor rather aromatic.
                                    12. *Chenopodium botrys*
    Plants sometimes with rank odor but not especially strong scented.
        Flower clusters axillary, shorter than leaves.
            Leaves pale green, oblong, evenly toothed, 1—5 cm. (⅖—2 in.)
                long.   10. *Chenopodium glaucum*
            Leaves dark green, triangular, 3—10 cm. long, little toothed.
                                    11. *Chenopodium rubrum*
        Flower clusters terminal and axillary, usually longer than leaves.
            Leaves nearly as wide as long with 2 or 3 large, triangular lobes
                on each side.   9. *Chenopodium gigantospermum*
            Leaves linear to ovate with small teeth, none, or only 2 toward
                base.
                Leaves linear to narrowly lanceolate, usually gray.
                                    7. *Chenopodium leptophyllum*
                Leaves ovate or broadly lanceolate, usually green or pale
                    green.
                    Leaves short ovate, 1—2 times as long as wide.
                        Leaves bright green, thin; plant of wooded places.
                                        8. *Chenopodium fremontii*
                        Leaves pale green, thick; plant of open ground.
                                        2. *Chenopodium berlandieri*
                    Leaves ovate, triangular or oblong, 3—5 times as long as
                        wide.
                        Leaves oblong or ovate, usually not toothed.
                            Leaves thick, blue green; stem purplish; in open ground.
                                        5. *Chenopodium strictum*

Leaves thin, light green; stem green; in shaded places.
6. *Chenopodium standleyanum*
Leaves triangular, ovate to lanceolate, more or less
toothed.
Seeds 1 mm. wide; petiole usually shorter than blade.
1. *Chenopodium album*
Seed 1.5 mm. wide; petiole often longer than blade.
4. *Chenopodium bushianum*

1. *Chenopodium album* L. LAMB'S QUARTERS. (Fig. 188A,C). Stout, annual, 3—15 dm. (1—5 ft.) high, stem often striped or spotted with red; leaves ovate, rhombic or lanceolate, usually with several teeth along each side; leaves bright green, quite mealy below; seed 1 mm. (1/25 in.) wide, little flattened on upper side, surface not pitted. Common weed, flowering all summer and fall. Plants grow 3—5 dm. high and little branched under poor conditions, becoming 1—2 m. high and widely branched under favorable conditions. Var. *stevensi* Aellen, has narrow, lanceolate, nearly entire leaves. I collected the type in my alley but cannot find it again. It probably is not worthy of recognition. Various specimens were labeled by Prof. Aellen as ssp. *dacoticum* Aellen, and ssp. *fallax* Aellen, but there are so many intergrading forms that separation is difficult. *Chenopodium album* is usually considered introduced from Europe. It is so common and widespread that it seems almost native. John Josselyn, in 1672 listed it as one of the plants which had appeared in America since white men came. Audubon mentioned it as very abundant around the Mandan villages in 1834, but he could easily have mistaken the following species for it.

2. *Chenopodium berlandieri* Moq. PITSEED GOOSEFOOT. Much like *album*, but leaves frequently not over 1 cm. long, abruptly sharp tipped; seeds distinctly flattened on upper side and surface pitted. This has long been misunderstood. It is common but more especially in native habitats, whereas *C. album* is usually more common in fields or other disturbed ground. Both species are valuable for greens, the tender tips and leaves, cooked alone or with other edible plants.

3. *Chenopodium berlandieri*, var. *farinosum* (Ludwig) Aellen. A low, compact form with especially nauseous odor. We have only one specimen from Billings County (W. Whitman in 1937). I looked for it in a nearby locality without success. The name *C. dacoticum* Standley was given to this but Prof. Aellen considers it a variety of *berlandieri*.

4. *Chenopodium bushianum* Aellen. This species had often but incorrectly been called *C. paganum* Reich. It is difficult to separate from *C. album*. The seeds are distinctly larger. Small plants often have very heavy, thick fruiting branches. Tall ones are more like *C. album* except for elongated petioles. We have specimens from edges of woods at Fargo, Valley City and Jamestown. Var. *acutidentatum* Aellen, based in part on Fargo specimens, seems only an extreme form with large, much toothed leaves.

5. *Chenopodium strictum* Roth. Resembles *C. album* but is finer, 4—8 dm. (16-32 in.) high, with rather distinctive olive green color, the stems usually red striped or splotched; leaves oblong or ovate-oblong, rounded at tip, sometimes with a few small teeth; seeds about 0.9 mm. wide. The typical form is Asiatic and is said to have toothed leaves. The latest disposition by Aellen and Just (Am. Mid. Nat. 30:67, 1943) is to call our plants ssp. *glaucophyllum* Aellen.

This is a recently introduced plant with an interesting history. It certainly was not here in 1917 when I collected 26 specimens of the *C. album* group at Fargo and sent to the U. S. National Herbarium. These were later examined by Prof. Aellen of Basel, Switzerland, in his study of all American species of the genus (Fedde, Rept. Spec. Nov. 26:31-64, 119-160, 1929). By 1931, when I again collected material for Prof. Aellen, it was rather common about the streets of Fargo. In 1942 I collected specimens (Stevens 668) from one of the experiment station plots where it had become well established. We have it from 11 counties west to Minot and Bismarck, mostly as a street and lawn weed. One striking feature of the plant is its late blooming, beginning regularly about August 20. In a lawn it continues to produce short branches after cutting, much like *Atriplex patula.*

6. *Chenopodium standleyanum* Aellen. Resembles *C. album* but has very slender branches, the tips often drooping; leaves thin, pale green, ovate to lanceolate, with a few small teeth; seeds very shining black as in *Amaranthus retroflexus.* Frequent in woods, at least at Fargo. We have it from Richland and Barnes Counties also. It was previously put in *C. boscianum* Moq., which Professor Aellen found to be a different species of southern U. S.

7. *Chenopodium leptophyllum* Nutt. NARROW-LEAVED GOOSEFOOT. Slender, 3—8 dm. (1—3 ft.) high, moderately to densely mealy; leaves linear or lanceolate, entire or with 1 or 2 teeth toward base; seed shining black. Common in dry soils. Aellen has consided our plants *C. pratericola* Rydb., and not the same as *leptophyllum.*

7a. *Chenopodium subglabrum* (S. Wats.) A. Nels. Leaves narrow as in *leptophyllum,* but green. Immature plants were collected on a sandbar of the Little Missouri River in Slope County, July 5, 1946 (Stevens 887). They were mostly small, 1—3 dm. high. Flower clusters are well separated in a widely branched top, the seeds 1.5 mm. wide.

8. *Chenopodium fremontii* S. Wats. Slender, 3—6 dm. high, often with widely spreading branches; leaves bright green, rather small and thin, short, broadly triangular or rhombic, sinuate-dentate, 1—5 cm. long, nearly as wide; seed black, shining. Frequent, usually in rather dry woods throughout the State.

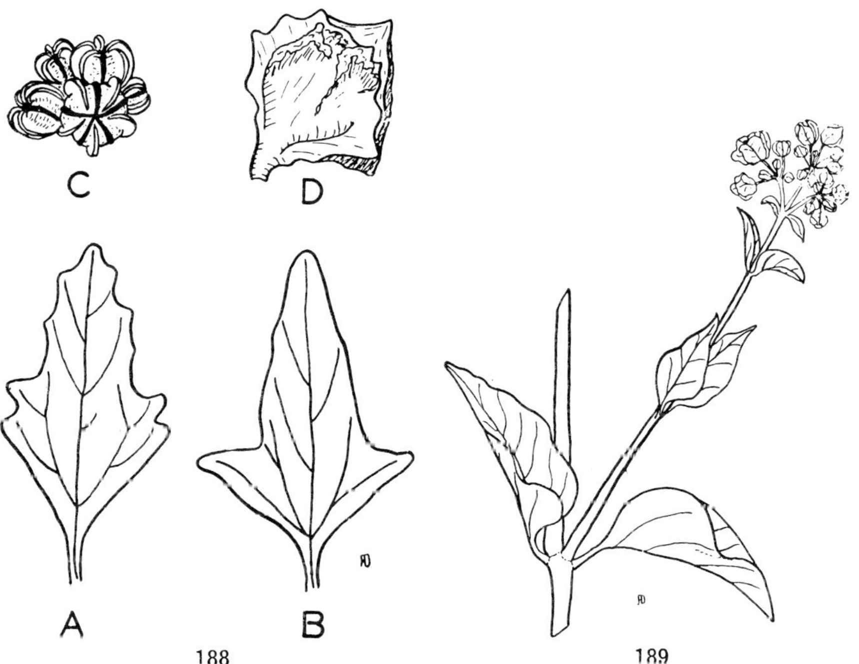

188. Leaves and fruits of. A, C. Lamb's Quarters (*Chenopodium album*); B, D. Saltbush (*Atriplex patula*).
189. Wild Four-o'clock (*Allionia nyctaginea*).

9. *Chenopodium gigantospermum* Aellen. MAPLE-LEAVED GOOSEFOOT. Coarse, rank scented annual; stem sharply angled, 1—2 m. (3—7 ft.) high; leaves broadly ovate, angular, 1—2 dm. (4—8 in.) long; seeds dull black, 1.5 mm. ($\frac{1}{16}$ in.) wide. Frequent in woods or about groves. Formerly placed in the European *C. hybridum* L., which Prof. Aellen considers a distinct species.

10. *Chenopodium glaucum* L. PALE GOOSEFOOT. Stems fleshy, spreading or forming a bushy plant 1—2 dm. high; leaves pale green, oblong or ovate, evenly small toothed, 1—5 cm. long; flowers in slender clusters at leaf bases; sepals thick, not keeled; seeds flat, 1 mm. wide. Frequent or common on bare saline soil. Our records are mostly from southern part of State. This plant has a pronounced bitter taste.

11. *Chenopodium rubrum* L. RED GOOSEFOOT. Erect, stout, fleshy plant, 3—8 dm. high; leaves 2.5—10 cm. long, often about as wide as long, thick, coarsely toothed; flowers very small in clusters at upper leaf bases; sepals thick, not keeled; seed flat, nearly black, smooth, 0.75 mm. wide. Frequent, sometimes very abundant on dried out pond beds; also along ditches and stream banks. During the drouth years 1934 and 1936, this attracted considerable attention as a forage plant. It was found quite palatable but on account of its fleshy nature was difficult to handle. It blooms in August and turns black when frosted.

12. *Chenopodium botrys* L. JERUSALEM OAK. Erect, branching annual, 1—6 dm. high; leaves 1—5 cm. long, deeply lobed; small flowers in axillary clusters; strongly scented but more aromatic than most species. A rare introduction, collected only along railroad tracks at Fargo in 1910 and some years later.

13. *Chenopodium foliosum* (Moench) Aschers. STRAWBERRY BLITE. Erect or spreading annual, 3—6 dm. high; leaves fleshy, broadly triangular, toothed, 3—10 cm. long; flowers at upper leaf bases, forming bright red clusters 5—10 mm. wide in fruit; seeds black, 0.75 mm. wide. In woods. We have only specimens received from Arthur Welford in 1925 from Neche, Pembina County, who wrote that it was in "brush land broken this spring". This plant had formerly been called *Blitum virgatum* L. I called *C. capitatum* (L). Aschers., a plant collected in 1932 in woods near Frazee, Minnesota, about 60 miles east of Fargo. It is similar to *C. foliosum,* but the flower clusters are in terminal, leafless spikes.

Atriplex SALTBUSH

Herbaceous or somewhat shrubby plants, some much resembling *Chenopodium;* flowers monoecious or dioecious, the staminate in clusters and having a calyx, the pistillate single or clustered, usually without calyx; fruit as in *Chenopodium,* enclosed by a pair of bracts which are usually united at base (fig. 188D); bracts often thick and hard, usually bearing tubercles on the surface; seed erect, attached at one edge. Many species in the Great Plains region, chiefly on saline to strongly saline soil. Some books use "orach", others "scale" as a general name.

Key to Species

Annual; stem often hard but not truly woody.
 Plant green or red or somewhat mealy when young, not gray.
  Stem 1—2 dm. (4—8 in.) high, brittle; leaves entire; plant of clay
   buttes.   17. *Atriplex dioica*
  Stem 3—10 dm. high, or low, spreading; leaves often 2-toothed;
   in ordinary soils.
   Bracts ovate, entire; plant often red; cultivated or escaped.
              14. *Atriplex hortensis*
   Bracts triangular or rhombic, margins usually toothed.
    Plants stout, erect; leaves thick; bracts broad at base.
              15. *Atriplex hastata*

    Plants slender, lower branches at least, low, spreading; bracts
       narrowed at base.   16. *Atriplex patula*
Plants gray; stem and leaves covered with scales; bracts hard.
    Stems rounded; leaves oblong or ovate, sharply toothed.
                    18. *Atriplex rosea*
    Stems angled; leaves triangular, little toothed.   19. *Atriplex argentea*
Perennial; stem woody at base; leaves entire, silvery.
    Leaves ovate to obovate.   21. *Atriplex confertifolia*
    Leaves oblong to linear.   20. *Atriplex nuttallii*

14. *Atriplex hortensis* L. GARDEN ORACH. FRENCH SPINACH. Erect annual,
    1—2 m. high; leaves ovate, 4—10 cm. long; flowers scattered along
    upper branches; bracts rounded, thin, 1 cm. wide. Frequently grown
    as a vegetable, also in a bright red form as an ornamental. Ordinary
    spinach is *Spinacea oleracea,* a smaller, related plant which does not
    have the bracts.

15. *Atriplex hastata* L. (Fig. 188B,D). Variable, but commonly stout, 5—10
    dm. (1½—3 ft.) high, with thick, fleshy leaves and dense flower clus-
    ters forming interrupted terminal clusters 1—2 dm. long; leaves ovate
    to lanceolate, 3—7 cm. (1½—3 in.) long, the larger ones nearly as wide,
    usually with 1 prominent, triangular tooth on each side below the
    middle, petiole shorter than blade; fruiting bracts triangular-ovate,
    thin, with a few short processes on the backs; seed 1—1.5 mm. (1/20 in.)
    wide, flat, black or brownish. Common in slightly saline soil, in wet
    or poorly drained places. This plant resembles lamb's quarters, but
    the two triangular teeth on the leaves are fairly distinctive. The seed
    has flat, instead of convex sides. It frequently makes a dense growth
    on dried out pond beds and sloughs.

16. *Atriplex patula* L. Slender, widely branched, lower branches lying on
    the ground, 4—8 dm. long; leaves thin, ovate to lanceolate; flower
    clusters quite scattered; otherwise similar to preceding. I reported
    this as found first in the State in 1919 (Bull. Torr. Bot. Cl. 49:97, 1922).
    Since then it has become common and troublesome at Fargo as a garden,
    lawn and grove weed. It seems to thrive in partial shade and that
    condition is probably responsible for the thin leaves. In the open it
    is not always distinct from *A. hastata.*

17. *Atriplex dioica* (Nutt.) Macbr. RILLSCALE. Erect, slender, or bushy,
    rounded plant 1—3 dm. high, the branches brittle; leaves ovate to
    lanceolate, smooth, 1—2.5 cm. long; bracts ovate, acute, 2 mm. long. On
    bare, clay buttes. We have specimens from Valley City, where clay
    exposures are local; otherwise it occurs from Missouri River westward.
    When in bloom (specimens June 27-Aug. 11), staminate flowers are
    in conspicuous, pinkish, terminal clusters. It has also been called *A.
    ovata* and *A. suckleyana.* Rydberg separated it from *Atriplex* as
    Endolepis dioica.

18. *Atriplex rosea* L. REDSCALE. Stout, bushy plant, 3—10 dm. high; leaves
    ovate, sharply, almost spiny toothed, very gray, rough, sessile, 1—6
    cm. long; bracts 5 mm. long, toothed near base, the sides with several
    green appendages. This has come in from farther west and is now
    occasional to frequent all over the State, especially as a street and
    roadside weed. It matures late and is decidedly coarse and persistent.
    Our first record was at Fargo in 1923. The name "red" is not appro-
    priate. Standardized Plant Names calls it Tumbling Orach, but it
    hardly breaks off easily enough to tumble well.

19. *Atriplex argentea* Nutt. SILVERSCALE. Stem stout, widely branched,
    1—6 dm. high; leaves ovate or triangular, 2—5 cm. long, rough with
    gray scales; bracts 4—8 mm. long, edges toothed and sides with prom-
    inent protuberances. Local in nearly bare, saline clay soil; specimens
    from Valley City, Bottineau area and the Bad Lands. A specimen
    collected in the middle of a railroad track at Ellendale in 1921 was iden-
    tified by Mr. E. C. Leonard of the U. S. National Museum as *A. expansa*
    Wats., but this is regarded by Clements and Hall (Carneg. Inst. Pub.

326:285, 1923) as a variety of *argentea* and limited to the far southwest. The Ellendale specimen is more slender and greener than is usual for *argentea*.

20. *Atriplex nuttallii* S. Wats. SALT SAGE. MOUNDSCALE. Perennial from a woody base, erect or spreading, 2—5 dm. (4—20 in.) high; leaves 1—5 cm. (⅖—2 in.) long, very gray, ovate to oblanceolate, margins entire but often wavy, petioles short or none; staminate flowers in dense, slender, yellow spikes or panicles; bracts 4—7 mm. (¼ in.) long, prominently toothed and long tubercled, forming a rounded, almost spiny fruit. Common on clay slopes. We have it from a butte near Walhalla, Pembina County, otherwise only from Missouri River westward. The name "moundscale" was applied to it because the soil between plants washes away, leaving them on small mounds. Seeds make up a relatively large part of the mature plant. Often it produces a large tuft of spreading stems. It is an important forage plant for sheep in the plains area.

21. *Atriplex confertifolia* (Torr.) S. Wats. SPINY SALTBUSH. Stiff, spiny, shrub, 2—8 dm. high; leaves ovate to elliptic, 1—2 cm. long, silvery gray; bracts rounded, similar to leaves, 5—10 mm. long. Local in Bad Lands.

22. *Suckleya suckleyana* (Torr.) Rydb. Annual; stems prostrate or nearly, 1—4 dm. long; leaves petioled, blades rounded or ovate, 1—3 cm. long, with few, short, sharp teeth; bracts 5—6 mm. long, toothed. Our only record of this is a specimen which I collected along a road near Belfield, Stark County, in 1914. The plant has been found quite poisonous in western states on account of prussic acid content.

23. *Monolepis nuttalliana* (Schultes) Greene. Tufted, fleshy winter annual; stem 1—3 dm. long; leaves oblong or triangular, 1—5 cm. long, somewhat toothed; flowers densely clustered at upper leaf bases; sepal 1, stamen 1; seed circular, flat, 0.75 mm. wide. Frequent on bare soil, gardens or roadsides. Flowers early in May, matures in July. A good plant for green vegetable.

24. *Cycloloma atriplicifolium* (Spreng.) Coult. TUMBLEWEED. WINGED PIGWEED. Widely branched, bushy annual, 3—6 dm. (1—2 ft.) high; leaves oblong, 1—5 cm. (⅖—2 in.) long, coarsely toothed, rather rough; fruit flattened, enclosed by calyx, each sepal developing a wing, entire calyx 5 mm. (⅕ in.) wide. Very sandy soil. Reported only in Richland-Ransom County area, Kidder and McHenry Counties (fig. 17); accidental along railroad at Fargo. Matures late.

Corispermum   BUGSEED

Annuals, resembling *Cycloloma*, leaves narrow, branches slender or widely spreading; fruit oblong, erect, flattened, slightly or broadly winged, projecting beyond calyx. The status of the species is not yet clear, but the following forms have been found. (See Bull. Torr. Bot. Cl. 49:97, 1920.).

Key to Species

Fruit with a distinct wing, 0.5 mm. wide; plant smooth or nearly so.
  Flower spike dense; bracts usually wider than fruit.
                    25. *Corispermum hyssopifolium*
  Flower spike slender; bracts narrower than fruit.
                    26. *Corispermum nitidum*
Fruit acute margined, scarcely winged; plant hairy.
                    27. *Corispermum villosum*

25. *Corispermum hyssopifolium* L. Only one specimen from Mandan (Wright 610) is placed here.

26. *Corispermum nitidum* Kit. This usually develops long, very slender, terminal branches. Specimens from Richland, Ransom, Cass and Towner Counties.

27. *Corispermum villosum* Rydb. Plant gray hairy; lower branches spreading, the whole plant bushy. Specimens from Richland, Ransom, Burleigh, Grant, Stark, Golden Valley, Williams, McHenry and Pierce Counties. It has been found at Fargo, introduced with gravel on railroad grades. It is one of the first plants to grow on shifting sand dunes. This seems to be our most common form. Prof. Aellen wrote in 1932 that he considered it the same as No. 26, but not identical with *nitidum* Kit.

28. *Kochia scoparia* L. BURNING BUSH. SUMMER CYPRESS. Erect, bushy annual, 6—20 dm. (2—7 ft.) high; stem somewhat hairy; leaves lanceolate to linear, 5—10 cm. (2—4 in.) long; calyx much as in *Chenopodium*, becoming narrowly winged in fruit. It is the first thing to germinate in spring but late in maturing. Introduced as an ornamental on account of its dense, conical form and often bright red color in fall. Now very common as a weed, mainly along roadsides, on dump heaps and other neglected ground, but also persistent in gardens. The plants vary greatly in color but I have been unable to separate distinct forms. It is usually regarded as a dry soil plant but in neglected, rich soil at Fargo in the wet year of 1943, it reached a height of 2 m.

29. *Axyris amaranthoides* L. RUSSIAN PIGWEED. Erect annual, 3—8 dm. high, often little branched; leaves ovate to lanceolate, thin, green but slightly gray from scattered, star shaped hairs; staminate flowers in terminal, interrupted clusters or mixed with pistillate ones in leaf axils; bracts thin, green, separate; sepals present in pistillate flowers, membranous; seed oblong, 2—3 mm. long, nearly black. First found at Pembina in 1889. We have specimens from Cass, Ransom, Steele, Griggs, Pierce, Rolette, Ward and Burke Counties, but it seems largely confined to Pembina and adjoining counties, growing chiefly in partial shade.

30. *Eurotia lanata* (Pursh) Moq. WINTER FAT. Tufted, white hairy perennial, 2—6 dm. high, more or less woody at base; leaves linear, 2—5 cm. long, 2—5 mm. wide; flowers either perfect, monoecious or dioecious, in small, axillary clusters; pistillate flowers with 2 united bracts with sharp, spreading tips, this fruiting body conical, 4—6 mm. long, covered with long hairs. Dry prairies and hills, west of Missouri river. This is regarded as a valuable forage plant, especially on winter ranges.

31. *Salicornia rubra* A. Nels. GLASSWORT. Fleshy, brittle annual, appearing leafless; stems erect or widely branched and spreading, flowers embedded in the stem in terminal spikes 2—4 cm. long. In strongly saline soil, sometimes forming a mat around ponds, the plants crackling when walked upon. We have it from Richland, LaMoure, Barnes, Ramsey, Pembina and Divide Counties.

32. *Suaeda depressa* (Pursh) S. Wats. SEA BLITE. Erect or spreading annual with crowded, fleshy, cylindrical leaves; stems 2—10 dm. (8—40 in.) long; leaves 1—4 cm. (½—1½ in.) long; flowers much as in *Chenopodium*, solitary or clustered at upper leaf bases; sepals keeled or narrowly winged. Saline soil throughout State. The plants suggest a Russian thistle, but are softer, leaves not spiny. They mature late and turn black when frosted. I have not been able to distinguish *S. erecta* (S. Wats) A. Nels. as a distinct form. The smaller plants may be upright, but weight of the longer branches seems to bend over the larger ones.

33. *Suaeda fruticosa* Forsk. Much like preceding, but growing from a woody base which is usually prostrate. Clay buttes in Bad Lands. Apparently Bergman overlooked this plant. Bell 509, 886, L. R. Waldron 1190, and Bergman (Medora, June 19, 1910), belong to it. Rydberg (Fl. Pl. Pr.) also failed to include it.

34. *Salsola kali* L. RUSSIAN THISTLE. Bushy, spiny annual, 1—10 dm. high, leaves slender, cylindrical, slightly flattened, 2—7 cm. long; flowers solitary at leaf bases, between 2 bracts which resemble short leaves; sepals thin and membranous, developing a horizontal wing 2 mm. wide in fruit; seed obconical, the embryo long and coiled. Very common, chiefly in dry soils. The plants are soft at first. The lower

leaves are 5—7 cm. long, but fall off as the plant develops. The upper ones are gradually shorter, those on the branch tips much resembling the bracts and becoming stiff and spiny. For further details, see N. D. Exp. Sta. Bull. 326.

35. *Sarcobatus vermiculatus* (Hook.) Torr. GREASEWOOD. Spiny, widely branched shrub, 1—3 m. high; leaves fleshy, cylindrical, 1—4 cm. long, pale green; staminate flowers in spikes 0.5—3 cm. long at ends of branches; pistillate flowers solitary at leaf bases; stamens 3; fruiting calyx 6 mm. long, with a wing margin 2—6 mm. wide. Local, chiefly on bare clay of Bad Land buttes; Grant, Billings, Golden Valley and McKenzie Counties. The pale green color of the leaves is rather striking as compared to rabbitbrush which is similar in general appearance.

PIGWEED FAMILY Amaranthaceae

Weedy annuals with alternate, not fleshy leaves; flowers perfect or dioecious, small, green, at leaf bases or in terminal clusters which are prickly because of small, spiny bracts below each flower; sepals 2—5, sometimes 0, usually with membranous edges; petals 0; stamens 5; pistil 1; fruit a 1-seeded, membranous utricle, which usually splits transversely, the seed falling free, rarely enclosed by wall and never by sepals; seed black, shiny.

Key to Species

Flowers perfect, calyx present.
  Flowers in dense, terminal clusters.
    Flower clusters 10—15 mm. (½ in.) wide; stems rough.
                                             1. *Amaranthus retroflexus*
    Flower clusters 4—6 mm. wide; stem smooth.
                                     2. *Amaranthus hybridus*
  Flowers in small, axillary clusters.
    Stems prostrate; seeds 1 mm. (1/25 in.) wide.
                               3. *Amaranthus graecizans*
    Stems erect, bushy branched; seeds 0.5 mm. wide.
                                  4. *Amaranthus albus*
Flowers dioecious, no calyx in pistillate ones.   5. *Acnida altissima*

1. *Amaranthus retroflexus* L. ROUGH PIGWEED. Annual with a red root; slender or widely branched, 2—10 dm. (8—40 in.) high; leaves ovate, long petioled, rather rough; flowers in dense, prickly, terminal clusters 5—10 cm. (2—4 in.) long. Very common weed, especially in rich soil.

2. *Amaranthus hybridus* L. SLENDER PIGWEED. Similar to last, taller and more slender; much like *Acnida* but has perfect flowers. I collected a single specimen at Fargo in 1921.

3. *Amaranthus graecizans* L. CREEPING PIGWEED. Stem red, prostrate, 2—10 dm. long; leaves spatulate or obovate, 1—3 cm. long, entire, long petioled; flowers in small, axillary clusters, hardly spiny. Very common weed, especially in gardens and along roadsides. Fernald has recently shown (Rhodora 47:139, 1945) that this name should be used for the plant formerly called *A. blitoides*.

4. *Amaranthus albus* L. TUMBLING PIGWEED. Much branched, forming a dense, rounded plant 3—6 dm. high; leaves oblong or spatulate, 1—4 cm. long; flowers in small, axillary clusters all along the branches, quite prickly. Common in dry soil, becoming a tumbleweed when dry. Formerly called *A. graecizans*.

5. *Acnida altissma* Riddell. WATER-HEMP. Slender or much branched, 3—15 dm. high; leaves lanceolate to ovate, 2—8 cm. long; flowers in long, slender, terminal clusters 1—3 dm. long; fruit not splitting open, seed 0.75 mm. long. Ditches and low spots. We have it chiefly from Cass, also Foster, Ransom and Slope Counties. It is a common weed farther south but this seems to be its northern limit. Quite variable in appearance, often having dense, reddish flower clusters. Late August.

FOUR-O'CLOCK FAMILY Nyctaginaceae

Annual or perennial herbs with opposite, entire leaves; corolla absent but calyx colored, resembling a corolla, funnel shaped; several flowers in a cluster surrounded by a calyx-like involucre; fruit 1-seeded, ribbed or winged.

Key to Species

Annual, 1 dm. (4 in.) high; bracts separate; fruit winged.
1. Abronia micrantha

Perennial, usually 3—8 dm. high; bracts united; fruit oblong, ribbed.
  Plant quite hairy. 2. Allionia hirsuta
  Plant not hairy. ·
    Leaves ovate, 2.5—5 cm. (1—2 in.) wide.   3. Allionia nyctaginea
    Leaves linear to lanceolate, 3—15 mm. (⅛—⅗ in.) wide.
4. Allionia albida

1. *Abronia micrantha* Torr. SAND VERBENA. Somewhat fleshy annual, 5—10 cm. high; leaves elliptic, 2—4 cm. long; flower 8 mm. long, with a narrow tube, flaring at tip; bracts ovate, 4—5 mm. long; fruit pink, 5 mm. long with 2—4 wings which are 3—4 mm. wide. Occasional in sandy soil, Bad Lands and along Little Missouri River. Late May.

2. *Allionia hirsuta* Pursh. HAIRY FOUR-O'CLOCK. Stems stout, little branched, rough hairy, 3—10 dm. (1—3 ft.) high; leaves ovate to oblong, 2.5—5 cm. (1—2 in.) long, with scattered, coarse hairs; flowers in a terminal panicle, but in scattered, small clusters; calyx pink, funnel shaped, 8 mm. (⅓ in.) wide; fruit oblong, strongly few ribbed, 5 mm. long. Common in dry, sandy soil. Flowers quite showy when open in early evening. June, July.

3. *Allionia nyctaginea* Michx. WILD FOUR-O'CLOCK. (Fig. 189). Erect or spreading, stems widely forked, 3—8 dm. high; leaves ovate, very flat and smooth, 3—8 cm. long, nearly as wide; flowers as in *hirsuta*, the involucre saucer-like, 5-angled, 1—1.5 cm. wide. Common weed from a very thick, black root. Blooms mostly in June and is very showy in fields or along roadsides in evening or early morning.

4. *Allionia albida* Sweet. Stems slender, 3—8 dm. high; leaves narrow and smooth.

These plants are also known as umbrella-worts but the name wild four-o'clock seems fitting, since they belong to that family and the flowers open in the evening. Some authors use the name *Oxybaphus* and Standley (Field Mus. Publ. 294:304-7, 1931) has placed them in *Mirabilis* with the garden four-o'clocks. The narrow leaved forms are puzzling. Several species have been described. Bergman discussed them at some length and separated *A. linearis* as having leaves less than 5 mm. wide. I have not been able to study them further.

CARPET-WEED FAMILY Aizoaceae

This includes chiefly fleshy plants of Africa, especially many species of *Mesembryanthemum*, some of which are known as ice plant, fig-marigold, etc. Another, frequently cultivated here, is New Zealand spinach (*Tetragonia expansa*). We have only one wild species.

1. *Mollugo verticillata* L. CARPET-WEED. (Fig. 88). Annual, stems prostrate, forming a mat 1—3 dm. (3—12 in.) wide; leaves spatulate or linear-oblong, 1—3 (⅖—1⅕ in.) long, in whorls of 4—8 at each node; flowers in few flowered, axillary clusters, greenish white, 3 mm. (⅛ in.) wide; sepals 4 or 5; petals 0; capsule rounded, 4—5 mm. long; seeds many, red. First collected along railroad at Fargo in 1910. It still grows there and has not been collected elsewhere. A common weed farther south.

PURSLANE FAMILY Portulacaceae

Succulent herbs; sepals usually 2; petals 4—5; stamens 4—many; fruit a capsule.

Key to Species

Tufted, low perennial; leaves cylindrical; flowers pinkish.
1. Talinum parviflorum

Prostrate annual; leaves thick but flat; flowers yellow.
2. Portulaca oleracea

1. *Talinum parviflorum* Nutt. FAME FLOWER. Leaves basal, 1—5 cm. (½—2 in.) long; flowers few in a forking cluster 5—10 cm. high; flowers 10—15 mm. (½ in.) wide; capsule 3—4 mm. long. We have a single record from Grant County by Bell in 1909.

2. *Portulaca oleracea* L. PURSLANE. Stems red, very fleshy, 1—5 dm. long; leaves thick, fleshy, wedge shaped or obovate, 1—3 cm. long; flowers at leaf bases, 5 mm. wide; capsule 3 mm. long, splitting transversely; seeds many, black. A troublesome weed in gardens, the plants remaining fresh for days after being cut off. Flowers open in bright sunshine. July-Sept.

PINK FAMILY Caryophyllaceae

Annual or perennial herbs with opposite, entire leaves; flowers usually in terminal cymes; sepals 5; petals 5, often 2-lobed at tip; stamens usually 10; pistil of 3—5 carpels, ovary 1, usually with only 1 chamber, styles separate; fruit a capsule, splitting down the sides or opening only at top; seeds usually many, flattened, wedge shaped or rounded, often marked with eccentric rows of small protuberances. Many species are weeds.

Key to Species or Genera

Sepals united into a cylindrical or ovate calyx; flowers usually showy.
  Calyx 10-ribbed.
    Calyx lobes 1 cm. (⅖ in.) long; coarse hairy plant; flowers purple,
      2 cm. wide.   1. *Agrostemma githago*
    Calyx lobes 1—5 mm. (1/25—⅕ in.) long; flowers white or pink.
      Styles usually 3; capsule divided inside at base.   2—5. *Silene*
      Styles 5; capsule not at all divided.
        Flowers dioecious, 1—2 cm. wide; coarse weed.   6. *Lychnis alba*
        Flowers perfect, 5 mm. wide; slender prairie plant.
7. Lychnis drummondii

  Calyx 5-ribbed.
    Calyx strongly 5-angled; flowers pink, 1 cm. wide.
8. Saponaria vaccaria

    Calyx not angled; flowers white or rose colored.
      Stems stout, flower cluster dense; flowers rose colored to white,
        15—25 mm. wide.   9. *Saponaria officinalis*
      Stems widely branched above; flowers white, 5—8 mm. wide.
10. Gypsophila paniculata

Sepals separate; flowers usually small.
  Leaves very narrow, in whorls of 6—8; flowers 3—5 mm. wide.
11. Spergula arvensis

  Leaves opposite.
    Stems very short; leaves not over 5 mm. (⅕ in.) long.
12. Paronychia sessiliflora

    Stems elongated; leaves 1—3 cm. (⅖—1⅕ in.) long.
      Leaves fleshy; petals pink.   13. *Spergularia rubra*
      Leaves not fleshy; petals white.
        Styles 3; capsule splitting down sides.
          Petals not notched.   18. *Arenaria lateriflora*
          Petals deeply notched.   14-17. *Stellaria*
        Styles 5; capsule opening at top only.   19-22. *Cerastium*

1. *Agrostemma githago* L. CORN COCKLE. PURPLE COCKLE. ROUGH COCKLE.
Annual, erect, often little branched, 3—10 dm. (1—3 ft.) high; stem and
leaves thinly covered with coarse hairs lying against surface; leaves
5—10 cm. (2—4 in.) long, 3—6 mm. (⅛—¼ in.) wide; flowers solitary
at ends of branches, dark red, 2 cm. wide; capsule oblong, 1—2 cm.
long; seeds black, wedge shaped, rough, 3—5 mm. long. Occasional
in grain fields or waste ground. It seems to have become less rather
than more common. Seeds poisonous.

Silene CATCHFLY

Smooth or hairy plants, mostly weeds; flowers usually opening
in the evening, rather showy. The name comes from the fact that
many species have sticky areas on stems in which small insects
become entangled. Of ours, only *Silene antirrhina* shows this
clearly.

Key to Species

Flowers 2—3 mm. ($\frac{1}{12}$—⅛ in.) wide, very inconspicuous; slender plant.
                                      2. *Silene antirrhina*
Flowers 1—2 cm. (⅖—⅘ in.) wide; stout plants.
  Leaves smooth and waxy; calyx tightly enclosing capsule.
                                      3. *Silene cserei*
  Leaves covered with fine, sticky, short hairs; calyx loose.
    Flowering branches slender, many flowers along one side.
                                      4. *Silene dichotoma*
    Flowering branches widely forked with few flowers.
                                      5. *Silene noctiflora*

2. *Silene antirrhina* L. SLEEPY CATCHFLY. Annual; stems erect, slender,
3—8 dm. (1—2½ ft.) high, a brown sticky area 1—2 cm. (⅖—⅘ in.)
long on each of upper internodes; calyx oblong, ribbed, 8—10 mm. (⅜
in.) long; petals white or pinkish, scarcely projecting beyond calyx;
seeds black, 0.5 mm. wide. Frequent in fields or waste ground. June,
July.

3. *Silene cserei* Baum. SMOOTH CATCHFLY. Biennial; stems stout below,
often widely branched, 4—8 dm. high; leaves oblong, ovate or lanceolate, thick, smooth and waxy, 2—5 cm. long; flowers somewhat clustered on elongated, leafless, terminal branches; corolla white, 1 cm.
wide; capsule oblong, 1 cm. long, closely covered by thin calyx, seeds
similar to No. 5 but smaller.

   This recently introduced species was first detected in specimens
collected at Oakes, Dickey County, in 1919. It was identified by Dr.
Paul C. Standley, who wrote that it seemed to be the first record for
North America. I found later that a specimen collected by Dr. J. F.
Brenckle at Kulm, LaMoure County, in 1916, had been incorrectly labeled *Saponaria vaccaria*, which has similar leaves and general growth
habits. Specimens were soon found in other localities and are now on
file from 13 counties. It seems to have spread along railroads and highways where it grows vigorously in gravel rather than in fields. Previously called *S. fabaria* Sibth. & Sm.

4. *Silene dichotoma* Ehrh. FORKED CATCHFLY. Biennial; stems densely
hairy, somewhat sticky, 3—6 dm. high; leaves lanceolate or oblanceolate, 5—8 cm. long; flowers on a 2—3 dm. long, slender, terminal
branch; calyx hairy, 5-ribbed; corolla 1 cm. wide, white; capsule oblong,
1 cm. long; seeds blackish gray, kidney shaped, 1 mm. wide, the tubercles well spaced and prominent. We have this only from Cass, Barnes,
LaMoure, Cavalier and Stark Counties. It seems not well established.
Plants were grown in the garden and the biennial nature determined.

5. *Silene noctiflora* L. NIGHT-FLOWERING CATCHFLY. (Fig. 190.) Annual,
3—10 dm. (1—3 ft.) high, erect or widely branched, stems densely
covered with short, fine, sticky hairs; leaves oblong or oblanceolate,

5—12 cm. (2—5 in.) long; flowers in widely branched, terminal cymes (first flower at end of stem, 2 branches arising below it and repeating the process); calyx bell shaped, loose, sticky, 1.5—2 cm. long; corolla white, fading pink, 1.5 cm. wide; capsule 1.5 cm. long; seeds rounded, gray, tubercles small and closely placed. This is our common catchfly, frequently abundant in old fields. Smooth catchfly, forked catchfly and white cockle are also night-flowering in habit.

6. *Lychnis alba* Mill. WHITE COCKLE. Perennial, much resembling *Silene noctiflora*, flowers somewhat longer, calyx more inflated, especially in staminate flowers. We have specimens from Cass, Ramsey, Bottineau. McKenzie and Divide Counties. So far as I have observed, it appears only as a stray plant, except in Turtle Mts., where it is common.

7. *Lychnis drummondii* (Hook.) S. Wats. Perennial; stems erect, little branched, 2—6 dm. high; stems and leaves somewhat gray or sticky hairy; leaves oblanceolate to linear, 5—10 cm. long; flowers few in a terminal cyme; calyx oblong, 10—12 mm. long, 5-toothed, 10-nerved; petals white or purplish, little longer than calyx; capsule narrowly oblong, 10 mm. long. Occasional on prairie, especially on sandy soil. June, July. Specimens from Richland, Ransom, Barnes, Pembina and Pierce Counties. Probably more widely distributed but it is inconspicuous and easily overlooked.

8. *Saponaria vaccaria* L. PINK COCKLE. COW COCKLE. (Fig. 191). Annual, widely branched, 3—6 dm. high; leaves ovate to lanceolate, sessile or sometimes united at base, 2—8 cm. long, rather thick, smooth and waxy; flowers numerous, pink, 1 cm. wide in a widely branched top; capsule 1 cm. long, strongly 5-angled, narrowed and opening slightly at top; seeds black, nearly spherical, 3 mm. wide, granular with minute protuberances. Common in fields, especially westward; frequently grown as an ornamental. The seeds are a little larger than those of mustard, a little smaller than those of vetch. Other names used for it are *Vaccaria pyramidata* and *Vaccaria vulgaris*.

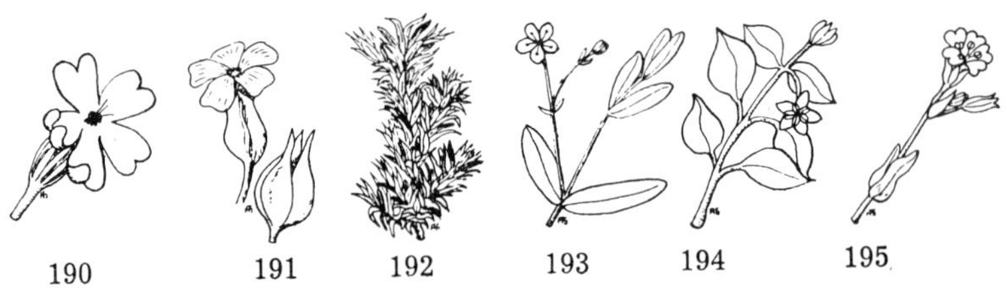

190   191   192   193   194   195

9. *Saponaria officinalis* L. BOUNCING BET Perennial; stems stout, little branched, 3—6 dm. high; flowers white to rose colored, 15—25 mm. wide in a dense, terminal cluster. A common escape from gardens farther east but not often grown here. In 1947 I found plants in a pasture near an old building site at Hankinson, Richland County.

10. *Gypsophila paniculata* L. PERENNIAL BABY'S BREATH. Perennial from a large tap root; stems slender, widely branched, 3—6 dm. (1—2 ft.) high; leaves oblanceolate or narrowly oblong, 2—5 cm. (1—2 in.) long; flowers numerous, white or pinkish, 3 mm. wide. Frequently grown as an ornamental. I collected a specimen along the railroad at Oakes, Dickey County, in 1919. A root was sent in by one man who thought it might be a bad weed, but it seems not to persist long, even in gardens. Found well established at Garrison and Dickinson in 1949.

11. *Spergula arvensis* L. SPURRY. Annual; stems slender, spreading, rather fleshy, 2—5 dm. high; leaves very slender, 3—5 cm. long, in whorls or clusters; flowers numerous, white, 3—4 mm. long, petals hardly longer than sepals; capsules on drooping stalks, rounded, 3—4 mm. long, splitting down the sides; seeds black, plump with a narrow wing around the middle. This is grown in Europe as a forage plant. Specimens were collected at Fargo in 1892 and 1903.

12. *Paronychia sessiliflora* Nutt. WHITLOW-WORT. (Fig. 192). Perennial, forming a dense tuft, 2—5 cm. high; leaves moss-like, 4—6 mm. long; flowers yellowish, very inconspicuous; sepals 3 mm. long; petals 0; fruit 1-seeded. On rocky ledges or hills, west of Missouri River; one specimen from Minot, Ward County. June.

13. *Spergularia rubra* (L.) J. & C. Presl. SAND SPURRY. Annual; stems branched, spreading from base, 0.5—1 dm. high; leaves fleshy, linear, 8—12 mm. long; flowers pink, 4 mm. wide, in slender, terminal racemes. We have found this only on strongly saline soil at edges of ponds or on mud of dried up ponds, in LaMoure, Barnes and Walsh Counties.

Stellaria CHICKWEED.   STITCHWORT

Low, slender annuals or perennials, with small, white flowers; capsules rounded, splitting down the sides; seeds red, flattened, circular or wedge shaped. The name *Alsine* is used by some writers for this genus.

Key to Species

Leaves ovate; the lower petioled. 14. *Stellaria media*
Leaves linear to oblong, not petioled.
  Bracts of flower cluster leaf-like. 15. *Stellaria crassifolia*
  Bracts of flower cluster white-membranous.
    Leaves widest at middle, narrowed toward base.
      16. *Stellaria longifolia*
    Leaves widest at base. 17. *Stellaria longipes*

14. *Stellaria media* (L.) Cyrill. COMMON CHICKWEED. (Fig. 194). Annual, stems prostrate, 1—3 dm. (3—12 in.) long; leaves ovate, 5—10 mm. (⅕—⅖ in.) long; flowers 3—4 mm. wide in terminal clusters or appearing axillary on slender, drooping stalks; petals shorter than sepals; capsule rounded, 3—4 mm. long. Lawns and woods, common. One of our most troublesome lawn weeds. June-Sept. This species is introduced from Europe, the three following are native.

15. *Stellaria crassifolia* Ehrh. Annual, stems spreading, very slender, 1—3 dm. long; leaves elliptic, lanceolate or oblong, 1—2 cm. long, upper similar to lower; flowers few at end of stem. Wet, usually shaded places. We have it only from Barnes, Stutsman, Grand Forks, Pembina and Williams Counties.

16. *Stellaria longifolia* Muhl. LONG-LEAVED STITCHWORT. Annual, stems very slender, 1—3 dm. high, usually supported by other vegetation; leaves linear or nearly so, 2—5 cm. long; flowers 3 mm. wide on very long, slender, upper branches. Frequent in woods. We have it west to Stutsman and Rolette Counties.

17. *Stellaria longipes* Goldie. Stems usually erect, 1—2 dm. high; leaves narrowly lanceolate, widest at base, sharply pointed at tip, 1—2 cm. long; uppermost leaves at base of flower stalks 2—5 mm. long, their edges thin and white. Low places on prairie or other open ground. We have it from Cass, Stutsman, Benson and Cavalier Counties.

18. *Arenaria lateriflora* L. BROAD-LEAVED STITCHWORT. (Fig. 193) Native perennial, forming patches; stems slender, erect or spreading, 1—2 dm. high; leaves ovate or oblong, rounded at tip, 1—3 cm. long; flowers white, 5—8 mm. wide; capsule ovoid, 5 mm. long, opening at tip by 3 parts which are 2-lobed. Dry woods, thickets, sometimes in open ground; quite common eastern part of State, west to Dickey, Stutsman and Ramsey Counties. June. Nearly as showy as *Cerastium arvense*, more so than species of *Stellaria*.

Cerastium MOUSE-EAR CHICKWEED

Similar to *Stellaria* but with sticky hairs; capsule cylindrical, opening only at tip.

Key to Species

Petals hardly longer than sepals; perennial. 19. *Cerastium vulgatum*
Petals 1½ to 2 times as long as sepals.
  Annuals; flowers 4—6 mm. (⅙—¼ in.) wide.
    Pedicels not much longer than sepals. 20. *Cerastium brachypodum*
    Pedicels 3—5 times as long as sepals. 21. *Cerastium nutans*
  Perennial; flowers 10—15 mm. (½ in.) wide. 22. *Cerastium arvense*

19. *Cerastium vulgatum* L. Introduced perennial forming patches; stems mostly prostrate, 1—2 dm. (4—8 in.) long; leaves ovate, oblong or spatulate, 1—2 cm. (⅖—⅘ in.) long; flowers 5—7 mm. (¼ in.) wide in terminal cymes; capsule 1 cm. long; seeds wedge shaped. We have this only from Fargo where it has survived at least some winters. It is a well known lawn weed.

20. *Cerastium brachypodum* (Engelm.) Robins. Native annual; stem erect, 2—4 dm. high; leaves oblong or oblanceolate, 5—20 mm. long; flowers few in a rather dense cluster at top of stem. Quite common on low, often saline prairie west of Missouri River.

21. *Cerastium nutans* Raf. Native annual; stem erect or often mostly prostrate, 2—5 dm. long. In wet, often shaded places, especially gravelly edges of streams or lakes. We have it from Richland, Stutsman, Pembina and Bottineau Counties.

22. *Cerastium arvense* L. PRAIRIE CHICKWEED. (Fig. 195). Native perennial, forming dense mats often 1—2 m. (3—7 ft.) wide; stems creeping, branches erect, hardly over 1 dm. high; leaves linear or narrowly lanceolate, 1—2 cm. long; flowers showy, 1 cm. wide, petals deeply divided. Very common on prairie. May-June.

WATER LILY FAMILY Nymphaeaceae

Large, perennial water plants with short, fleshy stems from which floating leaves and flowers arise; sepals 5—7, leathery; petals 10-20, narrow; pistils several, mostly united, stigma disk shaped; fruit tough, rounded, with many seeds.

1. *Nuphar advenum* Ait. YELLOW POND LILY. Leaves ovate, 1.5—3 dm. (6—12 in.) long, tip broadly rounded; flowers yellow, 2—3 cm. (1 in.) wide; fruit 4 cm. long. Local in ponds or slow streams; Richland, Ransom, Stutsman, Steele, Grand Forks, Rolette and Bottineau Counties (fig. 14). Some authors put it in *Nymphaea*, others in *Nymphozanthus*.

HORNWORT FAMILY Ceratophyllaceae

Small, water plants with finely divided leaves; flowers monoecious, very small, greenish, at leaf bases, sepals 6—12, petals 0; stamens 10—24, pistil 1; fruit nut-like.

1. *Ceratophyllum demersum* L. HORNWORT. (Fig. 65). Stems floating, 3—12 dm. long; leaves 6—9 in a whorl, each forking 1—3 times, segments very narrow, spiny toothed; fruit 5 mm. long with persistent style and 2 appendages at base. Ponds. Records from Steele, Rolette and Ward Counties. Probably more common but easily overlooked or confused with *Potamogeton* and other water plants.

BUTTERCUP FAMILY Ranunculaceae

Herbs or vines; leaves mostly alternate, without stipules, often compound or lobed; sepals separate, 3—15, often colored and petals lacking; petals, if present, as many as sepals; stamens usually many; pistils 3—many, each producing an achene or follicle. Plants often with acrid juice or poisonous, usually flowering early; leaves often ternate (divided once or more into 3 parts).

Key to Species or Genera

Pistil with several ovules; fruit a follicle, rarely fleshy.
  Flowers regular; sepals not spurred.
    Leaves simple, cordate, mostly basal.   1. *Caltha palustris*
    Leaves all ternately compound.
      Petals 0; sepals 3—5, petal-like; fruit fleshy.   2. *Actaea rubra*
      Petals 5, cornucopia shaped; sepals colored; fruit 5 follicles.
                  3. *Aquilegia canadensis*
  Flowers irregular; upper sepal prolonged into a spur 1 cm. long.
    Flowers deep blue; plant 2—4 dm. (8—16 in.) high, western.
                  4. *Delphinium bicolor*
    Flowers mostly white; plant 3—10 dm. (1—3 ft.) high.
                  5. *Delphinum virescens*

Pistils 1-ovuled, ripening into achenes.
  Leaves basal, except for some which form an involucre some
        distance below the flower or flowers.   6-11. *Anemone*
  Stem leaves not forming a pronounced involucre.
    Leaves opposite, with 3—7 leaflets; achenes with long, feathery
        tips.
      High climbing vines; flowers white.
        Leaflets 3; eastern.   12. *Clematis virginiana*
        Leaflets 5—7; western.   13. *Clematis ligusticifolia*
      Low plant with creeping stem; flowers blue.
             14. *Clematis pseudoalpina*
    Leaves alternate or only basal.
      Leaves simple, very narrow and basal only. 15. *Myosurus minimus*
      Leaves broad, usually lobed or compound.
        Petals present, usually yellow.   16-28. *Ranunculus*
        Petals absent; flowers usually dioecious; leaves large,
          compound.
          Stem leaves petioled; plant 3—10 dm. (1—3 ft.) high.
              29. *Thalictrum venulosum*
          Stem leaves sessile; plant 10—20 dm. high.
              30. *Thalictrum dasycarpum*

196.   Canada Anemone (*Anemone canadensis*).

1. *Caltha palustris* L. MARSH MARIGOLD. Perennial; stems hollow, erect, 2—6 dm. (8—24 in.) high; leaves dark green, smooth, mostly basal on long petioles, blades rounded, 5—20 cm. (2—8 in.) wide, with short, rounded teeth, the upper leaves narrower, sometimes elliptic; flowers yellow, 2—5 cm. (1—2 in.) wide, at ends of the few stem branches;

sepals 5—15, golden yellow; petals 0; fruit 5—15 flattened follicles, 1—2 cm. long. Local in swamps. We have it from Richland, Ransom, Benson and Pembina Counties (fig. 14). May, June. The leaves are used for greens, this being one of the few species of the family which is edible. It is also called "cowslip", a name used for *Dodecatheon* and other plants. "Marigold" is no better, since it is usually applied to plants of the aster family.

2. *Actaea rubra* (Ait.) Willd. BANEBERRY. Perennial with many stems 3—6 dm. high, and large, much divided leaves; leaflets ovate, 3—5 lobed, sharply serrate, 2—5 cm. long (fig. 198); flowers white, 3—5 mm. wide in a short, terminal raceme; fruit oblong, 8—10 mm. long, fleshy, bright red or white. Frequent in woods, west to Stutsman and Bottineau Counties. Flowers late May, fruit mature July, remaining all summer. This is an attractive ornamental. I have a plant which was brought from the woods over 20 years ago. The berries are reputed to be poisonous but the evidence is not clear. The red fruited form seems more frequent than the white fruited one.

3. *Aquilegia canadensis* L. COLUMBINE. Perennial with few stems 4—8 dm. high and much divided leaves; leaflets 1—3 cm. long, broader than long, partly divided into 3 blunt lobes (fig. 197); flowers 3—5 cm. wide, scarlet and yellowish, pendant on slender, upper branches; sepals lanceolate, horizontal; petals conical, 2 cm. long, the blunt tips pointing upward and curving together; fruit of 5 follicles 2 cm. long; seeds shining black. Frequent in woods, west to Barnes, Benson and Bottineau Counties (fig. 9). Late May and June. A handsome plant which should be protected against free picking for the plants are easily destroyed. Columbine leaflets are somewhat larger, wider and more deeply divided than those of meadowrue, which is much more common. It is readily grown in partial shade. The garden columbines are mostly derived from the European *A. vulgaris*, but other species are used.

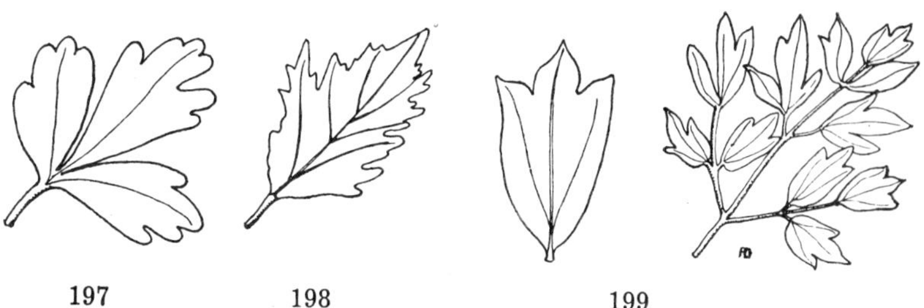

197   198   199

4. *Delphinium bicolor* Nutt. BLUE LARKSPUR. Perennial with a single stem 2—4 dm. (8—16 in.) high; leaves few, 2—4 cm. (⅘—1½ in.) wide, palmately divided into several narrow segments; flowers deep blue, 2 cm. long, scattered along the upper stem; fruit of 3 broad, widely spreading follicles, 1—2 cm. long. Only in extreme western part of State (fig. 9), but sometimes locally common. May. It is a showy plant, but one of the "low larkspurs", which are poisonous to livestock.

5. *Delphinium virescens* Nutt. WHITE LARKSPUR. Perennial with a single stem 4—12 dm. high; leaves 3—5 cm. wide, rounded, palmately divided into narrow lobes; flowers white with some purple inside, 2—3 cm. long, on the terminal 2—6 dm. of the stem; fruit 3 follicles, slender and nearly erect, 2 cm. long. Frequent on low prairie in the southeast to Traill and Stutsman Counties. Single records from Emmons and Walsh Counties. July. This plant has been called at different times, *D. carolinianum, D. penardi* and *D. albescens*.

Anemone   WIND-FLOWER

Perennials from crowns; proper leaves all basal, palmately lobed or divided; flowers at ends of 1 or more slender stalks; petals 0; sepals 5, usually white or purple, often showy; pistils many, each maturing into an achene.

Key to Species

Flowers lavender, 3—7 cm (1—3 in.) wide; achenes with long,
   feathery tips.   6. *Anemone patens*
Flowers white or greenish, 1—3 cm. wide; achenes not feathery tipped.
  Plants 1—2 dm. (4—8 in.) high, not hairy; flower solitary.
       7. *Anemone quinquefolia*
  Plants 3—8 dm. (1—2½ ft.) high, hairy; flowers usually 2 or more.
   Flowers showy, white, 2—5 cm. (⅘—2 in.) wide.
      8. *Anemone canadensis*
   Flowers inconspicuous, greenish, 1 cm. wide.
    Prairie plants 2—4 dm. high; leaf lobes 2—5 mm. ($\frac{1}{12}$—⅕ in.)
     wide.
      Flowers greenish; fruiting heads 2—4 cm. long; involucral
        leaves petioled.   9. *Anemone cylindrica*
      Flowers reddish; fruiting heads 1—2 cm. long; involucral
        leaves sessile.   10. *Anemone hudsoniana*
    Woodland plants 4—10 dm. high; leaf lobes 5—10 mm. wide.
      11. *Anemone virginiana*

6. *Anemone patens* L. Pasque-flower. (Fig. 200). Flowers dark lavender to nearly white, cup shaped, 3—7 cm. (1—3 in.) wide, solitary on stems only 5—10 cm. (2—4 in.) long at flowering time, these elongating to 2—3 dm. (8—12 in.) in fruit; leaves 5—10 cm. wide, divided into narrow lobes, not present at flowering time; whole plant quite hairy, especially flowering stems and a row of narrow bracts 2—3 cm. long below flower. April 15 to May 1. Common on higher prairie (fig. 16) and one of our most noted spring flowers. Commonly called "crocus" to which the flowers have only superficial resemblance. *Anemone patens* is the European plant. Ours is usually regarded as var. *wolfgangiana* (Bess.) Koch, or var. *nuttalliana* A. Gray. By some authors it is considered a distinct species, *A. ludoviciana* Nutt. or *Pulsatilla ludoviciana* (Nutt.) Heller.

7. *Anemone quinquefolia* L. Wood Anemone. Perennial; stem 1—2 dm. high with a single white flower 1 cm. wide, one basal leaf divided into 5 wedge shaped lobes and 3 smaller, 3-divided leaves near top of stem; achenes smooth, oblong with short tip. Late April, the basal leaf developing after flowering. Rare and local in aspen woods. We have it only from Fargo where it may now be extinct.

8. *Anemone canadensis* L. Canada Anemone. (Fig. 196). Stems 3—6 dm. high; basal leaves on petioles nearly as long as flowering stem; leaves rounded in general outline, 5—15 cm. wide, deeply 3—5-lobed, the lobes 1—3 cm. wide and usually 3-cleft; on upper part of stem a group of leaves similar to basal ones forms an involucre, at which point the stem often divides into 2 or 3 stalks, each bearing a single white flower above a second involucre; achenes flattened, 5 mm. wide, somewhat hairy, sharp pointed. Common in woods, coulees, and low prairie (fig. 16), forming large beds. June.

9. *Anemone cylindrica* A. Gray. Cottonweed. Basal leaves 3—5 cm. (1⅕—2 in.) wide, divided into lobes 4—5 mm. (¼ in.) wide; stem 3—6 dm. (1—2 ft.) high with an involucre of leaves at about two-thirds the height, above which are 1—5 flowers on single stalks; flowers greenish white, inconspicuous, 10—15 mm. (½ in.) wide; fruiting heads cylindrical, 2—4 cm. long. July. Common on prairie (fig. 16). The fruit heads remain all winter breaking up in spring, each achene covered with a fine tuft of crinkly hairs. This loose, cottony mass suggested the name "cottonweed".

200.   Pasque-flower (*Anemone patens*).

10. *Anemone hudsoniana* Richards. Cut-leaved Anemone. Similar to last but leaf divisions 2—3 mm. ($\frac{1}{12}$—⅛ in.) wide; involucral leaves without petioles; sepals reddish; fruiting heads rounded, 1—2 cm. long. We have this only from Rolette and Benson Counties.

11. *Anemone virginiana* L. Tall Anemone. Stem 6—10 dm. high; basal leaves 7—15 cm. wide, deeply divided into segments 1—2 cm. wide; flowers 1—2 cm. wide, greenish white; fruiting heads rounded, 1—2 cm. long; achenes hairy. Occasional in woods. We have records for Cass, Richland, Pembina, Benson, Bottineau and Dunn Counties. June-Aug.

12. *Clematis virginiana* L. Virgin's Bower. Herbaceous vine, climbing to 5 m. (17 ft.) or more. Leaves opposite, compound, of 3 leaflets which are ovate, toothed and 3-lobed, 2—5 cm. (⅘—2 in.) long (fig. 45); flowers white, 1—1.5 cm. wide in axillary panicles; achenes many from each flower, with feathery tips 3—5 cm. long. Flowers in August, quite showy. The plant climbs by means of petioles twisting around other stems. Common in woods or thickets in eastern part of State, but we do not have it west of Ransom and Walsh Counties (fig. 9).

13. *Clematis ligusticifolia* Nutt. Western Virgin's Bower. Very similar to the preceding; leaflets usually 5 and narrower. Common in woods along streams and in brushy coulees from Missouri River west (fig. 9).

14. *Clematis pseudoalpina* (Kuntze) A. Nels. Rocky Mountain Clematis. Perennial, stem very short from creeping base; leaves twice divided into 3 parts; petioles 1 dm. long, widely spreading; leaflets ovate or lanceolate, 1—3 cm. long, 3-lobed; flower nodding on stalk 1—2 dm. high from leaf base; sepals blue or purple, lanceolate, 2—4 cm. long. I found this first in 1935 at Killdeer Mts. where it is fairly common on upper, steep slopes and rocky ridges. June. Fruits much like those of pasque-flower.

15. *Myosurus minimus* L. Mousetail. Small, tufted annual, 5—10 cm. (2—4 in.) high; leaves very narrow, 2—5 cm. long; flowers yellow, 5 mm. (⅕ in.) wide. Frequent in fields or prairie. June. We have records for six widely distributed counties but believe it is found all over the State. The center of the flower becomes 3 or 4 cm. long as the fruits develop and this suggests the name mousetail.

Ranunculus BUTTERCUP. CROWFOOT

Annual or perennial, sometimes growing in water; leaves usually palmately 3-lobed or divided, the lower often not lobed; flowers usually yellow; stamens many; pistils many; fruits many seed-like achenes, usually flattened. Mostly of no importance, some species weedy, some reported poisonous. They might be confused with cinquefoils which have similar flowers and fruit but cinquefoil leaves have stipules and many species have pinnately compound leaves.

Key to Species

Stem entirely prostrate, in water or on wet ground.
    Flowers white, stems in water, leaves very finely divided.
              16. *Ranunculus aquatilis*
    Flowers yellow, stems creeping on wet ground, leaves rounded.
              17. *Ranunculus cymbalaria*
Stems upright, sometimes spreading or floating.
    Plants usually in water, sometimes on mud after water recedes.
        Flowers 15—25 mm. (⅗—1 in.) wide; achenes with thick, corky
           edge.   18. *Ranunculus flabellaris*
        Flowers 7—15 mm. wide; achenes without thickened edge.
            Stems upright, rather stout and fleshy, 2—6 dm. (8—24 in.) high.
              23. *Ranunculus sceleratus*
            Stems floating or creeping.   19. *Ranunculus gmelini*

Plants usually not in water; stems upright or spreading.
Plants smooth or nearly so, 1—3 dm. high.
Plants about 1 dm. high; flowers 1—2 cm. (⅗ in.) wide in early
spring.
Lower leaves elliptic; western plant. 20. *Ranunculus glaberrimus*
Lower leaves crenate, rounded or ovate; common on prairie.
21. *Ranunculus rhomboideus*
Plants usually 3—8 dm. high; flowers 4—8 mm. (¼ in.) wide in
summer.
Leaves all deeply lobed; plants growing in open places.
23. *Ranunculus sceleratus*
Lower leaves rounded, not divided; plants in woods.
22. *Ranunculus abortivus*
Plants rough hairy, 3—8 dm. high, erect or spreading.
Stems somewhat creeping, often rooting at lower joints.
24. *Ranunculus septentrionalis*
Stems upright or sometimes spreading.
Flowers 1.5—2 cm. wide; rare introduced weed.
25. *Ranunculus acris*
Flowers 6—12 mm. wide; native plants, mostly common.
Fruiting head oblong; stem erect, stout.
26. *Ranunculus pennsylvanicus*
Fruiting head rounded; stem widely branched or weak.
Beak of achene strongly hooked; rare plant in woods.
28. *Ranunculus recurvatus*
Beak of achene straight; common plant in meadows and
ditches. 27. *Ranunculus macounii*

16. *Ranunculus aquatilis* L. WHITE WATER CROWFOOT. Stems extensively
branching through the water; leaves divided into very narrow segments; flowers white, 2—3 cm. (1 in.) wide, projecting just above
surface of water. Frequent in ponds or slow streams throughout State.
June, July. Quite showy because of the large numbers of flowers. Our
plant is var. *capillaceus* (Thuill.) DC.

They had been called *Ranunculus trichophyllus* Chaix or *Batra-
chium trichophyllum*. According to the revision by Drew (Rhodora
38:1-47, 1936) our specimens seem to be R. *subrigidus* Drew except
Bell 353 from Grant County, which is R. *longirostris* Godron. This has
achenes with beaks about 1 mm. long as compared with 0.2—0.5 mm.
for R. *subrigidus*. Benson (Amer. Midl. Nat. 40:240-245, 1948) considers *subrigidus* a variety of R. *circinatus* Sibth. and *longirostris* a
distinct species. He cites a specimen of the last, collected by Lunell at
Leeds, Benson County.

17. *Ranunculus cymbalaria* Pursh. SEASIDE BUTTERCUP. Perennial with
fine, creeping stems, rooting at joints; flower stems 5—20 cm. high;
leaves smooth, rounded, 1—3 cm. wide, usually with only short, rounded
teeth; flowers yellow, 6—8 mm. wide; head of fruit oblong, 6—10 mm.
long. Very common in low places or edges of ponds, flowering all
summer. Some botanists put this in a separate genus, *Oxygraphis* or
Halerpestes.

18. *Ranunculus flabellaris* Raf. YELLOW WATER CROWFOOT. Stems usually in water, up to 2 m. long, sometimes erect on mud, 2—4 dm.
high; leaves in water divided into very narrow segments (fig. 201);
leaves above water 2—5 cm. wide, coarsely or finely lobed (fig. 203),
sometimes hairy; flowers bright yellow, 1—3 cm. wide. June, July.
Very showy, usually in dense masses. Previously known as R. *delphini-
folius* Torr.

19. *Ranunculus gmelini* var. *terrestris* (Ledeb.) L. Benson. Stems weak,
floating in water or creeping on mud, 2—6 dm. long; leaves 1—3 cm.
wide, smooth, variously cut into wedge shaped segments; flowers yellow, 4—10 mm. wide. Rare in marshes. Records from Ransom, Benson
and Bottineau Counties. Previously called R. *purshii* Richards.

20. *Ranunculus glaberrimus* Hook. Stems 1 dm. or less high; leaves thick, smooth, lower ones rounded or elliptic, sometimes lobed, 1—3 cm. long, upper 3-lobed; flowers yellow, 1.5—2.5 cm. wide. Wet places on prairie or along streams in extreme western part of State. Apr., May.

21. *Ranunculus rhomboideus* Goldie. PRAIRIE BUTTERCUP. Stems 1—3 dm. (4—12 in.) high; upper leaves smooth or somewhat hairy, divided into 3—5 lobes, the lower leaves rounded or ovate, 1—2 cm. (⅗ in.) long; flowers bright yellow, 2—3 cm. wide. Common on prairie. This is one of the earliest flowers, beginning to bloom the middle of April. Some authors use *R. ovalis* Raf. for this plant.

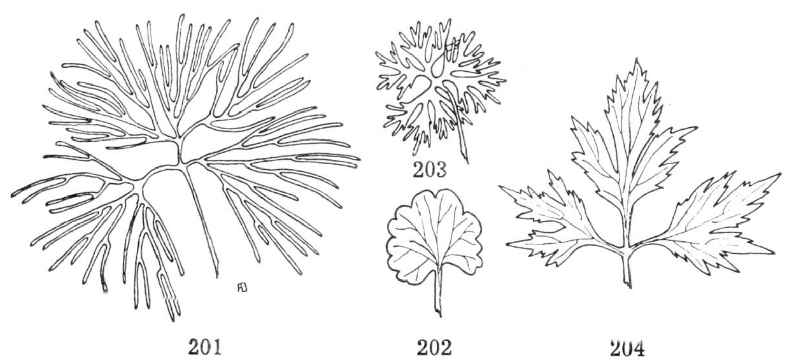

201   202   204

22. *Ranunculus abortivus* L. EARLY WOOD BUTTERCUP. Stems 2—6 dm. high; lower leaves rounded, often wider than long, 1—5 cm. long with rounded teeth (fig. 202); upper leaves divided into several lobes; flowers yellow, 4—6 mm. wide, numerous in large, branching top. Very common in woods. May, June.

23. *Ranunculus sceleratus* L. DITCH BUTTERCUP. Stems smooth, hollow, widely branched, 3—6 dm. high, erect or sometimes partly floating; leaves thick, smooth or sometimes hairy, 3—5 cm. long, deeply 3-lobed and these segments less deeply lobed; flowers yellow, 6—8 mm. wide. Common in shallow water or wet places. June-Aug. This is reported poisonous. It is sometimes very abundant at edges of ponds.

24. *Ranunculus septentrionalis* Poir. MARSH BUTTERCUP. Stems 3—10 dm. long, more or less creeping; leaves far apart, long petioled, blades 8—15 cm. long, divided into 3 leaflets and these again deeply and sharply lobed; flowers bright yellow, 20—25 mm. wide, at ends of long branches. June. Frequent in swamps or low meadows. Records chiefly from southeast and one from Morton County. One specimen from Valley City was verified by Benson, another was referred to *R. hispidus* Michx. The two species are similar. *R. septentrionalis* "produces long stolons very soon after the beginning of flowering" (*hispidus* does not) and has "broader and often apically free stipular leaf bases".

25 *Ranunculus acris* L. TALL BUTTERCUP. Annual; stem erect, 6—10 dm. high, the elongated flowering branches nearly leafless; leaf blades 5—10 cm. long, deeply lobed, the lower ones long petioled; flowers yellow, 2 cm. wide. Introduced weed collected in Cass, Richland, McLean and Billings Counties.

26. *Ranunculus pennsylvanicus* L. f. BRISTLY BUTTERCUP. Annual; stems stiff, erect, 3—10 dm. (1—3 ft.) high, very rough hairy; leaf blades 5—10 cm. (2—4 in.) long, divided into 3 leaflets and these deeply lobed and sharply toothed; flowers yellow, 6—8 mm. (¼ in.) wide; fruiting heads oblong, 1—2 cm. long. Aug. Frequent in wet places. We have it from only eastern part of State.

27. *Ranunculus macounii* Britton. MACOUN'S BUTTERCUP. Annual; stems 3—6 dm. high, rough hairy, usually widely branched and spreading; leaves 5—10 cm. long, divided into 3 leaflets which are more or less lobed and toothed (fig. 204); flowers yellow, 10—15 mm. wide; fruiting

heads rounded, 8—10 mm. long. June, July. Common in low meadows and wet places. This species is much like the preceding but has larger flowers, weak instead of stiff stems and blooms earlier.

28. *Ranunculus recurvatus* Poir. HOOKED BUTTERCUP. Stem upright, 2—6 dm. high, hairy but not as coarsely so as in the two preceding species; leaves 5—8 cm. wide, deeply divided into 3 or 5 lobes which have short, rounded teeth; flowers pale yellow, 8—10 mm. wide; fruiting heads rounded, 8—15 mm. wide. We have this only from swampy woods in northern Richland County.

29. *Thalictrum venulosum* Trel. EARLY MEADOWRUE. Perennial, smooth, erect, 5—8 dm. (20—32 in.) high; leaves few, 1—3 dm. long, divided into 25—100 oblong, rounded or wedge shaped leaflets 1—2 cm. (⅜ in.) long, which may be 3-lobed and toothed at tip (fig. 199); flowers numerous, in a large, terminal cluster; achenes oblong, 5 mm. (⅕ in.) long, somewhat flattened, strongly ridged. Common, especially along edges of woods or brush. May, June. Staminate plants are quite conspicuous from the yellow anthers. Stigmas of pistillate flowers are purplish. Our plants have often been called *T. dioicum* L., but the most recent authority (Boivin, Rhodora, 1944) considers that a different species.

30. *Thalictrum dasycarpum* Fisch. & Lall. TALL MEADOWRUE. Stems stout, 1—2 m. (3—7 ft.) high; leaves few, 2—4 dm. long; leaflets oblong or obovate, 3-toothed at tip, 1—4 cm. long. A stout plant, taller than the preceding. Common in woods. July. Leaves of meadowrues are often confused with columbine when plants are not in bloom. Columbine leaflets are often wider than long, rounded at base, deeply lobed. Meadowrue leaflets are often narrowed at base with 3 quite short teeth at tip. The basal leaves may be very large, 3—4 dm. long, the entire leaf having up to 100 leaflets.

BARBERRY FAMILY   Berberidaceae

Herbs or shrubs with various leaves; flowers small; sepals and petals usually 6 each; stamens 6; pistil one; fruit a berry, 1—several-seeded.

Key to Species

Shrub with spiny stipules: leaves small, simple, often clustered.
1. *Berberis vulgaris*

Perennial herb; leaves large, compound.   2. *Caulophyllum thalictroides*

1. *Berberis vulgaris* L. COMMON BARBERRY. Shrub 1—2 m. (3—7 ft.) high with many stems as in gooseberry; leaves obovate or oblong, 2—5 cm. (¾—2 in.) long, edges sharply toothed, alternate on new branches, apparently clustered on very short branches on old stems; stipules spiny, 3—5-pointed; flowers yellow, 5—6 mm. (¼ in.) wide in loose, drooping racemes 3—8 cm. long; berries bright red, oblong, 6—8 mm. long. June. The common barberry was quite widely planted before 1914, when a campaign of eradication was begun because the plant bears the cluster cup stage of black stem rust. Most of the plants have been destroyed but occasional ones are still found around old farm yards. Very few have been found growing naturally where seeds have been scattered, Japanese Barberry (*Berberis thunbergii*), which does not carry rust, is planted chiefly for a low hedge. It grows 3—8 dm. high, has rounded leaves which are not toothed, spines 1—3-pointed and flowers in an umbel.

2. *Caulophyllum thalictroides* (L.) Michx. BLUE COHOSH. Perennial, 4—6 dm. high from a short, thick rhizome; leaves usually 2—3 dm. long, compound, with many leaflets which are oblong, 2—5 cm. long, 3-toothed at tip; flowers yellowish green, 8—10 mm. wide, rather densely clustered at end of stem; berries 1 cm. wide, dark blue. Late May. Rather rare in woods. Records from Cass and Richland Counties.

MOONSEED FAMILY Menispermaceae

1. *Menispermum canadense* L. MOONSEED. Slender vine, 1—2 m. (3—7 ft.) high from long roots; leaves alternate, rounded or angular, 5—15 cm. (2—6 in.) wide; flowers white, 3—4 mm. (⅛ in.) wide in axillary racemes 3—5 cm. long; fruit a blue black berry 5—8 mm. wide. June. Frequent in woods, eastern part of State, not recorded west of Barnes County. The berry contains a flat, circular seed with a small notch at base.

POPPY FAMILY Papaveraceae

Herbs, usually with milky juice; leaves alternate, large, coarsely toothed or lobed; flowers large, showy, each at end of an upright stem; sepals 2, petals 4 or more; stamens many; fruit a capsule with many seeds.

206. Bloodroot. (*Sanguinaria canadensis*).

Key to Species

Perennial from a thick, red rhizome; no leafy stem.

          1. *Sanguinaria canadensis*

Annual, leafy stemmed plants, 3—10 dm. (1—3 ft.) high.

 Stout, prickly plant; flowers white.   2. *Argemone alba*

 Sometimes hairy but not prickly; flowers red. 3. *Papaver rhoeas*

1. *Sanguinaria canadensis* L. BLOODROOT. (Fig. 205). Perennial from a short, thick rhizome with orange red juice, leaves and flowers coming directly from rhizome; leaves 1—2 dm. (4—8 in.) wide, rounded, often wider than long, usually with rounded, oblong lobes; flowers white, 3—5 cm. (1—2 in.) wide; petals 8, narrowly oblong; sepals 2, whitish; capsule narrowly oblong, pointed, 4—8 cm. long, splitting lengthwise. Flowers about May 1. The petals fall the first day and sepals as soon as flower is well opened. Abundant locally in woods along Red River

and a short distance up the main tributaries (fig. 10). This is a popular wild flower and may be destroyed by picking because the rhizomes pull up easily. It was abundant in many places but use of woods for pasture has destroyed much of it.

2. *Argemone alba* Lestib. WHITE PRICKLY POPPY. Stout annual, leaves ovate, 1—2 dm. long, coarsely toothed; flowers white, 10—15 cm. wide; capsule rounded, 2—3 cm. long. A specimen of this was received in 1938 from Ralph K. Welch, county extension agent in McKenzie County. I have seen it a little farther west in Montana. It is found in South Dakota and quite commonly farther south. Sometimes it is planted for an ornamental and may have escaped from cultivation.

3. *Papaver rhoeas* L. FIELD POPPY. Erect annual, 3—8 dm. high with coarse but not prickly hairs; leaves oblong, 10—15 cm. long, pinnately deeply lobed; flowers 8—15 cm. wide, usually scarlet with dark center; capsule oblong, 1—2 cm. long, opening by several small holes just below the flat top. Not established as a wild plant, but occasionally escaped from cultivation along streets or yards.

206. Fumitory (*Fumaria officinalis*).

FUMITORY FAMILY Fumariaceae

Herbs with alternate, finely divided leaves and small flowers; sepals 2; petals 4, narrow; stamens 6; pistil 1; fruit a slender pod with 1 to several seeds.

Some authors combine this family with the Poppy Family though the flowers are quite different in shape.

Key to Species

Only one outer petal with basal projection.
  Fruit 1-seeded, not opening; flowers pink.   1. *Fumaria officinalis*
  Fruit several-seeded, splitting lengthwise; flowers yellow.
                                                    2. *Corydalis aurea*
Both outer petals with a basal projection; flowers white or pinkish.
                                                    3. *Dicentra cucullaria*

1. *Fumaria officinalis* L. FUMITORY. (Fig. 206). Annual with slender, spreading branches, 2—4 dm. (8—16 in.) high; flowers pale pink, 8—10 mm. (⅜ in.) long in slender racemes 3—5 cm. (1—2 in.) long; fruit green, rounded, 3—4 mm. wide. Flowers in June, seeds mature July 1. Earliest record of this was from Barnes County in 1923. Since found in Griggs County and abundant in a few localities. Plants dry up at maturity and fruit falls off as soon as ripe.

2. *Corydalis aurea* Willd. GOLDEN CORYDALIS. Spreading, much branched annual, forming tufts or mats 2—6 dm. wide; flowers bright yellow, 10—15 mm. long (fig. 102); fruits nearly cylindrical, 1.5—2.5 cm. long; seeds black, very shiny, 1 mm. wide. Common native plant, especially in open woods. May, June. This plant and the preceding have leaves much like those of California Poppy.

3. *Dicentra cucullaria* (L). Bernh. DUTCHMAN'S BREECHES. Perennial, leaves and flower stalks coming from a thick rootstalk; flower stalks 1 1.5 dm. (4—6 in.) long, curving, flowers hanging downward; flattened outer petals widely spreading and projecting at base; fruit 1—2 cm. long. Late May. We have only one record of this from Enderlin, Ransom County. It is a popular spring flower in eastern U. S., growing on wooded banks. Bleeding Heart (*Dicentra spectabilis*), frequently grown as an ornamental, has similar, larger flowers.

MUSTARD FAMILY Brassicaceae. Cruciferae

Annual or perennial herbs with alternate leaves and usually pungent taste; flowers usually small, yellow or white; sepals 4; petals 4; stamens 6, 2 shorter than others; pistil 1; seeds usually many in two chambers; fruit elongated or short, opening by the sides falling away, leaving seeds attached to central partition. Many species are introduced weeds.

Key to Species or Genera

Flowers purple, purplish or white, 2 cm. (⅘ in.) wide.
  Pods slender, splitting open.   53. *Hesperis matronalis*
  Pods stout, not splitting open.   23-24. *Raphanus*
Flowers white, yellow or yellowish, usually smaller.
  Flowers white.
    Leaves very large, 1—4 dm. (4—16 in.) long; pods usually not
      formed.   26. *Rorippa armoracia*
    Leaves ordinary, 5—20 cm. (2—8 in.) long; pods common.
      Pods much elongated, 2—15 cm. long.
        Pods sharply 4 angled, 8—15 cm. long.   54. *Conringia orientulis*
        Pods rounded, not 4-angled, 3—10 cm. long
          Pods 3—5 cm. long; rare swamp plants.
            Some leaves deeply pinnately lobed.
                                        32. *Cardamine pennsylvanica*
            All leaves rounded, not deeply lobed. 33. *Cardamine bulbosa*
          Pods 5—10 cm. long; woodland or prairie plants. 46-49. *Arabis*
      Pods short, oblong or rounded, 3—20 mm. long, usually
        flattened.
        Pods 3—8 times as long as wide.   43. *Draba micrantha*
        Pods hardly longer than wide.
          Pods scarcely flattened, oblong or rounded.
            Pods oblong, 5 mm. (⅕ in.) long; seeds thin and flat.
                                                    45. *Berteroa incana*

Pods rounded, 3—5 mm. long; seeds only slightly flattened.
3-4. *Cardaria*
Pods much flattened.
  Pods triangular, widest at tip. 37. *Capsella bursa-pastoris*
  Pods circular.
    Pods 2—5 mm. long; 2-seeded. 5-9. *Lepidium*
    Pods 1—2 cm. long; about 12-seeded. 10. *Thlaspi arvense*
Flowers yellow.
  Pods 5—20 mm. long, 1—5 times as long as wide.
    Pods 8—20 mm. long, 4—5 times as long as wide.
      Pods much flattened; leaves not lobed. 44. *Draba nemorosa*
      Pods little flattened; leaves lobed or divided.
        Pods constricted between large seeds; leaves coarsely lobed.
24. *Raphanus raphanistrum*
        Pods not constricted, seeds small; leaves finely divided.
14-16. *Descurainia*
    Pods 5—8 mm. long, 1—2 times as long as wide.
      Pods much flattened, notched at tip; rare western perennial.
38. *Physaria brassicoides*
      Pods not flattened, oblong or ovate.
        Pods oblong, thin walled.
          Seeds many, very small. 26-31. *Rorippa*
          Seeds 10—20, medium sized. 39-41. *Camelina*
        Pods rounded.
          Pods thick walled, 1-seeded; annual weed.
42. *Neslia paniculata*
          Pods thin walled, several seeded; prairie perennials.
34-36. *Lesquerella*
  Pods 5—15 cm. (2—6 in.) long, more than 5 times as long as wide.
    Pods hanging downward; tall plant in Bad Lands.
1. *Stanleya pinnata*
    Pods erect or spreading.
      Pods 3—5 mm. (⅙ in.) wide, 2—8 cm. long.
        Petals with purple veins; flowers 2 cm. wide. 17. *Eruca sativa*
        Petals without purple veins; flowers 0.5—2 cm. wide.
          Pods angular; leaves with rounded lobes; rare plant.
25. *Barbarea vulgaris*
          Pods rounded; leaves with irregular lobes and teeth.
18-21. *Brassica*
      Pods 1—2 mm. wide, slender, usually long.
        Leaves narrow, not divided. 50-52. *Erysimum*
        Leaves deeply lobed or divided.
          Pods 2—5 cm. long, erect. 2. *Thelypodium lilacinum*
          Pods 5—15 cm. long, 1—2 mm. wide, spreading.
            Pods 4-angled, 1 mm. wide; leaves with sharp teeth.
11-13. *Sisymbrium*
            Pods rounded, 2 mm. wide; leaves with rounded lobes
              or teeth. 22. *Erucastrum gallicum*

1. *Stanleya pinnata* (Pursh) Britton. PRINCE'S PLUME. Perennial, 4—10
dm. (16—40 in.) high, smooth, waxy; flowers 1 cm. (⅖ in.) wide,
bright yellow, in a long, terminal cluster; leaves 1—2 dm. long, pinnately deeply lobed; fruit 4—8 cm. long, curving downward. Late
June. Recorded from clay buttes in Billings County only.

2. *Thelypodium lilacinum* Greene. Biennial, 4—8 dm. high; leaves smooth,
entire, the lower sometimes toothed or lobed, 2—5 cm. long; flowers
4—5 mm. (⅕ in.) wide, white or nearly so, in a rather dense, terminal,
branching cluster; pods narrow, 2—3 cm. long. Saline, rather wet soils.
June, July. Records from Kidder, Burleigh and Ward Counties. This
is now considered distinct from *T. integrifolium* of Washington and
Oregon.

Cardaria PERENNIAL PEPPERGRASS. WHITETOP

Perennial by deep, vertical, and by horizontal roots; leaves sessile, broad, gray hairy, entire or toothed; flowers white, in a dense, terminal cluster; pods rounded or flattened, several seeded. These recently introduced plants are among our most persistent weeds. They have commonly been included in *Lepidium*.

Key to Species

Pods circular, flattened.   3. *Cardaria draba*, var. *repens*
Pods rounded, not much flattened.
  Pods heart shaped, not hairy.   3. *Cardaria draba*
  Pods rounded or longer than wide, very hairy.   4. *Cardaria pubescens*

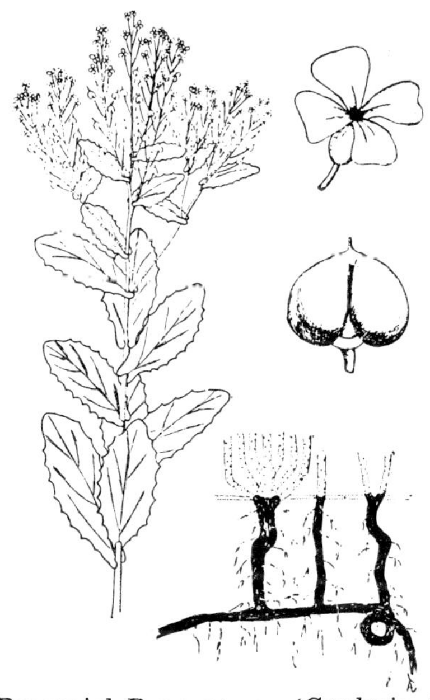

207.   Perennial Peppergrass (*Cardaria draba*).

3. *Cardaria draba* (L.) Desv. PERENNIAL PEPPERGRASS. (Fig. 207). Perennial by deep and by long, horizontal roots; stems 1—4 dm. (4—16 in.) high; leaves lanceolate or oblong, sharply toothed or nearly entire, 3—10 cm. (1—4 in.) long, gray with fine hairs; flowers white, 3 mm. (⅛ in.) wide in a large, terminal, flat-topped cluster; pods ovate or heart shaped, only slightly flattened, 2—4 mm. long, 3—5 mm. wide. Late May. Our first record of this came from Williams County in 1918. It may occur in every county although the local extent of it is usually limited.

Var. *repens* (Schrenk) Schulz differs chiefly in shape of pod which is distinctly flattened and evenly rounded, 4—6 mm. long. We have specimens from Cass, Grand Forks, Richland and Golden Valley Counties.

4. *Cardaria pubescens* (Meyer) Rollins. Much like preceding except the fruiting cluster becomes more elongated, pods oblong, densely covered with fine hairs. We have this from Fargo in 1940 (Stevens 460) and McKenzie County in 1941. Formerly known as *Hymenophysa pubescens*.

Lepidium   PEPPERGRASS

Annual or biennial; flowers small, white or greenish; pods rounded, much flattened, 2-seeded. The outer seed coat becomes mucilaginous when wet so the seeds cling to passing objects.

Key to Species

Upper leaves entire, with broad, clasping bases.   5. *Lepidium perfoliatum*
Leaves toothed or lobed, the upper narrow; bases not clasping.
   Pods circular, 3 mm. (⅛ in.) long, narrowly winged.
       8. *Lepidium virginicum*
   Pods elliptic, ovate or obovate, 2.5—5 mm. long.
    Pods elliptic, not winged, 2.5 mm. long.   7. *Lepidium ramosissimum*
    Pods obovate, winged at tip.
     Pods 2.5—3 mm. long, narrowly winged; common weed.
       6. *Lepidium densiflorum*
     Pods 5 mm. long, strongly winged; garden escape.
       9. *Lepidium sativum*

5. *Lepidium perfoliatum* L. Erect, branching above, 2—4 dm. (8—16 in.) high; lower leaves divided into narrow segments, upper leaves entire, 2—4 cm. (1—2 in.) long, rounded, with clasping bases; flowers yellowish; pods 4—5 mm. (⅕ in.) long. This introduced species was first collected at Fargo in 1932, later at Mott, Hettinger County.

6. *Lepidium densiflorum* Schrad. PEPPERGRASS. Erect, 1—4 dm. high, not branched on lower part, several, spreading, flowering branches on upper part; leaves oblong, 2—4 cm. long with short, sharp teeth or sometimes deeply cut; flowers greenish, 1 mm. wide, stamens 2; pods 2.5—3 mm. long; seeds reddish, oblong, twice as long as wide. Very common weed in fields, pastures and roadsides. Usually grows as a winter annual, the leaves forming rosettes 3—10 cm. wide in fall. June.

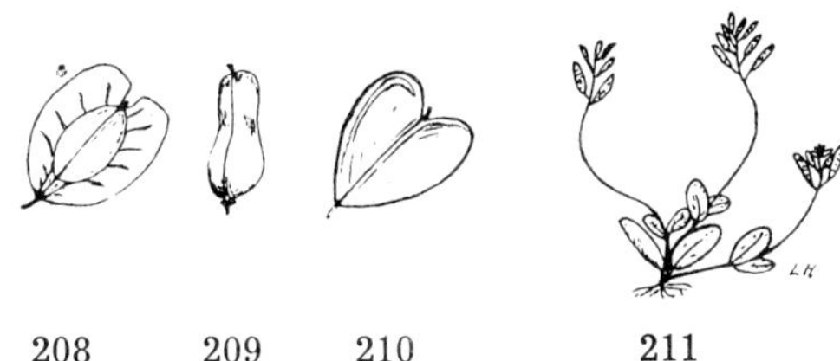

208   209   210   211

7. *Lepidium ramosissimum* A. Nels. BRANCHED PEPPERGRASS. Biennial, 2—4 dm. high, much branched over entire length of stem; leaves pinnately divided into narrow segments; pods elliptic, not winged at tip. This plant appears definitely a native biennial. It blooms a little later than *L. densiflorum* and well developed first year rosettes are found during summer. It seems most common in central part but has been collected in western part of State. Stray plants have been found at Fargo and there is a specimen from Bathgate, Pembina County, in 1892.

8. *Lepidium virginicum* L. Similar to *L. densiflorum* except for the distinctly whitish flowers and difference in structure of seeds which are about three-fourths as wide as long. It is a common weed farther south and east. Recorded at Fargo in 1892 and 1916. Two specimens from McKenzie and Grant Counties seem to be this species. It may be more common than we have supposed.

9. *Lepidium sativum* L. GARDEN CRESS. Annual, 6 dm. high; lower leaves much divided, upper entire or lobed; pods 4—5 mm. long, distinctly winged all around. Bergman cited records for Fargo and Bathgate, Pembina County, but I find no specimens. The plant is sometimes grown in gardens and may be found wild but hardly established.

10. *Thlaspi arvense* L. FRENCHWEED. PENNYCRESS. Annual or winter annual, 1—6 dm. (4—24 in.) high with flowering branches from upper half of stem; leaves 3—6 cm. (1½—2½ in.) long, oblong, usually somewhat toothed, clasping at base, rather thick and smooth; flowers many on elongated branches, white, 3—4 mm. (⅛ in.) wide; pods circular, much flattened, 1—1.5 cm. long, broadly winged all around (fig. 208); seeds about 12, oblong, flattened, 2 mm. long, dark reddish brown, marked with eccentric ridges. Very common weed, especially eastern part of State. Seeds germinate freely in early fall and seedlings winter in stages from 2—3 cm. rosettes to 2 dm. high and flowering. Plants with flower buds begin to bloom about April 20. It is often called "garlic" because of the pronounced odor caused by a substance that is also present in onions.

Sisymbrium   HEDGE MUSTARD

Tall, branching, introduced annuals with broad, coarsely lobed leaves, small yellow flowers and elongated pods with many, small, angular seeds.

Key to Species

Pods 2 cm. (¾ in.) long, erect, lying close to stem.
                                                  12. *Sisymbrium officinale*
Pods 3—10 cm. long, spreading.
  Leaves rather regularly and finely lobed.   11. *Sisymbrium altissimum*
  Leaves coarsely, quite irregularly lobed.   13. *Sisymbrium loeslii*

11. *Sisymbrium altissimum* L. TUMBLING MUSTARD. (Fig. 212). Annual or winter annual, 3—10 dm. high, widely branched above; leaves 1—3 dm. long, somewhat hairy, thin and soft, pinnately divided into narrow segments; flowers 4—5 mm. wide, pale yellow; pods 6—10 cm. long, 1 mm. wide, 4-angled; seeds very numerous, yellow or greenish, irregularly angled, 0.75 mm. long. June-Sept. One of the commonest of all mustards. The old plant breaks off and is blown by wind, scattering the seeds.

12. *Sisymbrium officinale* (L.) Scop. HEDGE MUSTARD. Erect or spreading annual, 3—6 dm. high; flowering branches stiffly erect or widely spreading, leafless, 2—4 dm. long; leaves 3—8 cm. long, coarsely lobed, the lobes somewhat toothed, sometimes with a few, narrow, basal lobes and a single terminal portion, or with 4—10 lobes; flowers yellow, 3 mm. wide; pods about 2 cm. long, narrowed to a sharp point and standing upright close to stem; seeds much as in tumbling mustard. Uncommon, chiefly in southeastern part of State.

13. *Sisymbrium loeselii* L. TALL HEDGE MUSTARD. Erect annual, 1—2 m. high, widely branched; leaves 1—2 dm. long, with 2—5 irregular, large lobes; flowers bright yellow, 6—8 mm. wide; pods 3—5 cm. long, on slender, spreading stalks; seeds much as in tumbling mustard. This has been found at several places since 1918, sometimes abundant locally but it seems restricted to roadsides and waste ground.

Descurainia   TANSY MUSTARD

Annuals or winter annuals with finely divided leaves, small yellow flowers, moderately short pods with many small seeds. This group is united by many authors with *Sisymbrium*. It has also been called *Sophia* but *Descurainia* is retained by present rules.

Key to Species

Pods spreading, slightly larger at upper end.   14. *Descurainia pinnata*
Pods erect, not enlarged at upper end
  Leaves finely divided, strong scented; pods 2 cm. (⅘ in.) long.
                                                  15. *Descurainia sophia*
  Leaves pinnately lobed, not strong scented; pods 1 cm. long.
                                                  16. *Descurainia richardsonii*

14. *Descurainia pinnata* (Walt.) Britt. Annual, 1—8 dm. (4—32 in.) high, usually slender, with few, long branches; leaves gray or bright yellowish green, 3—10 cm. (2—4 in.) long, pinnately divided into lobes 2—4 mm. (⅛ in.) wide; pods 1—1.5 cm. long, widely spreading on stalks of equal length; seeds dark reddish brown, narrowly oblong. June. This is a native plant found chiefly along roadsides, in prairies, etc. Large plants sometimes develop in fields where not disturbed, but it is not very troublesome as a weed. Detling (Am. Mid. Nat. 22:481—520) calls our plant subspecies *brachycarpa* (Richards.) Detling. It has also been called *Sisymbrium canescens* and *Sophia intermedia*.

212

213

212.　Tumbling Mustard (*Sisymbrium altissimum*).

213.　Hare's-ear Mustard (*Conringia orientalis*).

15. *Descurainia sophia* (L.) Webb. FLIXWEED. (Fig. 214). Annual, 3—8 dm. high, widely branched above; leaves twice divided into very narrow segments, gray with fine hairs and strongly scented; pods upright, 2 cm. long. June. This was introduced about 1910 and has become extremely common in the last 15 years especially in waste ground about towns and small pastures.

16. *Descurainia richardsonii* (Sweet) O. E. Schulz. Erect, slender, 3—8 dm. high; leaves gray as in preceding, but not so finely divided nor strongly scented; pods erect, 1—1.5 cm. long. June, July. Frequent, chiefly around edges of woods, thickets or groves. Our records show only Grant County and eastern third of State. We have only the typical subspecies according to Detling. Formerly called *Sisymbrium hartwegianum*.

17. *Eruca sativa* Mill. GARDEN ROCKET. Coarse, branching annual, 3—5 dm. high; leaves 1—2 dm. long, pinnately divided into irregular lobes; flowers 1—2 cm. wide, yellow with purple veins; pods 2—3 cm. long, stout, with a flattened beak; seeds oblong, smooth 2—3 mm. long. Collected in Cass County in 1902 and 1912, Rolette 1911, Pierce 1912, but probably not established.

Brassica   MUSTARD

Annuals with broad leaves and large, yellow flowers; pods cylindrical, seeds spherical or nearly so.

Key to Species

Leaves clasping the stem by a broad base.   21.   *Brassica campestris*
Leaves not clasping at base, usually slender petioled.
  Pods erect, 1—1.5 cm. (½ in.) long; rare weed.   20.   *Brassica nigra*
  Pods spreading, 3—5 cm. long; common weeds.
    Stalk of pod stout, thick, 5—10 mm. (⅕—⅖ in.) long; plant rough.
                18.   *Brassica arvensis*
    Stalk of pod slender, 10—20 mm. long; plant smooth, waxy.
                19.   *Brassica juncea*

18. *Brassica arvensis* (L.) Rabenh. FIELD MUSTARD. YELLOW MUSTARD. CHARLOCK. Stout, widely spreading, 3—8 dm. (12—32 in.) high; leaves 1—2 dm. long, broadly ovate, more or less deeply lobed at base; flowers yellow, 1—1.5 cm. (½ in.) wide; pods 3—5 cm. long, stiff, usually curved; seeds spherical, black, almost smooth, 1 mm. wide. Very common weed, especially eastern part of State. *B. kaber* (DC.) Wheeler has been adopted by many recent authors for this plant.

19. *Brassica juncea* (L.) Cosson. INDIAN MUSTARD. Rather slender, upright, 6—10 dm. (2—3 ft.) high; leaves and stems with a waxy coating; flowers deep yellow, 1 cm. wide; seeds reddish brown, irregularly rounded, covered with fine network of ridges. Common weed especially in western part of State. Both this and preceding are common, sometimes found growing together and often not distinguished from each other. Indian mustard is an upright, smooth, gray plant as compared with the dark green, widely spreading, field mustard, which is somewhat rough with scattered, coarse hairs. The seeds of Indian mustard have been used commercially to produce ordinary table mustard, and it has been grown under the name of "brown" mustard. The seeds of field mustard are not used. White mustard (*B. alba* (L.) Rabenh. or *B. hirta* Moench.) has been collected a time or two but only escaped from cultivation. It has paler flowers, hairy, oblong leaves, rather evenly divided into several triangular lobes; seeds light yellow. This and *B. arvensis* are often placed in a separate genus *Sinapis,* distinguished from *Brassica* by the flattened, empty tip of pod.

20. *Brassica nigra* (L.) Koch. BLACK MUSTARD. 1—2 m. high, widely branched above; leaves 1—2 dm. long with a few, large lobes, somewhat rough hairy; flowers yellow, 1 cm. wide; pods upright, lying against stem, 1—1.5 cm. long; seeds oblong, dark reddish brown covered with network of ridges. Collected in Cass and Cavalier Counties only. It is a common weed in central U. S. Seeds of this and of white mustard are the ones chiefly used for condiment.

21. *Brassica campestris* L. WILD TURNIP. Erect, smooth and waxy, 4—8 dm. high; leaves ovate with broad, clasping base; flowers yellow, 1 cm. wide; pods 4—6 cm. long; seeds dark brown or black, nearly smooth. Occasionally found in waste ground. Plants at least similar to this sometimes develop in rape fields because of seeds having been mixed with rape seed. Several related species are involved and the identity of them is confusing. None of them are frequent here as weeds.

22. *Erucastrum gallicum* (Willd.) O. E. Schulz. DOG MUSTARD. (Fig. 215). Stems spreading, 2—4 dm. (8—16 in.) high; leaves 5—15 cm. (2—6 in.) long, dark green, with coarse, rounded lobes; flowers yellow, 4—5

mm. (⅕ in.) wide along the terminal branches; pods cylindrical, 3—6 cm. (1½—2½ in.) long; seeds pale brown, oblong, a little flattened, 1—1.25 mm. long. First collected along a railroad at Fargo in 1909. At that time it had not yet been recorded in North America. It spread rapidly and became well established along railroads but is still rare in fields except in the Red River Valley from Grand Forks northward, where it is quite abundant. It flowers late in fall as well as in summer.

214 215

214. Flixweed (*Descurainia sophia*).
215. Dog Mustard (*Erucastrum gallicum*).

23. *Raphanus sativus* L. WILD RADISH. Coarse stemmed annual, 3—6 dm. high; leaves oblong, 1—1.5 dm. long, pinnately lobed or divided, with one large, terminal lobe and 2 or 3 pairs below; flowers pink or white, 2 cm wide, scattered along leafy branches; pods 3—5 cm. long, rounded, about 8 mm. thick at base, tapering to a point; seeds oblong, brownish, 3 mm. long. A wild form of common radish was introduced in flax in Pembina County in 1943 but has not become naturalized as far as known. We have specimens from Richland and Barnes Counties, both collected in 1912, probably only garden escapes.

24. *Raphanus raphanistrum* L. Similar to last but flowers yellow, pods slender, constricted between seeds. Bergman reported it from Barnes County.

25. *Barbarea vulgaris* R. Br. YELLOW ROCKET. Smooth biennial, 3—6 dm. high; leaves 5—10 cm. long with a terminal, large, rounded leaflet and 2 or 3 pairs of quite small ones below; flowers yellow, 6—8 mm. wide; pods 4-angled, 2—3 cm. long; seeds rounded and flattened. Collected at Fargo in 1910. It is a rather common weed farther east.

Rorippa YELLOW CRESS

Annual, perennial or biennial, stems often spreading; leaves usually pinnately lobed; flowers small, usually yellow; pods short, oblong, 2 or 3 times as long as wide; seeds numerous, very small.

Key to Species

Flowers white; leaves 2—4 dm. (8—16 in.) long, 1 dm. wide; escaped
  garden plant.   26. *Rorippa armoracia*
Flowers yellow, leaves usually 5—10 cm. (2—4 in.) long.
  Annuals or biennials.
    Style of pod 0.5 mm. long; surface of seed pitted.
                                                  27. *Rorippa islandica*
    Style of pod 1 mm. long; surface of seed tubercled.
                                                  28. *Rorippa obtusa*
  Perennials.
    Stems erect; leaves with many, fine, short teeth. 29. *Rorippa austriaca*
    Stems spreading; leaves deeply toothed or divided.
      Leaves 2—3 cm. (1 in.) wide, deeply cut into narrow lobes.
                                                  31. *Rorippa sylvestris*
      Leaves 5—15 mm. (⅕—⅗ in.) wide, coarsely cut about half way.
                                                  30. *Rorippa sinuata*

26. *Rorippa armoracia* (L.) Hitchc. HORSERADISH. Perennial from a thick
    root; stems 4—8 dm. (16—32 in.) high; leaves 2—4 dm. long, sometimes
    deeply lobed; flowers white, 4—6 mm. (¼ in.) wide, in large, terminal
    clusters. June. Frequently persisting in old gardens; usually not pro-
    ducing seeds but spreading by pieces of roots. This plant is sometimes
    separated from *Rorippa* and called *Armoracia rusticana* Gaertn.

27. *Rorippa islandica* (Oeder) Borbas. MARSH YELLOW CRESS. Biennial:
    stems 3—6 dm. high; leaves oblong, 5—15 cm. long, usually deeply
    pinnately lobed; flowers yellow, 4—5 mm. wide, in clusters 5—10 cm.
    long from upper leaf bases; pods 5—6 mm. long, half as wide (fig. 209).
    Very common in wet places. June-Sept. The plants are commonly
    without hairs but some are hairy and these have been called var.
    hispida (Desf.) Butters and Abbe. The American plant is now called
    var. *fernaldiana* Butters and Abbe. Formerly referred to R. *palustris*
    (L.) Bess.

28. *Rorippa obtusa* (Nutt.) Britt. Similar to preceding and not easily
    distinguished from it. Records from Cass, Pembina, Bottineau and
    Williams Counties. It may be common but is not conspicuous. Most
    of our specimens are low, almost tufted plants, but two from Turtle
    Mts. are 4—6 dm. high.

29. *Rorippa austriaca* (Jacq.) Crantz. AUSTRIAN FIELD CRESS. Stems 4—6
    dm. (16—24 in.) high; leaves narrowly oblong or oblanceolate, 5—10
    cm. (2—5 in.) long, with many, small teeth; flowers yellow, 2—3 mm.
    (¹⁄₁₀ in.) wide, in long, terminal clusters. First recorded at Edinburg,
    Pembina County in 1935 and since in Richland, Cass and Barnes Coun-
    ties. A very persistent weed.

30. *Rorippa sinuata* (Nutt.) Hitchc. SPREADING YELLOW CRESS. Perennial
    with prostrate stems 3—5 dm. long; leaves 3—5 cm. long, pinnately
    lobed; flowers yellow, 4 mm. wide on short branches; pods slightly
    curved, 6—10 mm. long. Frequent in low ground west of Missouri
    River. June, July.

31. *Rorippa sylvestris* (L.) Besser. Much like preceding but with finer,
    taller stems and more finely divided leaves. This was found at Fargo,
    apparently well established in 1922 but has not been seen since. The
    location was probably destroyed by building operations.

32. *Cardamine pennsylvanica* Muhl. BITTER CRESS. Slender annual, 2—6
    dm. (8—24 in.) high; leaves 5—10 cm. (2—4 in.) long, pinnately deeply
    lobed; flowers white, 4 mm. (⅙ in.) wide; pods very narrow, 1.5—3 cm.
    long. Rare, chiefly in cold springs or streams, Richland, Ransom and
    Pembina Counties. June, July.

33. *Cardamine bulbosa* (Schreb.) BSP. BULBOUS CRESS. Perennial from
    short, tuberous base; stems slender, 2—6 dm. high; lowest leaves round-
    ed, upper ones oblong to lanceolate, often with short, broad teeth; flow-
    ers and pods as in preceding. Cold, wet places, Ransom County.

Lesquerella BLADDERPOD

Low, native perennials with golden yellow flowers and rounded pods; stems and leaves rough and gray with short, star shaped hairs.

Key to Species

Pods oblong or ovate; plants tufted, 5—10 cm. (2—4 in.) high.
                                       34. *Lesquerella alpina*
Pods rounded, stems spreading, 1—3 dm. (4—12 in.) long.
    Stems slender, pods all hanging to one side.   35. *Lesquerella arenosa*
    Stems stout, pods arranged rather equally.   36. *Lesquerella argentea*

34. *Lesquerella alpina* (Nutt.) Wats. Stem very short; leaves oblanceolate or spatulate, 1—2 cm. long; pod narrowed toward tip, slightly flattened, 4 mm. long. Dry hills, southwestern part of State (fig. 10). June.

35. *Lesquerella arenosa* (Richards.) Rydb. Stems spreading or sometimes erect, 2—3 dm. long; pods rounded, 4—5 mm. long. Quite common on prairie (fig. 10). Early May and June.

36. *Lesquerella argentea* (Pursh) MacM. This has been maintained as a separate species by Payson (Ann. Mo. Bot. Gard. 8:150, 1921), though one is inclined to think it is just an early stage of development.

37. *Capsella bursa-pastoris* (L.) Medic. SHEPHERD'S PURSE. Winter annual; stems erect, 2—6 dm. (8—24 in.) high, somewhat branched and flowering along most of the length of the branches; larger leaves near the ground 5—10 cm. (2—5 in.) long, toothed or deeply pinnately lobed; flowers white, 2—3 mm. wide; pods triangular, 4—8 mm. (¼ in.) long (fig. 210). Common introduced weed, blooming April to June.

38. *Physaria brassicoides* Rydb. DOUBLE BLADDERPOD. Tufted, silvery perennial with low flowering branches; leaves ovate, oblong or spatulate, 1—3 cm. long, sometimes with 2 or 4 short teeth; flowering stems 5—10 cm. high; flowers yellow, 1—2 cm. wide; pods inflated, partly divided at tip, 8—10 mm. long. Recorded only from buttes in Billings County (fig. 10). June. This was formerly referred to *P. didymocarpa* but Rollins (*Rhodora* 41: 410, 1939) considers our plants distinct from that species.

Camelina FALSE FLAX

Erect, narrow leaved, introduced, weedy annuals; flowers pale yellow; pods obovate, slightly flattened; seeds several, yellowish or reddish, oblong or flattened.

Key to Species

Plant slender, quite hairy; pods 4—6 mm. (⅕ in.) long.
                                       39. *Camelina microcarpa*
Plants widely branched, smooth or nearly so; pods 6—8 mm. long.
    Pods slightly wider at tip; seeds oblong, not flattened, 2 mm. long.
                                       40. *Camelina sativa*
    Pods much wider at tip; seeds flattened, 3 mm. long.
                                       41. *Camelina dentata*

39. *Camelina microcarpa* Andrz. SMALL-SEEDED FALSE FLAX. Slender winter annual, 3—6 dm. (1—2 ft.) high; leaves quite hairy, lanceolate, base clasping; flowers 5 mm. (⅕ in.) wide, yellow, on long branches; pods oblong, slightly flattened, 4—6 mm. long; seeds oblong, 1 mm. long, reddish with pitted surface. June, July. This plant has much the appearance of Ball Mustard from which it differs in the several seeded pods. It was first collected at New England, Hettinger County, in 1935. It seems to have worked in from the west and has since been collected in Pembina, Emmons, Burleigh, Mountrail and Williams Counties.

40. *Camelina sativa* (L.) Crantz. Erect annual, 3—6 dm. high, spreading branches on upper half; leaves oblong or lanceolate, only slightly

rough, 3—6 cm. long; pods slightly wider toward tip, 6—8 mm. long; seeds oblong, reddish yellow, 2 mm. long.

41. *Camelina dentata* Pers. LARGE-SEEDED FALSE FLAX. Quite similar to last but lower leaves often coarsely toothed; pods pear shaped, 6—10 mm. long; seeds reddish yellow, oblong, 2.5 mm. long, not flattened, or much flattened and nearly as wide as long. These plants are called false flax because the general habit of growth is quite similar to that of flax. The branching stems and pods are easily overlooked in flax fields. Seeds of large-seeded false flax are not readily removed from flax seed and they contain oils which make them a serious impurity. The plants do not seem to be common nor to have become more common in recent years.

42. *Neslia paniculata* (L.) Desv. BALL MUSTARD. Slender annual, 3—6 dm. (1—2 ft.) high; leaves oblong-lanceolate, slightly hairy, clasping at base, 2—7 cm. (1—3 in.) long; flowers bright yellow, 3 mm. (⅛ in.) wide, in long, terminal clusters; pod 1-seeded, rounded, 3—4 mm. long, rough, not opening; seed yellowish. Occasional, especially northeast part of State. An introduced species which does not seem to become generally distributed.

43. *Draba micrantha* Nutt. WHITE WHITLOWWORT. (Fig. 211). Small, tufted, native, winter annual; leaves rounded, 8—15 mm. long, rough hairy; flowers white, 2 mm. wide, in close clusters on 1 or more stalks 5—10 cm. long; pods flattened, 8—15 mm. long. Dry or sandy prairie, chiefly western part of State; also Richland County. May. This has usually been regarded as a variety of *D. caroliniana* and in recent years *D. reptans* (Lam.) Fern. has been substituted for that.

44. *Draba nemorosa* L. YELLOW WHITLOWWORT. Slender, branching annual, 1—3 dm. (4—12 in.) high; leaves ovate or oblong, somewhat toothed, 6—12 mm. (¼—½ in.) long; flowers yellow, 1 mm. wide, scattered along upper branches; pods flattened, 8—15 mm. long. Very common on prairie, forming yellow patches in late May and early June. Some authors separate *D. lutea* Gilib. as a related species without hairs on pods. Most of our specimens have no hairs. A few do and one or two collections have both kinds. It is native to both Europe and America.

45. *Berteroa incana* (L.) DC. HOARY ALYSSUM. Perennial; stem usually widely branched from near ground, 4—8 dm. high; leaves lanceolate, entire, 4—8 cm. long; flowers white, 4—5 mm. wide, scattered along upper branches; pods 5—8 cm. long with flattened, reddish brown, nearly circular seeds. July, Aug. This introduced weed was first found in Emmons County about 1910, since then in 8 other counties. I have seen it chiefly in northwestern Cass and in Walsh County. It seems restricted to sandy soils but is a persistent weed. The stems are exceptionally tough and so rough with short, stiff hairs that it is dangerous to try to pull the plants without gloves.

Arabis ROCKCRESS

Tall, erect, native plants with white flowers and long flowering branches with few, small leaves; pods long, narrow, flattened; seeds angular or flat with thin winged edges.

Key to Species

Pods 1 mm. wide, stiffly erect.   46. *Arabis hirsuta*
Pods 2—2.5 mm. wide, spreading, drooping or erect.
  Pods hanging downward when mature; western prairie plant.
                                    47. *Arabis holboellii*
  Pods erect or spreading.
    Pods erect or spreading when mature.   48. *Arabis drummondii*
    Pods stiffly spreading.   49. *Arabis divaricarpa*

46. *Arabis hirsuta* (L.) Scop. Slender, erect, 3—9 dm. (1—3 ft.) high; leaves oblong to lanceolate, all except uppermost distinctly toothed, rather evenly short hairy, the upper ones less so; pods 1 mm. (1/25 in.) wide, in a thick, compact, terminal cluster. Common on prairie. June, July.

47. *Arabis holboellii* Hornem. Stems 3—6 dm. high, not much branched; lower leaves rounded, 1 cm. long, densely short hairy, the upper narrowly lanceolate, smooth and waxy; pods flattened, 2—2.5 mm. wide. Quite common on prairie west of Missouri River; collected also in Rolette County.

48. *Arabis drummondii* A. Gray. Erect, 6—10 dm. high; flowering branches usually standing close together; leaves lanceolate, clasping at base, lower ones slightly toothed and hairy, the upper narrow and smooth; pods 2 mm. wide, 8—15 cm. long. Occasional in open woods or near woods or thickets, mostly in sandy soil, northeastern part of State; one record from Grant County.

49. *Arabis divaricarpa* A. Nels. Erect, 4—8 dm. high, branches spreading; mature pods at about a right angle to stem. Occasional near edges of woods or thickets. Records from 6 scattered counties. Formerly called *A. brachycarpa,* a name found to have been used previously for a different species. In this plant and the preceding, the basal, winter leaves are hairy, the next year's stem leaves smooth or nearly so. In *A. holboellii* the lowest leaves are densely hairy, the upper ones nearly smooth. In *A. hirsuta* the leaves are more nearly evenly hairy.

Erysimum

Native biennials with narrow leaves, bright yellow flowers and slender pods; seeds oblong, yellow or reddish brown.

Key to Species

Flowers 1—2 cm. (²⁄₅—⅘ in.) wide; pods 1—1.5 dm. (4—6 in.) long.
50. *Erysimum asperum*
Flowers 5—8 mm. (⅕—⅓ in.) wide; pods 1—8 cm. long.
Plants gray, stiff; pods 4—8 cm. long. 51. *Erysimum parviflorum*
Plants green, not stiff; pods 1—2 cm. long. 52. *Erysimum chieranthoides*

50. *Erysimum asperum* DC. Western Wall-flower. Stiff, widely branched biennial, 2—8 dm. (8—32 in.) high; leaves 3—8 cm. (1—3 in.) long, oblong with a few teeth, rough with short hairs; fruiting branches thickly covered with stiff, widely spreading pods 1—1.5 dm. long. Common on prairie. June. This is a showy plant and is frequently grown as an ornamental.

51. *Erysimum parviflorum* Nutt. Small Erysimum. Erect, slender biennial, 3—6 dm. high; leaves narrow, 3—5 cm. long; pods stiff, erect or somewhat spreading, 3—6 cm. long. Frequent, sometimes common on prairie. June, July.

52. *Erysimum chieranthoides* L. Wormseed Mustard. Slender with short branches on upper part, 1—2 m. (3—7 ft.) high; leaves numerous, lanceolate, 3—8 cm. (1—3 in.) long; pods 1—2 cm. long, somewhat flattened; seeds narrow, twisted, reddish yellow. Common, usually in wooded places. July-Sept.

53. *Hesperis matronalis* L. Sweet Rocket. Coarse biennial, 6—15 dm. high; leaves numerous, narrowly oblong, 6—10 cm. long; flowers purple or white, 2 cm. wide; pods slender, 6—10 cm. long. June. This introduced plant is often grown as an ornamental and escapes to roadsides or remains about abandoned dwellings.

54. *Conringia orientalis* (L.) Dum. Hare's-ear Mustard. (Fig. 213). Slender annual; erect, or larger plants widely branched, 2—10 dm. high; leaves oblong-elliptic, 6—15 cm. long, clasping at base, surface waxy;

flowers yellowish white, 4—6 mm. wide in long, terminal clusters; pods 10—15 cm. long, stiff, sharply 4-angled, 3 mm. wide; seeds dark brown, oblong, 2—2.5 mm. long. June-Aug. Common introduced weed especially on drier soils. Frequently winters as a rosette of rounded leaves 5—10 mm. long.

CAPER FAMILY Capparidaceae

Upright herbs with alternate, compound leaves; sepals usually 4, stamens 6 or more, elongated; pistil 1; fruit a 2-celled pod resembling a legume.

Key to Species

Pods drooping; plant not sticky.   1. *Cleome serrulata*
Pods erect; plant sticky, rank smelling.   2. *Polanisia graveolens*

1. *Cleome serrulata* Pursh. BEE PLANT. (Fig. 216). Erect, branched annual, 1—2 m. (3—7 ft.) high; leaves of 3, narrowly oblong leaflets 2.5—10 cm. (1—4 in.) long; flowers lavender, in long, terminal clusters; pods 5—8 cm. long, cylindrical but constricted between seeds; seeds rounded, somewhat flattened, 3 mm. (⅛ in.) long. Common in sandy soils. July, Aug. Quite showy and frequently grown as an ornamental. Spider Flower (*C. spinosa*), sometimes seen in gardens, is similar but has 5 larger leaflets.

2. *Polanisia graveolens* Raf. CLAMMY WEED. Erect, somewhat branched annual, 3—8 dm. high; leaflets 3, narrowly to broadly oblong; flowers 4—5 mm. long; petals white, sepals purplish; pods somewhat flattened, 5—8 cm. long; seeds brown. Frequent, chiefly western part of State, usually in gravelly soil on lake shores and hillsides. July, Aug.

SUNDEW FAMILY Droseraceae

Perennial or biennial, small, sticky plants with only basal leaves and a short flower cluster; sepals 4—5, separate or partly united; petals 5; stamens 4—20; pistil 1; styles usually 3; fruit a few seeded capsule. Leaves of some species are long and slender, others rounded. The sticky hairs capture and digest insects.

1. *Drosera rotundifolia* L. ROUND-LEAVED SUNDEW. Leaf blades rounded, 5—15 mm. (⅕—⅗ in.) wide on slender stalks (fig. 107); flower stem 1—2 dm. (4—8 in.) high; flowers pinkish, 3—5 mm. wide. The plants grow in peat bogs and our only record is a specimen collected by Thomas Street near Boundary Butte, Turtle Mts., in 1939.

ORPINE FAMILY Crassulaceae

Mostly succulent plants with small flowers. Here belong hen-and-chickens and species of *Sedum* often grown in rock gardens. Our only native species is little succulent.

1. *Penthorum sedoides* L. DITCH STONECROP. Perennial from a short, thick rootstock; stems erect or spreading, 2—4 dm. (8—16 in.) high; leaves alternate, elliptic, sharply toothed, 3—8 cm. (1—3 in.) long; flowers greenish white, 4—8 mm. (¼ in.) wide, arranged along several spreading branches 3—10 cm. long, at top of stem; sepals 5; petals 0; stamens 10; fruit a thin walled, lobed capsule, 6—8 mm. wide with many, very small seeds. Wet ground, edges of ponds or streams. July, Aug. Richland, Cass and Grand Forks Counties.

SAXIFRAGE FAMILY Saxifragaceae

Herbs or shrubs with opposite or alternate leaves; flowers various; sepals 5, separate or united in a tube surrounding, or attached on top of ovary; petals 5 on tube of calyx; stamens 5—10; pistil 1; fruit a capsule or berry.

Key to Species or Genera

Herbs with rounded or shallowly lobed leaves; fruit a capsule.
 Flowers white or greenish, 1 cm. (⅖ in.) wide, solitary on stem
  1—3 dm. (3—12 in.) high.
   Leaves somewhat heart shaped, false stamens 5—15 at each petal,
    united at base.   1. *Parnassia palustris*
   Leaves rounded, false stamens 3—5 at each petal.

     2. *Parnassia glauca*
 Flowers white or purplish, 4—8 mm. (¼ in.) wide in slender
  clusters.
   Petals entire or nearly so; common prairie plant.

     3. *Heuchera richardsonii*
   Petals finely divided; rare plant in woods.   4. *Mitella nuda*
Shrubs; leaves usually sharply lobed; fruit a berry.   5-10. *Ribes*

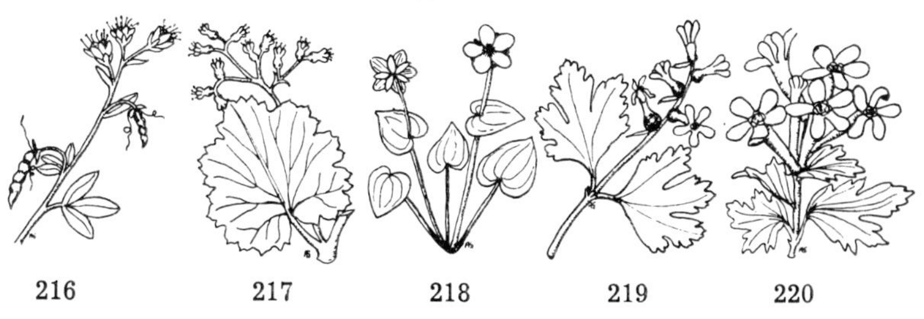

216   217   218   219   220

1. *Parnassia palustris* L. NORTHERN GRASS OF PARNASSUS. (Fig. 218). Perennial from short, upright base; leaves smooth, 1—3 cm. (⅖—1⅕ in.) long, mostly basal, 1 or 2 on lower part of flowering stem; flower stalks upright, 1—3 dm. (4—12 in.) high with a white flower at top. Frequent in wet meadows. July, Aug. The plants usually grow in patches and the leaves are suggestive of a blue violet. Showy, false stamens consist of slender stalks with rounded, yellowish tips.

2. *Parnassia glauca* Raf. CAROLINA GRASS OF PARNASSUS. This species seems rare. Several specimens from Richland, Ransom, McHenry and Pembina Counties, the last without flowers. Formerly called *P. caroliniana* Michx.

3. *Heuchera richardsonii* var. *grayana* Rosendahl, Butters and Lakela. ALUM ROOT. (Fig. 217). Perennial from a stout, thick base; leaves all basal, rounded, 5—15 cm. wide, rough hairy, edges coarsely lobed and finely toothed; flowering stem 3—6 dm. high with a dense cluster of flowers 1—2 dm. long; flowers yellowish and purplish; fruit a 2-pointed capsule, 5—8 mm. long. Common on prairies. June, July. The thick cluster of leaves suggests those of a geranium. Formerly included in *H. hispida* Pursh.

4. *Mitella nuda* L. BISHOP'S CAP. MITREWORT. Slender perennial with creeping stems; leaves rounded, somewhat hairy, 1—3 cm. wide; flower stalks 1—1.5 dm. high with 3—5 flowers and 1 or 2 small leaves below; pod rounded, 5—6 mm. wide; seeds black, shining. Several collections from swampy woods at Walhalla, Pembina County.

Ribes CURRANTS. GOOSEBERRIES

Shrubs with or without prickles; leaves more or less rounded, usually 3-lobed with basal segments partly lobed, edges more or less toothed; flowers in clusters of several on a short branch; calyx tube either short and flat or elongated tubular; fruit an edible, many seeded berry. Some authors separate the gooseberries as a genus *Grossularia*.

Key to Species

Stems without prickles; currants.
  Flowers bright yellow, 10 mm. (⅖ in.) long, slender. 5. *Ribes odoratum*
  Flowers pale yellow or purplish, 5—8 mm. long.
    Flowers yellowish, narrowly cup shaped.   6. *Ribes americanum*
    Flowers purplish, saucer shaped.   7. *Ribes triste*
Stems very spiny; gooseberries.
  Fruits smooth, not spiny.
    Stamens longer than petals; plant 1—2 m. (3—7 ft.) high.
        8. *Ribes missouriense*
    Stamens shorter than petals; plant 3—6 dm. (1—2 ft.) high.
        9. *Ribes setosa*
  Fruits covered with soft spines.   10. *Ribes cynosbati*

5. *Ribes odoratum* Wendl. Golden Currant. (Fig. 220). Shrub 1—2 m. (3—7 ft.) high, widely branched; leaves rounded or triangular, 3—5-lobed; flowers numerous at leaf bases; fruit black or yellowish, 8—12 mm. (⅓—½ in.) wide. Native along streams and hillsides, western part of State. Late May. Spreads rather freely by roots. This had been called *R. aureum* Pursh, which is regarded as a slightly different species of the west coast.

6. *Ribes americanum* Mill. Wild Black Currant. Erect or spreading, 1—1.5 m. high; leaves 3—8 cm. wide, deeply 3—5-lobed, with yellowish, glandular dots on lower surface; flowers in axillary racemes; calyx somewhat inflated, 6—8 mm. long; fruit black, 6—10 mm. wide. Common in woods especially where moist. Early June.

7. *Ribes triste* Pall. Swamp Currant. Stems creeping or ascending, 5—10 dm. long; leaves few, 3—8 cm. wide with 3—5 short lobes; flowers in axillary racemes, purplish, quite flat, 4—5 mm. wide; fruit bright red, 6—8 mm. wide. Rare in swampy woods in Pembina, Rolette and Bottineau Counties. Early June.

8. *Ribes missouriense* Nutt. Missouri Gooseberry. (Fig. 219). Slender, branched, spiny shrub, 1—2 m. (3—7 ft.) high; leaves 2—5 cm. (1—2 in.) wide, sharply 3—5-lobed and toothed; flower clusters about 3-flowered from short branches; flowers drooping, nearly white, 1—1.5 cm. long; petals and stamens erect; sepals narrow, 5 mm. long, at right angles to flower tube; fruit 8—12 mm. wide, purple or black when mature. Common in woods or open places. Mid-May.

9. *Ribes setosa* Lindl. Bristly Gooseberry. Widely spreading, 4—8 dm. high, more densely branched and stiffer than preceding; calyx tube cylindrical, 5—8 mm. long; fruit red or black, smooth or somewhat bristly, 5—9 mm. wide. Chiefly hillsides, western part of State and on rocky lake shores in Rolette and Ramsey Counties.

10. *Ribes cynosbati* L. Prickly Gooseberry. Erect, widely branched, 6—15 dm. high; leaves rounded, 2—5 cm. wide, 3—5-lobed; flowers 1—3 together, calyx tube 3—8 mm. long, stamens exserted; berry purplish, 6—8 mm. wide with numerous, soft prickles. We have this from wooded bank of Sheyenne River in Richland County near Kindred, and at Fort Ransom, Ransom County (Robert Kloubec in 1917).

ROSE FAMILY Rosaceae

Herbs or shrubs, with alternate, usually compound leaves and a pair of stipules at leaf base (fig. 222); sepals 5; petals 5; stamens 5 to many; pistils 1, several or many, separate or united; fruit dry, 1-seeded (achene, nutlet), pod-like (follicle) or fleshy (drupe, pome, aggregate). This family is usually recognized by its many stamens and by petals which fall off readily. In many species there are small bracts just below sepals which are easily mistaken for sepals. The calyx is somewhat extended at base of flower with petals and stamens attached on it (fig. 221). In strawberry the receptacle enlarges and becomes fleshy. In rose the same is true but receptacle is hollow and encloses stony fruits (nutlets). In the apple group, the fleshy receptacle is firmly united with the ovary. Some authors separate the apple group as another family, *Pomaceae*, and the plums as *Drupaceae*.

Key to Species or Genera

Fruit fleshy.
  Fruit with a single, stony seed.   38-41. *Prunus*
  Fruit with 3—many seeds.
    Shrubs or small trees with stout thorns; leaves simple.
      36-37. *Crataegus*
    Shrubs or herbs, with or without small spines.
      Tall shrub with simple, rounded leaves.   35. *Amelanchier alnifolia*
      Low shrubs or herbs; leaves compound.
        Stems 2—15 dm. (8—60 in.) high with small spines or prickles.
        Leaflets usually 3; fruit of several, rounded, 1-seeded parts.
          27-29. *Rubus*
        Leaflets usually 5—7; fruit enclosing nutlets.
          31-34. ·*Rosa*

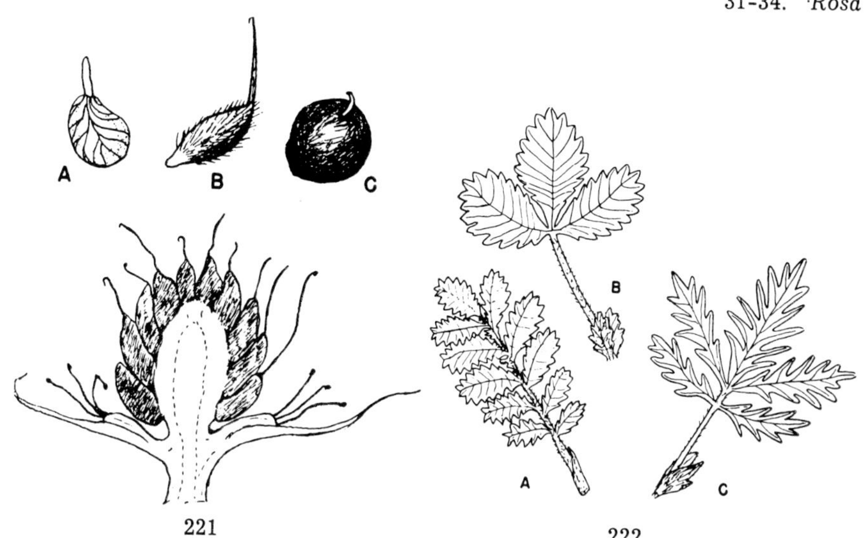

221. Diagram of flower in rose family and fruits of: A. Cinquefoil; B. Avens; C. Raspberry (one drupelet).

222. Cinquefoil leaves: A. *Potentilla anserina;* B. *P. norvegica;* C. *P. pennsylvanica.*

Stems about 1 cm. (½ in.) high, producing runners.
          20-21.   *Fragaria*
Fruit not fleshy.
  Fruit a bur with many hooked prickles.   30. *Agrimonia striata*
  Fruit not bur-like; pistils 5—many, separate.
    Fruit pod-like, each with 2—4 seeds; usually 5 from each flower.
          1. *Spiraea alba*
    Fruits seed-like, many from each flower.
      Fruits hooked or feathery; style persistent.   23-26. *Geum*
      Fruits very small, rounded; style not persistent.
        Leaves 1—2 cm. long, finely divided, closely tufted at base
          (fig. 225).   22. *Chamaerhodos erecta*
        Leaves 3—20 cm. long, not finely divided.   2-19. *Potentilla*

1. *Spiraea alba* DuRoi. Wild Spiraea. Fine stemmed shrub, 6—12 dm. (2—4 ft.) high; leaves oblong-lanceolate, finely toothed, 2—6 cm. (¾—2½ in.) long, flowers white, 5—8 mm. (¼ in.) wide in a branching cluster; fruit of 5 follicles, 3—4 mm. long, splitting on inner edge. June, July. Rather common in low, moist soil. This seems ornamental but is hardly satisfactory for cultivation. Some authors have considered our plant the same as *S. salicifolia* L. of the eastern hemisphere.

Potentilla CINQUEFOIL. FIVE-FINGER

Mostly biennial or perennial herbs (a few shrubs) with compound leaves and usually yellow flowers; stamens many or rarely only 5; pistils many, each developing into a very small, seed-like achene. Some groups are quite variable, the species difficult to separate. Some authors put *tridentata, fruticosa, anserina,* and *arguta* in separate genera, *Sibbaldiopsis, Dasiphora, Argentina* and *Drymocallis,* respectively. The French name "cinquefoil" means 5-leaved, but many species have only 3 or else more than 5 leaflets (fig. 222).

Key to Species

Stems woody.
  Stems horizontal, underground; flowers white; rare, eastern.
        2. *Potentilla tridentata*
  Stems upright, 3—10 dm. (1—3 ft.) high; flowers yellow; western.
        3. *Potentilla fruticosa*
Stems herbaceous.
  Stems creeping or spreading; leaves pinnate.
    Stems creeping, rooting at joints; leaves large, silvery below.
        4. *Potentilla anserina*
    Stems weakly spreading; leaves fine, green. 5. *Potentilla plattensis*
  Leaves digitately compound; leaflets 3 or more.
    Leaves green on both sides.
      Leaflets 3 or some lower leaves with 5.
        Stamens 10 or more; leaflets 3.
          Flowers 3—4 mm. (⅛ in.) wide; stamens 10; achenes
            smooth, pale.   7. *Potentilla millegrana*
          Flowers 7—10 mm. wide; stamens 15—20; achenes ridged,
            brown.   6. *Potentilla norvegica*
        Stamens 5; lower leaves with 5 leaflets. 8. *Potentilla pentandra*
      Leaflets 5—7, rarely 9.
        Stems slender; 1—2 dm. (4—8 in.) high; leaflets oblanceolate,
          much narrowed at base.   9. *Potentilla diversifolia*
        Stems stout, 3—6 dm. high; leaflets oblong, not strongly
          narrowed at base   10. *Potentilla nuttallii*
    Leaves densely white hairy on lower side, leaflets usually 5.
      Stems short and thick or slender and spreading
        Flowers 5—8 mm. (¼ in.) wide; stems slender.
          12. *Potentilla argentea*

Flowers 10—15 mm. wide; stems short, thick.
13. *Potentilla concinna*

Stems stout, erect, 3—8 dm. (1—2½ ft.) high.
11. *Potentilla viridescens*

Leaves, at least the lower, pinnately compound.
Leaves silvery or gray, at least on under side.
Leaves, especially upper ones, nearly palmate.
14. *Potentilla pennsylvanica*

Leaves strictly pinnate.
Flower cluster loose; leaves silvery.   15. *Potentilla hippiana*
Flower cluster dense, leaves dull gray.   16. *Potentilla arygyrea*
Leaves more or less hairy but not gray.
Stems very stout; plant densely glandular hairy; petals white.
17. *Potentilla arguta*

Stem medium; plant not conspicuously glandular; petals yellow.
Achene with a thickened, corky, upper edge.
Leaves all pinnate with 7—11 leaflets.   18. *Potentilla paradoxa*
Lower leaves with 5—7, upper with 3 leaflets.
19. *Potentilla nicolletii*

Achene without much thickened edge; leaflets 3—7.
19a. *Potentilla rivalis*

2. *Potentilla tridentata* Ait. THREE-TOOTHED CINQUEFOIL. Perennial from a woody, horizontal, underground stem; annual branches 1—2 dm. high; leaflets 3, obovate, 3-toothed at tip; flowers white, 1 cm. wide. Found in late bloom on shale outcrop, Pembina Mts., near Olga, Cavalier County, Aug. 4, 1937.

3. *Potentilla fruticosa* L. SHRUBBY CINQUEFOIL. Shrub with many stems, 3—6 dm. (1—2 ft.) high; leaflets 5 or 7, close together, linear-lanceolate, 1—2 cm. (⅖—⅘ in.) long, entire, soft hairy; flowers golden yellow, 1.5—2.5 cm. wide. Local on buttes, western part of State. A widely distributed plant, found all around the world. It is a desirable shrub for ornamental use, bearing flowers in June on old growth and again in August on new growth.

4. *Potentilla anserina* L. SILVERWEED. Perennial by long runners, no erect stem; leaves 1—3 dm. long, of 4—15 pairs of larger leaflets and other small ones (fig. 222A); larger leaflets 1—4 cm. long, oblong or oblanceolate, deeply and sharply toothed, white hairy below and sometimes above; flowers on single stalks from leaf clusters, golden yellow, 1—2 cm. wide. Common on low, especially saline ground. Late May-Aug. This plant also occurs all around the world.

5. *Potentilla plattensis* Nutt. Perennial; stems slender, spreading, 1—3 dm. long; basal leaves 5—15 cm. long, pinnately compound, 5—13 leaflets deeply divided into 5—9 slender lobes; stem leaves few, of only 3—5 slender lobes; flowers 1—1.5 cm. (½ in.) wide, few on slender branches. June, July. First found by Mrs. Alvina Halgrimson, McGregor, Williams County, where in 1946 we found it common at the edge of a prairie coulee. After considerable search we found it in two other places in adjoining counties.

6. *Potentilla norvegica* L. ROUGH CINQUEFOIL. Biennial; stem stout, erect, widely branched and leafy above, coarse hairy, 3—7 dm. (1—2½ ft.) high; leaflets 3, rounded, sharply toothed, 2—5 cm. (⅘—2 in.) long (fig. 222B); flowers yellow, 8—10 mm. (⅜ in.) wide; achenes 0.75 mm. long, brown, with fine, parallel, curved ridges (fig. 221A). Common weed in low places and neglected fields. The first year's growth closely resembles a strawberry plant, but produces no runners. Often called *P. monspeliensis* L. Occurs also in Europe and Asia.

7. *Potentilla millegrana* Engelm. Much like the last, but finer in all respects. We have it from eastern third of State only. Native to central and western U. S. This and the next species are regarded by some botanists as varieties of *P. rivalis* (see No. 19a.)

8. *Potentilla pentandra* Engelm. FIVE-STAMENED CINQUEFOIL. Much like *P. norvegica* except that the lower leaflets are deeply lobed or completely divided. Specimens from Richland, Ransom, LaMoure, Cass and Ward Counties. I have looked for it in recent years without success. Native to central U. S.

9. *Potentilla diversifolia* Lehm (?). Perennial from a thick crown; stem slender, erect or spreading, 1—2 dm. long; leaves mostly near the ground; leaflets 5—7, upper leaves with 3 oblanceolate leaflets, much narrowed at base, 1—2 cm. long, deeply but not sharply lobed; flowers few, 5—8 mm. wide, on slender branches. Specimens received in June, 1935, from Mrs. Ethel Atkinson at Vim, Slope County, I had referred to this western species, but Keck (Carneg. Inst. 520:137, 1940) indicates that it should not occur here. A specimen from Leeds, Benson County (Lunell, May 31, 1909), seems similar in habit but has rounded leaflets. It was labeled *concinna* by Lunell and var. *divisa* Rydb. by Bergman.

10. *Potentilla nuttallii* Lehm. Perennial; stem stout, erect, 3—6 dm. (1—2 ft.) high, the upper part widely branched in the flower cluster; leaflets usually 7, 2—4 cm. (¾—1½ in.) long, narrowly oblong, sharply toothed, moderately hairy but not gray; flowers 1.5—2 cm. wide. As in other prairie species, the leaves are mostly near the ground, the upper stems nearly leafless. We formerly had only two specimens from Grant County (Bell 502, 1174). The plant was originally found in the Mercer County area by Nuttall in 1811. In 1946, I found it in Logan and Mountrail Counties, locally common in prairie sloughs (*Zygadenus* zone), but not distinct from the next species or other variations.

11. *Potentilla viridescens* Rydb. Much like the preceding and perhaps only a form of it, differing in the densely hairy covering, which makes the plant appear quite silvery. Plants with large, obovate leaflets have been called *P. camporum* Rydb. Two specimens determined by Bergman as *P. pulcherrima* Lehm. seem better placed in *viridescens* and *argyrea.*

12. *Potentilla argentea* L. SILVERY CINQUEFOIL. Perennial; stems many, slender, 3—6 dm. high, about equally leafy to the short, flowering clusters; leaflets usually 5, obovate, 1—3 cm. long, much narrowed toward base, deeply toothed toward tip, silvery below, green above. This was collected in a bromegrass field in Stark County in 1914. It was reported by the Nicollet Expedition of 1839 at Devils Lake but I have been unable to verify that record. It occurs in sandy soil in many places in Minnesota.

13. *Potentilla concinna* Richards. Perennial; stems stout, spreading, hardly 1 dm. high; leaves closely crowded and silvery below; flowers 1—2 cm. wide. Frequent on hills or high prairie, western half of State. This is a conspicuous plant, beginning to bloom in early May, the leaves appearing very gray in their young, folded condition. Bergman 2662, collected Aug. 6, shows a mature condition, the flowering stems spreading, about 1 dm. long, slender, with few flowers in fruit. Similar specimens sent by Mrs. Atkinson were collected at the same date as those referred above to *P. diversifolia*, and one of July 28 referred by Bergman (2522) to *P. viridescens* is apparently *concinna.* *P. divisa* Rydb. is here included.

14. *Potentilla pennsylvanica* L. Perennial; stems several, 2—4 dm. (8—16 in.) high, erect but usually spreading at base; lower leaves long-petioled, usually a few leaves about half way up the stem; leaflets narrowly or broadly oblong, 2—8 cm. (¾—3 in.) long, deeply divided into narrow lobes (fig. 220C), silvery below; flower clusters compact, 3—8 cm. long; flowers 8—10 mm. (⅜ in.) wide. Common on prairie. July, Aug.

From this have been separated *P. bipinnatifida* Dougl. and *P. strigosa* Pall. among others. I am unable to see good distinctions on either form of leaf or pubescence. The leaflets are commonly distinctly pinnate, sometimes so close together as to be practically digitate. A specimen from McHenry County, which I determined as *P. atrovirens* Rydb., seems rather distinctive by its dark green leaves. Two broad leaved specimens had been labeled *P. platyloba* Rydb.

15. *Potentilla hippiana* Lehm. Perennial from a thick crown; stem erect, 3—6 dm. high with few leaves, widely branched in the loose flower cluster; leaflets 7—11, well separated, 2—6 cm. long, narrowly oblong, shallowly or deeply toothed, shining white hairy below; flowers 15—20 mm. wide. Two specimens are placed here, Stevens 575 and Bergman at Dickinson, June 21, 1910, Slope and Stark Counties respectively.

16. *Potentilla argyrea* Rydb. Very similar to last in leaf shape but with dull gray leaves and branches; flower clusters short, compact, the main one 5—8 cm. long. Specimens from Bottineau and Burke Counties, also Perrine 1198 from Valley City, which has a loose flower cluster 15 cm. long.

17. *Potentilla arguta* Pursh. TALL CINQUEFOIL. Perennial, the whole plant densely and coarsely glandular hairy; stem single, very stout, 6—12 dm. (2—4 ft.) high; leaves 1—3 dm. long, leaflets 7—11, lateral ones obliquely ovate; flower cluster dense; flowers 12—15 mm. (½ in.) wide, petals white. Common on prairie. July.

18. *Potentilla paradoxa* Nutt. Biennial; stems 3—8 dm. high, often coarse, at first single, later with wide spreading branches; leaves numerous, 5—20 cm. long, leaflets 9—11, obovate with rounded teeth; flowers many, 5—8 mm. wide; achenes dark brown with one side much thickened. Frequent in eastern part of State, especially on gravelly pond shores. The smaller branches resemble those of *P. millegrana*.

19. *Potentilla nicolletii* (S. Wats.) Sheld. Similar to the last but smaller, the flowers seeming in axillary clusters. First collected at Devils Lake by the Nicollet Expedition of 1839. We have only two specimens, Leeds in Benson County and Williston in Williams County.

19a. *Potentilla rivalis* Nutt. Biennial; stem leafy, erect, 4—6 dm. high, widely branched above and in large plants with spreading branches from base; leaflets usually 5, sometimes 7, often 3 on upper stems, obovate, 2—5 cm. long with coarse teeth. First recorded at Edgeley, LaMoure County (Stevens 845) in 1945. It grew in an old field with *P. norvegica* which it resembles in general appearance.

Fragaria STRAWBERRY

Perennials with thick, short stems and long runners; leaflets 3, ovate or wedge shaped, coarsely toothed; flowers white; sepals and petals 5, stamens and pistils many; receptacle becoming swollen and edible in fruit, each pistil developing into a tiny achene. There are 5 bracts, very similar to and just below the sepals. Leaves and flower cluster stalks come from a short stem which is mostly underground, erect or partly horizontal.

Key to Species

Fruit conical; achenes on surface.   20. *Fragaria vesca*
Fruit rounded; achenes in cavities.   21. *Fragaria virginiana*

20. *Fragaria vesca*, var. *americana* Porter. WOOD STRAWBERRY. Leaflets pale green, thin, 3—6 cm. (1—2 in.) long; flowers 8—10 mm. (⅜ in.) wide; fruit conical, 10—15 mm. long, usually half or less as wide, surface not pitted. Rather local in woods; chiefly eastern, also Killdeer Mts. and Little Missouri River, Slope County. Flowers early June, fruit ripe July 1. This is sometimes considered a distinct species, but more often as a variety of *F. vesca* L., the garden strawberry. It seems to produce fruit more freely than the next species.

21. *Fragaria virginiana* Duchesne. WILD STRAWBERRY. Leaflets dark green above, whitish below, rather thick, 2—10 cm. long; flowers 15—20 mm. wide; fruit rounded, hardly longer than wide, 8—15 mm. long, achenes set in pits. Common, both on low prairie and in woods, usually not fruiting freely.

22. *Chamaerhodos erecta* (L.) Bunge. LITTLE ROSE. (Fig. 225). Biennial; stem erect, 1—3 dm. (4—12 in.) high, branching above; leaves finely divided, 1—3 cm. (⅖—1⅕ in.) long and often as wide; flowers numerous, white, 2—3 mm. ($\frac{1}{10}$ in.) wide; achenes greenish, smooth. Frequent on gravelly hills, Barnes County westward. June, July. The first year it forms a tuft of leaves 2—5 cm. wide. Some authors regard our plant as distinct from the Asiatic one and call it *C. nuttallii* Pickering.

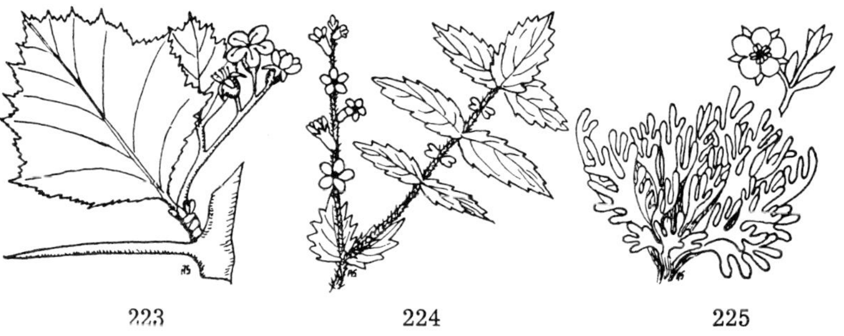

223   224   225

Geum   AVENS

Perennials from thick crowns or short rhizomes; leaves alternate, pinnately compound; flowers white, yellow or purple; achenes with long beaks, hooked or plumed in fruit. Torch Flower (*G. triflorum*), which has plumed achenes, is separated as *Sieversia* by some authors. The hooked fruits of the others are among the common "sticktights" which one finds in woods.

Key to Species

Sepals purple; achenes with long, feathery plumes.　　23.　*Geum triflorum*
Sepals green; achenes with hooked tips (fig. 221B).
　Petals white.　　24.　*Geum canadense*
　Petals yellow.
　　Terminal leaflet and petals wedge shaped.　　25.　*Geum strictum*
　　Terminal leaflet and petals rounded.　　26.　*Geum macrophyllum*

23. *Geum triflorum* Pursh. TORCH FLOWER. (Fig. 226). Perennial from a short, thick, horizontal stem just below ground; basal leaves 1—2 dm. (4—8 in.) long, leaflets many, wedge shaped, 3-lobed, soft hairy; stem leaves few, 1—3 cm. (⅖—1⅕ in.) long with narrow lobes; flowers drooping, 10—15 mm. (½ in.) long, often 3 on a stem; fruiting head erect, plumes 2—6 cm. long, delicate, bronze or purplish. Frequent to abundant on prairie. Begins to bloom late April or early May, continuing well into June. The petals are yellowish white but do not show much because the flowers droop and do not open widely. The narrow bracts are longer than the sepals. This is an attractive and popular wild flower. Also called Maidenhair, Old Man's Whiskers and Prairie Smoke.

24. *Geum canadense* Jacq. WHITE AVENS. Stem leafy, erect, 3—6 dm. high, leaves quite variable, pinnate to 3-lobed or simple; flowers white, 1—1.5 cm. wide, at tips of upper branches; achenes 25—75, rather inflated and bent back, 8—12 mm. long. Woods. June, July. Not com-

226.   Torch Flower (*Geum triflorum*).

mon but recorded for 8 counties west to McLean. Lowest leaves rounded, 3—5 cm. wide, not divided or 3-foliate; next ones with 5—9 leaflets, the terminal rounded; upper leaves 3-lobed or simple, lanceolate.

25. *Geum strictum* Soland. YELLOW AVENS. Stem leafy, erect, 5—10 dm. high; leaflets usually 5—9, wedge shaped or the terminal rounded; flowers golden yellow, 10—15 mm. wide; fruiting head obovate, 1—2 cm. long; achenes slender, not inflated, 6—8 mm. long. (fig. 221B). Woods, brushy or open ground in coulees. Fairly common and over entire State. July, Aug. Fernald has referred this to a variety of *G. aleppicum* Jacq.

26. *Geum macrophyllum* Willd. LARGELEAVED AVENS. Similar to last and not readily distinguished from it. Specimens from 6 eastern counties have been referred to this, but I am unable to recognize them as distinct from the preceding species.

Rubus   RASPBERRIES

Woody or herbaceous perennials, spiny or not; leaves alternate, of 3—5 leaflets; flowers white; sepals and petals 5, stamens and pistils many; each pistil becoming pulpy (drupelet), all more or less cohering into a rounded or oblong fruit.

Key to Species

Stems trailing, herbaceous, not spiny; drupelets 5—15, loosely
 cohering.   27. *Rubus pubescens*
Stems erect or arching, woody, spiny; drupelets 25—50, closely
 cohering.
 Fruit red; stems bristly but not prickly.   28. *Rubus idaeus*
 Fruit black; stems with stout spines.   29. *Rubus occidentalis*

27. *Rubus pubescens* Raf. DWARF RASPBERRY. Stems weak, trailing, 2—6 dm. (8—24 in.) long; leaflets 3, rarely 5, ovate or angular, 2.5—10 cm. (1—4 in.) long, sharply toothed; flowers 8—15 mm. wide; drupelets bright red, 3—4 mm. (⅛ in.) wide, readily separating. Swampy woods or open boggy ground; Cass, Richland, Ransom, Pembina, Ramsey and Bottineau Counties. Flowering in June, fruit ripe in July. The fruits are pleasantly acid. Formerly called *R. americanus* (Pers.) Britton.

28. *Rubus idaeus* L. RED RASPBERRY. Stems erect or nearly, 6—10 dm. high, new ones covered with weak bristles; leaflets 5, ovate, 5—10 cm. long, sharply toothed, lateral ones oblique at base; flowers 10—15 mm. wide on short branches from stems of year before; drupelets 2 mm. wide, closely cohering in a fruit 1.5—2 cm. wide and slightly longer. Frequent to common in woods or coulees. Flowering in June, fruit ripe in July. The American plant has often been called *R. strigosus* Michx., but many authors regard it as only a variety of the European which is the one cultivated. Hybrids of that with black raspberry are also grown.

29. *Rubus occidentalis* L. BLACK RASPBERRY. Stems 1—2 m. (3—7 ft.) long, curving to root at tip, quite spiny, leaflets 3 or 5, ovate, cordate or lanceolate, sharply toothed, white below, 5—8 cm. long; flowers 1.0—1.5 cm. wide. in a short, dense cluster; drupelets black or nearly so, cohering in a fruit 10—15 mm. wide, hardly as long. Plants have been found in Richland and Oliver Counties, probably introduced and not persisting more than a few years.

30. *Agrimonia striata* Michx. AGRIMONY. (Fig. 224). Perennial; stem erect, 6—10 dm. high; leaves pinnate, leaflets 5—13, lanceolate to obovate, 3—10 cm. long with rounded or sharp teeth and with small leaflets 3—15 mm. long between the large ones; flowers yellow, 6—10 mm. wide, along the slender, upper branches; fruit a bur, 6—8 mm wide, 2-seeded. Occasional in woods all over State. June, July.

Rosa ROSE

Shrubs with spiny stems and pinnate leaves of 5—11 oblong leaflets; flowers large, 1—several on short branches; stamens many; pistils many, each producing a hard nutlet, these enclosed by the fleshy, bright red, hollow receptacle; sepals persisting. Identification of species of roses is very difficult. The following treatment may not be technically complete but will distinguish our general kinds.

Key to Species

Stems bristly, not very spiny, slender, mostly dying back.
31. *Rosa arkansana*

Stems spiny or smooth, stout, living at least 2 years.
Older stems mostly smooth; eastern woodland plants.
32. *Rosa blanda*

Older stems with stout spines.
A stout spine below each leaf; western, common plant.
34. *Rosa woodsii*

No spine just below leaf; eastern, not common. 33. *Rosa acicularis*

31. *Rosa arkansana* Porter. PRAIRIE WILD ROSE. Stem slender, 3—6 dm. (1—2 ft.) high, not much branched, mostly dying back each year, covered with weak bristles; flowers dark red to white, several in a cluster at tip of stem or sometimes on a branch. Common on prairie, hills and fields. Begins to flower about June 15. The names *pratincola, heliophila* and *suffulta* have been used by some authors who consider our plants not the real *arkansana*.

32. *Rosa blanda* Ait. SMOOTH WILD ROSE. Stem stout, 1—2 m. (3—7 ft.) high, woody, widely branched above, these branches usually smooth; new stems covered with bristles; flowers uniformly pink. Common in woods west to Barnes and Ramsey Counties. Begins to bloom about June 10, lasting about 3 weeks.

33. *Rosa acicularis* Lindl. PRICKLY WILD ROSE. Stem stout, 6—10 dm. high, widely branched, very spiny. Specimens from Pembina and Turtle Mt. areas seem to be this species. The fruits are either rounded or elongated.

34. *Rosa woodsii* Lindl. WESTERN WILD ROSE. Stem moderately fine but woody, widely branched throughout, 6—20 dm. high, very spiny; flowers usually pale pink, sometimes dark. Common along streams and coulees, Missouri River westward. Begins to bloom about June 20. the name "western" is used here as applied to North Dakota. This is the common wild rose in the woods in that section. The name "Woods' Rose" generally used, refers to John Woods, British botanist, for whom the plant was named.

Some specimens from central and eastern parts of State have been referred to *woodsii*. E. W. Erlanson (Bot. Gaz. 96:227, 1934) stated: "*R. blanda* also gives fertile, vigorous hybrids with the diploid *R. woodsii*. An intergrading series between these forms makes it hard to classify the diploid roses of the Northern Great Plains region. In herbaria these are usually called *R. macounii*. **** *R. blanda* and *R. woodsii* both give partially sterile tetraploid F_1 hybrids with *R. acicularis*, which bear corymbs of flowers and fruits."

35. *Amelanchier alnifolia* Nutt. JUNEBERRY. SERVICE BERRY. SASKATOON. Slender shrub, 1—3 m. (3—10 ft.) high; leaves alternate, broadly oblong to rounded, 1—3 cm. (½—1½ in.) long, with a few, small teeth on upper part or nearly all around (fig. 60); flowers white, 1—2 cm. wide, 3—10 in a cluster 2—5 cm. long on branch tip; fruit a blue black berry 7—8 mm. (⅓ in.) wide. Common in woods, coulees or sometimes on open hillsides. Flowers early May, fruit ripe July 1. It seems best to recognize only one species. Several have been described, but separation of them is difficult and authorities do not agree.

Crataegus HAWTHORN. RED HAW

Shrubs or small trees, with broad leaves and stout, brown thorns 2—5 cm. (1—2 in.) long; flowers white, in large, flat clusters at ends of branches; fruit red, fleshy, with 3—5 large nutlets. These plants are closely related to apple. The flesh of the fruit is partly from the receptacle which surrounds, and is united to, the ovary.

Key to Species

Leaves broadest near base, soft hairy; fruit 15—20 mm. (⅗—⅘ in.)
  wide.   36. *Crataegus mollis*
Leaves broadest near middle, little hairy; fruit 8—10 mm. wide.
                                        37. *Crataegus rotundifolia*

36. *Crataegus mollis* (T. & G.) Scheele. Downy Hawthorn. Tree, 2—5 m. (6—17 ft.) high; leaves broadly ovate, truncate or rounded at base, doubly sharp toothed, soft hairy below, 5—12 cm. (2—5 in.) long; flowers 1.5—2 cm. wide. Flowers in early June; fruit ripening late Sept., remaining firm. A few trees in woods along the river at Fargo have been our only record, but it probably occurs, or did formerly, at other places along the Red River.

37. *Crataegus rotundifolia* (Ehrh.) Borckh. Round-leaved Hawthorn. (Fig. 223). Shrub or small tree 1—3 m. high; leaves obovate or rounded, sharply toothed, slightly hairy below, 3—5 cm. long; flowers 1—1.5 cm. wide. Flowering late May; fruit ripening late Aug. and becoming soft. Frequent in woods, coulees or open hillsides, all over State. This had been called *C. chrysocarpa* Ashe, but present authorities do not regard that as different from *C. rotundifolia*. Some specimens have been called *C. macrantha* (Lindl.) Lodd. (= *C. succulenta* Schrad.) but all seem better placed under the name *rotundifolia*.

Prunus PLUM. CHERRY

Shrubs or trees with alternate, simple, smooth leaves; sepals and petals 5; stamens 10—25; pistil 1; calyx cup thin, scarcely green, not persisting; fruit fleshy, with a 1-seeded stone.

Key to Species

Flowers 2—5 at leaf base before or as leaves develop.
  Flowers 1.5—2 cm. (¾ in.) wide; stone or fruit flattened.
                                        38. *Prunus americana*
    Flowers 8—10 mm. (⅜ in.) wide; stone rounded.
      Fruit red; slender shrub or small tree 2—5 m. (6—16 ft.) high.
                                        39. *Prunus pennsylvanica*
      Fruit black; spreading shrub 3—10 dm. (1—3 ft.) high.
                                        40. *Prunus pumila*
Flowers many, in clusters 5—15 cm. long, at ends of new branches.
                                        41. *Prunus virginiana*

38. *Prunus americana* Marsh. Wild Plum. Widely spreading, thorny tree, 1—3 m. high; leaves elliptic to oblanceolate, finely toothed, 4—10 cm. long; flowers 2—5 at old leaf base; fruit yellow or red, rounded or oblong, 1—2 cm. long. Frequent to common, edges of woods and open coulees. Flowers mid-May; fruit ripe Aug., Sept.

39. *Prunus pennsylvanica* L. f. Bird Cherry. Pin Cherry. Shrub or tree, 2—5 m. high, main trunk erect, the very slender branches spreading or drooping; leaves oblong—lanceolate, finely toothed, 5—10 cm. long; flowers 2—5 at old leaf base; fruit scarlet, rounded, 4—6 mm. wide. Woods, eastern part of State, locally west to Killdeer Mts. and Mountrail County. Flowers mid-May; fruit ripe July. This is a very characteristic small tree of eastern and northern America. The acid fruit is prized for jelly making.

40. *Prunus pumila* L. SAND CHERRY. Spreading shrub, 3—8 dm. high; leaves oblanceolate or elliptic, 3—6 cm. long, with some small teeth on at least upper half; fruit rounded or oblong, 1—1.5 cm. long. Sand hills and sometimes hillsides, chiefly southeast, south and southwestern counties. Flowers in mid-May; fruit ripe Aug. *P. besseyi* Bailey with larger fruits, is commonly recognized as the western form and most of our plants may belong to it, but the distinctions are slight.

41. *Prunus virginiana* L. CHOKE CHERRY. Shrub or tree 1—10 m. (3—35 ft.) high; leaves oblong-obovate, finely toothed, 5—15 cm. (2—6 in.) long; flowers 1 cm. wide, in a dense, narrow cluster 10—15 cm. long at end of young, leafy branch; fruit rounded or slightly oblong, black, reddish or rarely yellow, 8—10 mm. (⅜ in.) wide. Very common in woods and coulees. Flowers late May; fruit ripe Aug., Sept. The fruit is astringent, becoming sweet when fully ripe but does not jel readily. It has long been discussed whether our plants are *P. virginiana*, *P. demissa* (Nutt.) Walp. or *P. melanocarpa* (A. Nels.) Rydb. Thick leaves are given as one character of *P. melanocarpa*, a western form, but thick leaved specimens have been collected in Turtle Mts. where the eastern form should occur. It seems best to leave them all as *P. virginiana*.

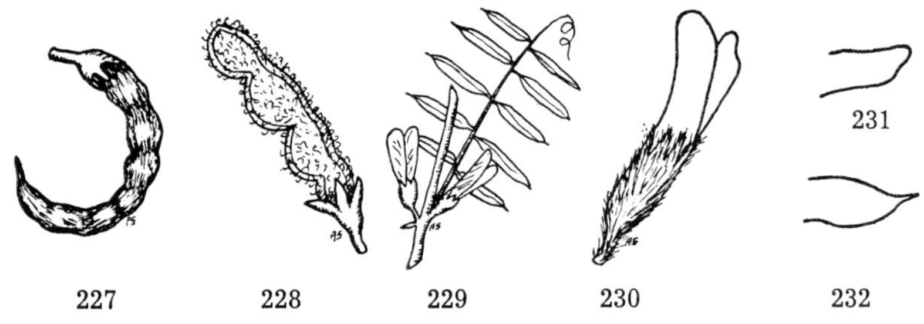

227   228   229   230   232

PEA FAMILY Fabaceae

Herbs, shrubs or trees; leaves alternate, usually pinnately or palmately compound; leaflets 3 to 25 or more; calyx somewhat cup shaped; corolla "pea-shaped"; petals 5: 1 large, upper (standard or banner), 2 narrow, lateral (wings) and 2 partly united, inner (keel); stamens 10 (rarely 5), usually 9 united by filaments halfway, rarely all united or all separate ; pistil 1, usually 1-celled, ovules 1—many; fruit a legume, splitting on two sides, in some species breaking into 1-seeded segments or not splitting.

One of the largest and most important groups of plants, many used for food and forage, few weedy or poisonous. Very important also because of nodule bacteria on roots which fix free nitrogen. Our species are nearly all herbs. Many woody ones occur in warmer regions and a few such are grown here as ornamentals.

Key to Species or Genera

Stamens separate, the filaments not grown together.
  Flowers in small heads; stamens 5, projecting beyond corolla.
               1. *Desmanthus illinoensis*
  Flowers large, golden yellow; stamens 10. 2. *Thermopsis rhombifolia*
Stamens partly united to form a sheath around ovary.
  Vines or vine-like; twining or with tendrils on leaves.
    Leaflets 3; twining or trailing vines.
      Flowers 12—15 mm. (½ in.) long; pods flattened, 1—3 cm.
        (⅖—1⅕ in.) long.   61. *Amphicarpa bracteata*

Flowers 5—6 mm. long; pods cylindrical, 3—6 cm. long.
                  62. *Strophostyles leiosperma*
Leaflets 5—14, usually a tendril in place of terminal leaflet.
Leaflets 5—9, no tendrils.   60. *Apios americana*
Leaflets 4—14, tendrils usually present (fig. 229).
  Style slender, with a tuft of hairs at tip.   63-65. *Vicia*
  Style flattened, hairy along upper side.   66-68. *Lathyrus*
Herbs or shrubs, not vine-like nor with tendrils.
Pods flat, breaking into 1-seeded segments.
  Leaflets 3; pods covered with short, hooked hairs (fig. 228).
                58-59. *Desmodium*
  Leaflets 9—15; pods with few, straight hairs.
             57. *Hedysarum boreale*
Pods not breaking into 1-seeded segments.
  Leaflets 3—9 (rarely 1), palmately arranged.
    Leaflets only 1 (terminal).
      Flowers blue, 8 mm. (⅓ in.) long; leaflet 1—3 cm.
      (⅖—1⅕ in.) long.   36. *Astragalus caespitosus*
      Flowers yellowish, 2 cm. long; leaflet 5—10 cm. long.
           33. *Astragalus longifolius*
    Leaflets 3—9.
      Leaflets 5—9.
        Leaflets mostly 3; flowers 5 mm. long.
           49. *Psoralea argophylla*
        Leaflets 5—9; flowers 8—15 mm. long.
        Leaflets 5; flowers 15 mm. long; plant coarsely hairy.
           48. *Psoralea esculenta*
        Leaflets 5—9; flowers 8—10 mm. long; hairs fine.
           3—4. *Lupinus*
    Leaflets 3.
      Leaflets entire.
        Annuals; flowers whitish, 5 mm. long, single at leaf
        base.   17. *Lotus americanus*
        Perennials; flowers blue or larger.
          Erect plants, flowers blue or bluish.   48-51. *Psoralea*
          Low, tufted plants.
            Flowers blue, 8 mm. long; leaves green.
             36. *Astragalus caespitosus*
            Flowers whitish, 15 mm. long; leaves silvery.
             34. *Astragalus triphyllus*
      Leaflets with fine teeth; flowers in long or rounded
        clusters.
        Flowers in slender clusters 5—10 cm. long.
           11-12. *Melilotus*
        Flowers in rounded heads or oblong clusters 2—3 cm. long.
          Pods evident, rounded, curved or coiled.
            13-15. *Medicago*
          Pods covered by calyx, straight, thin. 5-10. *Trifolium*
Leaflets mostly 5—25, pinnately arranged.
  Flowers in heads surrounded by small leaves; rare plant.
           16. *Anthyllia vulneraria*
  Flower clusters not surrounded by small leaves.
    Pod a bur, covered with hooked prickles.
           44. *Glycyrrhiza lepidota*
    Pod not a bur.
      Woody plants; only standard of corolla present.
          45-47. *Amorpha*
      Herbs; all petals present.
        Pods several—many-seeded, not covered by calyx.
          Tip of keel rounded (fig. 231). 18-38. *Astragalus*
          Tip of keel with sharp point (fig. 232).
            30 43. *Oxytropis*
        Pods thin, 1-seeded, covered by calyx.
         Stamens 5, partly attached to petals.
           54-56. *Petalostemum*

Stamens 10, not attached to petals.
Calyx with a conspicuous tuft of white hairs.
53.   *Dalea enneandra*
Calyx without such hairs.
52.   *Dalea alopecuroides*

1. *Desmanthus illinoensis* (Michx.) MacM. Prairie Mimosa. Perennial; several stout stems 3—10 dm. (1—3 ft.) high; leaves alternate, 5—15 cm. long, twice pinnate, leaflets many, 2—3.5 mm. ($\frac{1}{10}$ in.) long; flowers very small, in several, dense, rounded heads 1 cm. (⅖ in.) wide on stalks 5—10 cm. long from upper leaf bases; petals all alike, stamens 5, longer than petals; pods flat, 1.5—2 cm. long, curved and twisted into a head about 2.5 cm. wide. Gravelly soil, old pond or lake edges. Sargent, Emmons and Ramsey Counties. July. This is related to the many species of acacia of warmer regions. The tiny flowers in heads are characteristic of a group which is sometimes treated as the mimosa family.

2. *Thermopsis rhombifolia* (Nutt.) Richards. False Lupine. Perennial from long roots; stems erect, 2—4 dm. (8—16 in.) high, usually widely branched above; leaflets 3, obovate, 1.5—3 cm. (⅗—1⅕ in.) long, smooth, rather thick; flowers golden yellow, 1.5—2 cm. long, in clusters

233.   Silvery Lupine. (*Lupinus argenteus*).

3—10 cm. long, at ends of branches; pods 5—6 cm. long, flat, curved in a semi-circle (fig. 227). Frequent to common, Missouri River and Ward County westward (fig. 13), especially on clay soils on buttes or banks. A very showy plant. May.

3. *Lupinus argenteus* Pursh. SILVERY LUPINE. (Fig. 233). Perennial; stems several in old plants, erect, branches widely spreading, 3—8 dm. high; entire plant silvery with dense hairs; leaflets 6—8, narrowly oblanceolate, 2—5 cm. long; flowers blue or violet, 1 cm. long, in dense, terminal clusters 1—2 dm. long; pods 2—3 cm. long, flattened, seeds 5—6, oblong, somewhat flattened, dark yellow. Prairie, chiefly southwestern 4 counties (fig. 13); records also from Grant, Stark, Billings and one from Mountrail County (Stevens in 1945). A showy species. June. Poisonous, especially pods, for sheep.

4. *Lupinus pusillus* Pursh. SMALL LUPINE. Fleshy annual, 1—2 dm. high, branches widely spreading, coarsely hairy, not silvery; leaflets 5—8, oblong-oblanceolate, 2—3 cm. long, smooth above, hairy below; flowers 6 mm. long, keel white, wings and standard blue, in terminal clusters 2—5 cm. long; pods 1—3 cm. long; seeds 4—6, nearly circular, quite flat, nearly white. Frequent to common in sandy soil, Missouri River westward. June, July. The pods and stem tips are brown hairy.

Trifolium   CLOVER

Annual, biennial or perennial; leaves 3-foliate; flowers small, slender, usually in rounded heads; pods membranous, 1—several-seeded. Here belong the true clovers; ours are all introduced, planted, naturalized or casual. Species of *Medicago, Melilotus* and *Lotus* may be confused with them.

Key to Species

Flowers red, pink, purple or white.
  Slender or spreading annual; pods inflated, veiny.
  8. *Trifolium resupinatum*
  Biennial or perennial, stout to slender; pods not inflated nor veiny.
  Flowers not stalked, remaining erect; heads 2—3 cm. (1 in.) wide.
  5. *Trifolium pratense*
  Flowers slender stalked, soon drooping; heads 1—2 cm. wide.
  Main stem prostrate, only leaf and flower stalks upright.
  6. *Trifolium repens*
  Main stem erect or ascending, branches widely spreading.
  7. *Trifolium hybridum*
Flowers yellow; rare annuals.
  Heads 20—40-flowered, 8—15 mm. (½ in.) long.
  9. *Trifolium procumbens*
  Heads 6—15-flowered, 5—6 mm. long.   10. *Trifolium dubium*

5. *Trifolium pratense* L. RED CLOVER. Biennial; stout, erect or spreading, widely branched; leaflets rounded, ovate or obovate, 2—5 cm. (1—2 in.) long; flowers bright pink or purple, 10—15 mm. (½ in.) long; pod 1-seeded, splitting around the middle; seeds rounded—triangular, yellowish or purple. Frequent along roadsides, rarely planted.

6. *Trifolium repens* L. WHITE CLOVER. DUTCH CLOVER. Perennial; stems creeping, leaf and flower stalks 5—20 cm. high; leaflets obovate, ovate or rounded, 1—2 cm. long; heads 1—2 cm. wide; flowers white, 7—8 mm. long; pods 2—5-seeded, hardly splitting; seeds squarish, yellow. Commonly planted in lawns; frequent in various places, especially where moist.

7. *Trifolium hybridum* L. ALSIKE CLOVER. Perennial; stems usually much branched, spreading, 2—5 dm. high; leaflets ovate to obovate, 2—4 cm. long; flowers very similar to last but commonly pinkish; seeds greenish to black. Frequent in meadows or low places.

8. *Trifolium resupinatum* L. STRAWBERRY CLOVER. Annual; stems 1—2 dm. long, very slender when growing in grass, densely tufted when alone; leaflets obovate, 1—2 cm. long; flowers bright pink, 3—4 mm. long, in heads 5—10 mm. wide; calyx much enlarged, prominently veined, 2-beaked and turned backward in fruit. Abundant in a newly seeded lawn at Fargo in 1941 (Stevens 586). It has been promoted as a forage plant but probably will not be successful nor persistent in this region.

9. *Trifolium procumbens* L. LOW HOP CLOVER. Annual; stems spreading or erect, 1—3 dm. long; leaflets obovate, 5—10 mm. long; flowers bright yellow, 3—4 mm. long, in rounded or oblong heads; seeds oblong, yellowish, shining. Records from Richland and Benson Counties; lawns, fields and roadsides.

10. *Trifolium dubium* Sibth. Similar to last, more slender, heads few flowered. In lawn with No. 8.

Melilotus SWEET CLOVER

Biennials, resembling *Trifolium,* but tall plants with flowers in long, slender clusters; pods 1—2-seeded; seeds oblong, slightly flattened. The plants are smooth, strong scented, much planted for forage and found established in all sorts of places.

11. *Melilotus alba.* Desr. WHITE SWEET CLOVER. Stem much branched, 1—2.5 m. (3—8 ft.) high; leaflets 3, oblong or obovate, 1.5—2.5 cm. (⅗—1 in.) long; flowers white, 3 mm. (⅛ in.) long, in slender clusters 5—10 cm. long; seeds yellow (green when immature), slightly granular. Very common. July-Oct.

12. *Melilotus officinalis* (L.) Lam. YELLOW SWEET CLOVER. Similar to last but with yellow flowers; seeds often olive green or purple spotted.

Medicago MEDICK

Annuals or perennials similar to *Trifolium;* flower clusters oblong, rather loose; pods rounded and 1-seeded, curved or coiled and several seeded.

Key to Species

Flowers purple, sometimes green or yellowish; pod coiled.
           13. *Medicago sativa*
Flowers yellow; pod rounded or curved.
  Flowers 6—8 mm. (¼ in.) long; pod curved, several seeded.
           14. *Medicago falcata*
  Flowers 3 mm. long; pod rounded, 1-seeded.   15. *Medicago lupulina*

13. *Medicago sativa* L. ALFALFA. LUCERNE. Perennial from a thick crown; stems upright, 3—10 dm. (1—3 ft.) high; leaflets 5, narrowly obovate or oblanceolate, toothed toward tip, slightly hairy, 1—3 cm. (⅖—1⅕ in.) long; flowers 8—10 mm. (⅜ in.) long, in oblong clusters 2—3 cm. long; pods coiled 1—3 times, 6—8 mm. wide; seeds yellow, bean shaped or irregular. Commonly planted and found along roads, etc. June-Sept. Plants with variegated (green, yellowish, blackish) flowers are hybrids with the next species.

14. *Medicago falcata* L. SICKLE LUCERNE. Similar to last but less upright and flowers yellow; pod slightly curved, 1—1.5 cm. long. A plant was found along a railroad track at Oakes, Dickey County, in 1918.

15. *Medicago lupulina* L. BLACK MEDICK. YELLOW TREFOIL. Annual; stems slender, weak, 1—4 dm. long; leaflets 3, obovate or rounded, 5—10 mm. long; flowers yellow, 3 mm. long, in dense, oblong clusters 5—8 mm. long; pods rounded, ridged, black at maturity; seeds yellow, obovate. Frequent, chiefly in lawns. July.

16. *Anthyllis vulneraria* L. KIDNEY VETCH. Perennial; stems erect, 4—6 dm. high, slender but rather stiff; leaflets 5—13, narrowly obovate, 2—5 cm. long; flowers yellow, 10—15 mm. long, in heads at ends of long branches; calyx thickly covered with yellowish hairs. One specimen from an alfalfa field at Fargo by Clarence Waldron in 1909.

17. *Lotus americanus* (Nutt.) Bisch. PRAIRIE BIRD'S-FOOT TREFOIL. Annual; stem slender, erect, branched, 3—5 dm. (12—20 in.) high; leaflets 3, elliptic to lanceolate, entire, soft hairy, 1—2 cm. (¾ in.) long; flowers solitary from leaf base, yellowish white, 4 mm. (⅙ in.) long; pod slender, 2—3 cm. long; seeds oblong, greenish or mottled. Frequent, sometimes abundant, especially in sandy soils. July. The pod is borne on a slender stalk 1—3 cm. long with a single leaflet just below the flower. Sometimes placed in a separate genus, *Hosackia*, or called *Lotus purshianus* (Benth.) Clements.

Astragalus   MILKVETCH

Mostly perennials with pinnate leaves; flowers (fig. 230) showy, purple or yellowish, in short or long clusters; pods fleshy, leathery, woody or papery, few to many seeded. Mostly plants of no value, some poisonous to livestock. Rydberg divided the group into about 20 genera, differing in pods and other characters, but most authors regard it as only one genus.

Key to Species

Flowers purple (pink, red, violet, etc.).
  Plants tufted, leaves mostly basal.
    Plants 10—15 cm. (4—6 in.) high; flowers 2 cm. long; common.
                      35. *Astragalus missouriensis*
    Plants 5 cm. high, woody; flowers 8 mm. (⅓ in.) long; rare,
      western.   36. *Astragalus caespitosus*
  Plants with leafy stems 1—6 dm. (4—24 in.) long.
    Stems erect, stout, 3—6 dm. high; pods 2-grooved above.
                  26. *Astragalus bisulcatus*
    Stems spreading, slender or stout; pods not grooved.
      Pods thick, fleshy, 2 cm. long; stems stout, often in a thick
        bunch.   18. *Astragalus caryocarpus*
      Pods not fleshy, 5—20 mm. long; stems slender.
        Flowers 15 mm. (⅗ in.) long.
          Plants growing in patches 1—3 m. (3—10 ft.) wide.
              20. *Astragalus goniatus*
          Plants growing in dense clumps 2—6 dm. (8—24 in.) wide.
              21. *Astragalus striatus*
        Flowers 6—10 mm. (⅓ in.) long.
          Stems 3—10 dm. (1—3 ft.) long; common plants.
             27. *Astragalus flexuosus*
          Stems 1—3 dm. long; rare western plant.
             31. *Astragalus vexilliflexus*
Flowers yellowish or white, often with a dark spot.
  Stems erect, 3—10 dm. high.
    Stems slender; flowers 5—12 mm. (⅕—½ in.) long.
      Pods rounded, 5—7 mm. long; rare plant. 29. *Astragalus gracilis*
      Pods flattened, 8—20 mm. long.
        Pods moderately flattened, 15—20 mm. long.
          30. *Astragalus aboriginorum*
       Pods very flat, 8—12 mm. long.   28. *Astragalus tenellus*
    Stems stout; flowers 15—20 mm. long.
      Pods 3-angled, 2.5 cm. (1 in.) long   23. *Astragalus racemosus*
      Pods oblong, not angled.
        Plant covered with long hairs; pods 2—2.5 cm. long.
          25. *Astragalus drummondii*
        Plant with short or few hairs; pods 1—1.5 cm. long.

Stems in dense clumps; leaflets very narrow.
24. *Astragalus pectinatus*
Stems few; leaflets broad. 22. *Astragalus canadensis*
Stems very short, slender or creeping, hardly over 1 dm. high.
Stems slender from long rhizomes; leaflet only 1.
33. *Astragalus longifolius*
Stems tufted or creeping; leaflets 3—23.
Stems very short, plants tufted.
Leaflets 5, very narrow, spiny; flowers 5 mm. (⅕ in.) long.
32. *Astragalus viridis*
Leaflets 3 or more, not spiny; flowers 8—20 mm. long.
Leaflets 3, in dense, silvery tufts; flowers 2 cm. long.
34. *Astragalus triphyllus*
Leaflets 7—13; flowers 8—9 mm. long. 37. *Astragalus lotiflorus*
Stems 1—3 dm. long, creeping, coarse.
Pods rounded, 2-celled; hairs of plant very short.
19. *Astragalus plattensis*
Pods oblong, pointed, 1-celled; hairs long, silky.
38. *Astragalus purshii*

18. *Astragalus caryocarpus* Ker. GROUND PLUM. (Fig. 234). Stems 1—3 dm. (4—12 in.) long, few, or many in a dense clump 3—6 dm. wide, stout, spreading or ends upright; leaflets 13—27, oblong to linear, 8—20 mm. (⅓—⅘ in.) long; flowers 5—15 in dense clusters on stalks from upper leaf bases, pinkish to bluish purple, 1.5—2 cm. (¾ in.) long; pods rounded, smooth, partly red, very fleshy, 2 cm. long, 2-celled; seeds black. Common on prairie. May, June. A very showy plant.

19. *Astragalus plattensis* Nutt. Similar to last in habit but apparently not developing large bunches; flowers yellowish with a purple center; pods 1 cm. long, covered with short hairs. Apparently not common. Specimens from Emmons, Grant and McHenry Counties.

20. *Astragalus goniatus* Nutt. Stems slender, 1—2 dm. long, growing in large patches; leaflets 15—21, narrowly oblong, 5—15 mm. long; flowers purple, 1.5 cm. long, in rounded clusters 2—3 cm long; pods rounded, pointed, densely hairy, 8—10 mm. long. Very common on prairie. May, June. Some authors consider this the same as *A. hypoglottis* L. of Europe.

21. *Astragalus striatus* Nutt. Many stems forming dense clumps 2—6 dm. wide; leaflets 9—19, narrowly oblong, 1—2 cm. long; flowers bluish purple, 12—15 mm. long, in squarish or oblong heads, 2—5 cm. long; pod 10—20 mm. long, hairy. Common, especially on clay soils and slopes. June, July. Formerly called *A. nitidus* Dougl. or *A. adsurgens* Pall.

22. *Astragalus canadensis* L. LITTLE RATTLEPOD. Stems very stout, 8—15 dm. (32—60 in.) high; leaflets 15—25, oblong, 2—4 cm. (⅘—1⅗ in.) long; flowers pale yellow, 1.5 cm. long; seeds yellow. July, Aug. *A. carolinianus* L. is usually regarded as the same plant. The pod clusters remain stiffly upright all winter. Indian children used them for rattles.

23. *Astragalus racemosus* Pursh. Stems erect, coarse, 3—6 dm. high, usually in clumps; leaflets 15—31, oblong, 1.5—2.5 cm. long; flowers yellowish white, 15—20 mm. long, in stiff clusters 1—2 dm. long; pods 2—2.5 cm. long, 3-angled in cross section. Early June. The large clumps of flowering stems make this a showy plant. It is a poisonous species, growing on soils containing selenium. Quite local, Barnes (Valley City), Grant, Morton and Stark Counties (fig. 6), where deep clays are exposed.

24. *Astragalus pectinatus* Hook. Stems many, in erect or spreading clumps 2—6 dm. high; leaflets 15—21, very narrow, 2—5 cm. long; flowers yellowish white, 2 cm. long, in dense clusters 8—15 cm. long; pods oblong, 8—15 mm. long, very woody. June. Abundant in extreme western part of State; specimens as far east as Ward Co. (fig. 6). A very showy, late spring flower. Quite similar in general appearance to *racemosus* but usually lower, more spreading. It is a selenium absorber to a high degree.

234. Ground Plum (*Astragalus caryocarpus*).

25. *Astragalus drummondii* Dougl.   Entire plant covered with long, soft hairs, otherwise similar to two preceding; pods 2—2.5 mm. long, slender, cylindrical.   First collected near Buford, Williams County, in June, 1945 (Stevens 817).

26. *Astragalus bisulcatus* Hook.   TWO-GROOVED MILKVETCH.   Stems stout, often in large clumps, 3—10 dm. (1—3 ft.) high; leaflets 17—27, elliptic, 1—3 cm. (⅖—1⅕ in.) long; flowers red purple, 12—15 mm. (½ in.) long, in dense clusters 1—3 dm. long; pods hanging sharply downward, 2—3 cm. long, with 2 deep, longitudinal grooves on upper side.   June.   Common westward, local in central and eastern parts of State, chiefly where deep clays are exposed or near surface (fig. 6).   A handsome plant and another of the selenium absorbers.

27. *Astragalus flexuosus* Dougl.   SLENDER MILKVETCH.   Stems slender, trailing, 3—10 dm. long; leaflets 15—25, narrowly oblong, 5—15 mm. long; flowers pink or purplish, 8—10 mm. long, in slender, loose clusters 8—15 cm. long; pod cylindric, 1.5—2 cm. long.   Common on prairie and frequently showy with large bunches of flowering stems in June and July.   I call this species "Slender Milkvetch" because it is easily the most conspicuous one.   The name would be equally applicable to No. 28, which is abundant in some places, or to No. 29 which is rare.

28. *Astragalus tenellus* Pursh.   Stems slender, 2—5 dm. long, usually many in a dense, spreading clump; leaflets 15—21, narrow, 8—20 mm. long; flowers white, 8—9 mm. long, in slender clusters 5—8 cm. long; pods very flat, 10—15, mm. long.   Frequent on hillsides and old beaches, often abundant in limited areas.   June, July.

29. *Astragalus gracilis* Nutt.   Stems slender, wiry, erect, 3—5 dm. high; leaflets 11—21, narrow, 1—2 cm. long; flowers nearly white, 5 mm. long, in slender, loose clusters, 2—5 cm. long; pods oblong, 5—10 mm. long.   Collected only at Marmarth, Slope County.

30. *Astragalus aboriginorum* Richards.   Stems erect, slender, 2—4 dm. high; leaflets 9—15, linear, 1—2 cm. long; flowers cream colored, 9—11 mm. long, in loose clusters at ends of branches; pods flattened, 15—20 mm. long.   Sandy hillside, site of Heart River dam in Morton County (probably will be exterminated), Stevens 811 in 1945.   Determined by A. C. Cronquist as var. *glabriusculum* (Hook.) Rydb.

31. *Astragalus vexilliflexus* Sheld.   Stems slender, 1—3 dm. long, usually prostrate, in a gray, dense mat; leaflets 7—11, oblong, 5—10 mm. long; flowers purplish, 6—8 mm. long in groups of 2—5; pod oblong, flattened, hairy, 8—10 mm. long.   Collected only once on White Butte, Slope County (Stevens 141), in 1935.   Some authors regard it the same as *A. pauciflorus* Hook.

32. *Astragalus viridis* (Nutt.) Sheld.   PRICKLY MILKVETCH.   Stems 1—4 dm. long, low, spreading; leaflets 5, 8—15 mm. long, narrow, stiff, spiny; flowers yellowish, 5 mm. long, 2 or 3 together at leaf base; pods ovate, flattened, 1-seeded, 5—7 mm. long. One specimen from Sully's Springs, Billings County, in 1898 (L. R. Waldron 173).

33. *Astragalus longifolius* (Pursh) Rydb.   LONG-LEAVED MILKVETCH.   Stems slender, 1—3 dm. high, from a slender rhizome; leaflet usually only 1, very narrow, 5—10 cm. long; flowers yellowish, 8—9 mm. long, 1—5 in a cluster; pods inflated, oblong, 2—3 cm. long, very thin walled, mottled with purple.   Two records from very sandy soil, Slope and Billings Counties.   The names *A. pictus* and *A. ceramicus* have been used for this plant.   The papery pods are very striking and their spotted appearance has given rise to the name "Birdsegg Pea".

34. *Astragalus triphyllus* Pursh.   TUFTED MILKVETCH.   (Fig. 235).   Compact tufts, 5—20 cm. wide, 2—3 cm. high, from a stout root; leaves from short, thick stems; leaflets 3, elliptic, 0.5—2 cm. long, silvery hairy; flowers many among the leaves, 2—3 cm. long, yellowish white.   Common on hills and buttes west of Missouri River.   One record from Stutsman County.   An outstanding early spring flower.   The name *Orophaca caespitosa* has often been used for it.

35. *Astragalus missouriensis* Nutt. MISSOURI MILKVETCH. Tufted, leaves
and flower stalks 5—20 cm. (2—8 in.) long; leaflets 11—21, oblong or
elliptic, 5—15 mm. (⅕—⅗ in.) long; flowers dark bluish purple, 1.5—
2 cm. long in a rounded group at end of a slender stalk; pods fleshy,
oblong, pointed, 5—15 cm. long. Hills and prairies. One of the commonest species and quite showy because of the large, dark flowers.
May, June. The cylindrical calyx, 7—8 mm. long, will distinguish it
from *lotiflorus* which has a shorter (4—5 mm.), more spreading calyx.

36. *Astragalus caespitosus* (Nutt.) Gray. Tufted but with spreading stems
5—20 cm. long, leaves and slender flower stalks from a hard, thick
crown; leaflets 1, 3 or 5, linear-lanceolate, 1—3 cm. long; flower stalks
3—10 cm. long with 2—10 blue flowers 8 mm. long; pods slender, 1 cm.
long. Hills and buttes, extreme west. The flowers are small but striking because of the dark blue color.

235.   Tufted Milk-vetch (*Astragalus triphyllus*).

37. *Astragalus lotiflorus* Hook. Tufted or with stems 2—10 cm. long; leaflets 5—13, oblong, 5—15 mm. long, gray with short hairs; flowers yellow, 7—9 mm. long, early ones in a small cluster on a stalk 5—10 cm.
long, later ones close to ground; pods ovoid or curved, 1—2 cm. long.
Quite common on hills and prairies. May, June.

38. *Astragalus purshii* Dougl. Tufted or nearly so, stems stout, 5—20 cm.
long, leaflets 9—13, oblong, 1—2 cm. long, gray with long hairs; flowers
yellow with purple tip, 2—2.5 cm. long, close to stem or in small clusters on stalks 5—10 cm. long; pods ovoid, 2—3 cm. long, densely covered
with long hairs. This had been overlooked until 1944 when Dr. M. L.
Fernald identified a specimen. It seems to be not common and to occur
on sandy flats rather than hillsides The much larger flowers and long
hairs distinguish it from *A. lotiflorus*. It apparently blooms in late May.
The following sheets reported by Bergman as *lotiflorus* bear plants of
purshii also; Medora (Bergman June 19, 1910), Dickinson (Stevens
July 16, 1915), Schaller, Grant Co., (Bell 313). Recent collections are
from Dunn, Emmons and Slope Counties.

Oxytropis LOCO-WEED. POINT-VETCH

Perennials, mostly tufted; leaflets many; flowers usually in
spikes or head-like clusters; keel of flower with a sharp point; pods
oblong, pointed, leathery. These plants much resemble species of

Astragalus and are distinguished from them by the sharp tip of the flower keel (fig. 232). Some species are poisonous (loco-weeds), others apparently harmless.

Key to Species

Leafy stemmed; stems 3—6 dm. (1—2 ft.) long, weak.
43. *Oxytropis deflexa*
Tufted; stems short, thick.
  Leaflets in 4 or 5 rows, not in pairs.   42. *Oxytropis splendens*
  Leaflets in pairs.
    Leaflets silky with long hairs.
      Flowers pale, nearly pink.   40. *Oxytropis sericea*
      Flowers yellow.   41. *Oxytropis gracilis*
    Leaflets gray with short hairs; flowers usually purple.
39. *Oxytropis lambertii*

39. *Oxytropis lambertii* Pursh. PURPLE LOCO. (Fig. 236). Leaflets usually well separated, the whole leaf 2—3 dm. (8—12 in.) long; flower stalks a little taller than leaves; flowers red purple, occasionally yellow, 2 cm. (¾ in.) long, in short, compact or long, loose clusters. The most common species on prairie. June. Several forms have been described as distinct species but their status is still questionable and it seems best to leave them all under this name. This species is regarded as poisonous, but less so than some others and no trouble is known to occur in this State.

236   237

236. Purple Loco (*Oxytropis lambertii*).
237. Wild Licorice (*Glycyrrhiza lepidota*).

40. *Oxytropis sericea* Nutt. Very similar to preceding. It had not been recognized until 1941 when I noticed plants beginning to bloom May 20 with uniformly pale flowers. Abundance and distinctiveness is still uncertain.

41. *Oxytropis gracilis* (A. Nels.) K. Schum. SLENDER LOCO. Leaves erect, 5—15 cm. long; flower stalks 2—3 dm. high, much longer than leaves; flowers yellow, 15 mm. long, in dense clusters 2—5 cm. long. Rather local, usually on steep hillsides or buttes, western part of State. Determination of these three species is often difficult, but this one seems distinctive as a rule.

42. *Oxytropis splendens* Dougl. SHOWY LOCO. Leaves very gray hairy, 1—1.5 dm. long, spreading, about two rows of leaflets erect; flower stalks 1—2 dm. high; flowers blue, 2 cm. long. This is a striking species because of additional rows of leaflets and blue flowers contrasting with gray leaves. It blooms in July and occurs in all northern counties (fig. 21). Finley in Steele County is our farthest south.

43. *Oxytropis deflexa* (Pall.) DC. LEAFY LOCO. Stems slender, spreading, 3—6 dm. long; leaflets 25—35, ovate to lanceolate, 1—2 cm. long; calyx with black hairs, corolla yellowish 1—1.5 cm. long; flowers in clusters 5—10 cm. long at ends of branches, becoming 10—15 cm. in fruit; pods 2 cm. long, grooved on upper side, bent backward. July. Abundant along gravelly shores of Lake Metigoshe in Turtle Mts.; also recorded from Walhalla and Langdon. The plant is brownish hairy, especially the young leaves and stem tips. The crowded leaflets give an unusually flattened appearance to the leafy stems.

44. *Glycyrrhiza lepidota* (Nutt.) Pursh. WILD LICORICE. (Fig. 237). Stout perennial from rhizomes; stems 3—10 dm. (1—3 ft.) high; leaflets 11—19, lanceolate, 2—3 cm. (1 in.) long, pitted on lower surface; flowers yellowish white, 12—15 mm. (½ in.) long, in dense clusters 1 dm. long on stalks of equal length from leaf bases; pods oblong, 10—15 mm. long, covered with hooked prickles; seeds 3—5, rounded, dark green, 4 mm. long. A common, weedy plant on lake shores and in brushy places, less often on prairie. July, Aug. The large clusters of burs remain through winter and are characteristic. When without flowers or burs, the plant might be confused with *Astragalus canadensis*, leaflets of which are less pointed, not pitted below and stems are in clumps rather than patches. Licorice root used in medicine is from a related species of Asia.

Amorpha   FALSE INDIGO

Shrubs with pinnate leaves and finger-like clusters of small, purple flowers; only upper petal present, but flowers closely crowded, the orange colored anthers conspicuous; pods oblong, somewhat flattened, usually 1-seeded, not opening. Blooming in summer, flowers quite showy.

Key to Species

Tall shrub, 1—2 m. (3—7 ft.) high; leaflets 2—5 cm. (1—2 in.)
   long.   45.   *Amorpha fruticosa*
Low shrubs, 2—8 dm. (8—32 in.) high; leaflets 5—15 mm. (⅕—⅗
   in.) long.
   Leaves green; flower clusters usually single.   46.   *Amorpha nana*
   Leaves gray hairy; flower clusters several together.
                        47.   *Amorpha canescens*

45. *Amorpha fruticosa* L. FALSE INDIGO. Widely branched shrub 1—2 m. (3—7 ft.) high; leaflets 11—25, oblong, 2—5 cm. (1—2 in.) long, pitted below; flower clusters 5—20 cm. long; flowers dark purple; pods 7—10 mm. (⅓ in.) long, brown, persisting into winter. Pond or stream banks, west to Missouri River, not common but occasionally abundant. Late June, July. Planted in some windbreaks in recent years.

46. *Amorpha nana* Nutt. DWARF WILD INDIGO. Low shrub, usually 2—4 dm. high; leaflets 15—31, oblong, 5—10 mm. long, pitted below; flower clusters 3—6 cm. long; flowers rose purple; pod 5 mm. long, warty, sticky resinous when bruised. Widely distributed, usually on moist prairie or slopes, but sometimes on apparently dry banks. Early June.

47. *Amorpha canescens* Pursh. LEAD PLANT. Gray, spreading stems 3—8 dm. high, hardly appearing shrubby; leaflets 15—47, oblong, 10—20 mm. long, gray hairy, especially below; flower clusters 5—20 cm. long, usually several near ends of main stems; flowers dark purple; pods 4 mm. long. Sandy prairie, chiefly southeastern part of State. July.

Psoralea   SCURF PEA

Perennials; leaflets 3—5, usually pitted below; flowers purple, blue or whitish, in few or many clusters; pods oblong, 1-seeded, not splitting. Most of these plants have no well accepted common name. "Scurf pea" refers to the pitted or warty, rough appearance of stems, leaves and pods of some species.

Key to Species

Flowers 12—15 mm. (½ in.)long; leaves coarsely hairy.
                    48. *Psoralea esculenta*
Flowers 5—10 mm. long; leaves finely hairy or rough.
  Leaves silvery; flowers blue, in small, scattered groups.
                    49. *Psoralea argophylla*
  Leaves green, rough; flowers in rounded or loose clusters.
    Flowers bluish white; pods rounded, very rough.
                    50. *Psoralea lanceolata*
    Flowers purple; pods moderately flattened and rough.
                    51. *Psoralea tenuiflora*

48. *Psoralea esculenta* Pursh. TIPSIN. INDIAN BREADROOT. (Fig. 238). Bushy, hairy plant, 1—3 dm. (4—12 in.) high, one or more stems from a root with a tuber-like body 3—8 cm. (1—3 in.) long; leaflets 5, obovate to elliptic, 2—6 cm. long, smooth above, hairy below, not pitted; flowers bluish purple, 12—15 mm. (½ in.) long, in a dense cluster 5—10 cm. long, on a long stalk from leaf base; pod 5 mm. long, papery, with a slender tip 10—15 mm. long; seed brown, flattened, 5 mm. wide. Frequent to common on prairie. June. The starchy root was a main food for the Indians. The brown, leathery covering was peeled off and roots dried for storage.

49. *Psoralea argophylla* Pursh. SILVER LEAF. Stem erect, widely branched, 3—6 dm. high; leaflets 3, lower leaves often with 5, which are obovate to elliptic, 1.5—4 cm. long, densely covered with long, close-lying hairs; flowers blue, 7—8 mm. long, in small, widely separated groups along upper branches; seed and pod as in *esculenta* but beak of pod 5 mm. long. I had not seen the seeds until one of my students, Lyle Nelson, brought some from Ward County in 1947. He reported they seemed abundant on some plants, not on others. Very common on prairie. July. This is one of the most familiar plants of the prairie, yet it has lacked a good common name. "Silver Leaf", here proposed, would apply equally well to many other plants, but we feel it may be used for this plant which is so common and has not had a name.

50. *Psoralea lanceolata* Pursh. SCURF PEA. Bushy, 3—6 dm. high, with many fine branches; leaflets 3, narrowly oblanceolate or linear, black dotted, 1.5—3 cm. long; flowers bluish white, 5—6 mm. long, in numerous clusters 1—3 cm. long; pods plump, rounded, warty, 4 mm. long. River bars and sand dunes, Missouri River westward, also Logan and Kidder Counties. July. Spreads by long rhizomes.

51. *Psoralea tenuiflora* Pursh. Erect, much branched, 3—5 dm. high; leaflets 3, rarely 5, oblanceolate, 2—3 cm. long; flowers purple, 5 mm. long, in slender, loose clusters 2—6 cm. long; pod ovoid, somewhat flattened,

8 mm. long. There is an old record of this, "Bad Lands", G. B. Aiton, in 1892. I collected one plant on the railroad track near Bowman in 1918. It may have come there in gravel ballast. Many authors regard this plant as a variety of *P. floribunda* Nutt., which is common from Nebraska southward, where it often makes the prairie look like alfalfa fields.

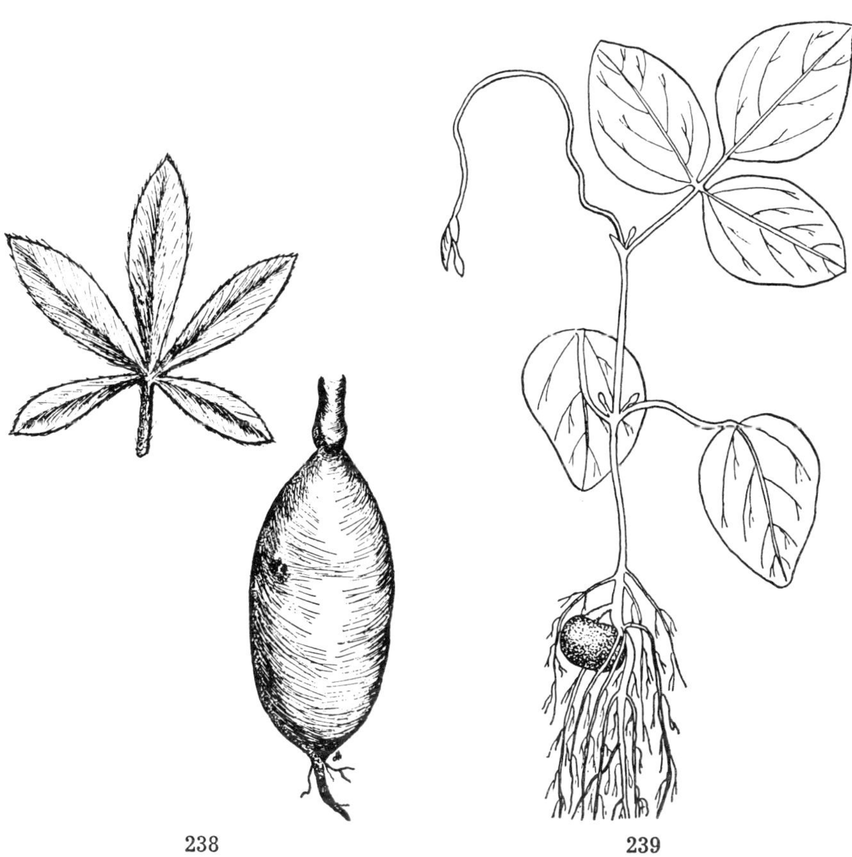

238   239

238. Tipsin (*Psoralea esculenta*).

239. Hog Peanut seedling (*Amphicarpa bracteata*).

52. *Dalea alopecuroides* Willd. Soft hairy annual; erect, branching above, 3—6 dm. high; leaflets 15—41, oblong, 3—8 mm. long; flowers pinkish white, 2.5—3 mm. long, in dense, cylindrical spikes 2—7 cm. long at ends of branches; pod thin, 1-seeded, covered by the soft hairy calyx. Local as if introduced: Fargo along railroad; Richland County in sandhills; Ransom, LaMoure and Stutsman Counties.

53. *Dalea enneandra* Nutt. Perennial, 6—10 dm. high, branching above, the branches long and slender; leaflets 5—11, linear or oblong, 5—15 mm. long, dark dotted; flowers 6 mm. long, in slender spikes 2—10 cm. long. We have this only from Morton and Grant Counties where it grows on clay hillsides. The hairy calyx is striking and below each is a scale-like, dotted bract, 4 mm. long.

Petalostemum PRAIRIE-CLOVER

Erect perennials with slender stems, pinnate, dotted leaves and small flowers in thimble-like spikes; keel petals not united; stamens 5, united in one group; pod thin, 1-seeded, covered by calyx.

Key to Species

Calyx not hairy, corolla white.   54. *Petalostemum candidum*
Calyx hairy, corolla pink or purplish.
  Leaflets usually 5, linear; corolla dark red.
                                      55. *Petalostemum purpureum*
  Leaflets 7—17, oblong; corolla pale rose.   56. *Petalostemum villosum*

54. *Petalostemum candidum* Michx. WHITE PRAIRIE-CLOVER. Stems 3—8 dm. (1—2½ ft.) high; leaflets 5—9, oblanceolate to linear, 1—2.5 cm. (⅖—1 in.) long; flower spikes 2—10 cm. long, on long, bare, stem tips. Common on prairies. July, Aug. Bergman referred most of our specimens to *P. oligophyllum* Torr., which had been regarded by many authors as a variety of *P. candidum* and recently listed as *P. occidentale* (Gray) Fern. I have been unable to distinguish the two forms satisfactorily.

55. *Petalostemum purpureum* Vent. PURPLE PRAIRIE-CLOVER. Stems 2—6 dm. high; leaflets 3—5, linear, 1—2 cm. long; flower spikes 1—5 cm. long. Common on prairie, especially hills. July, Aug. This plant has a pronounced, aromatic odor when bruised, which white prairie-clover does not have.

56. *Petalostemum villosum* Nutt. HAIRY PRAIRIE-CLOVER. Stems erect or spreading, 3—6 dm. high, entire plant gray with fine, soft hairs; leaflets 7—17, oblong, crowded, 6—12 mm. long; flower spikes 2.5—8 cm. long, corolla pale rose color. Found only on sand dunes, Richland, Ransom, Emmons, Grant, Pierce and McHenry Counties (fig. 17).

57. *Hedysarum boreale* Nutt. SWEET VETCH. Perennial; stems sometimes clustered, erect or spreading, 3—6 dm. (1—2 ft.) high; leaflets 9—15, oblong, 1—2 cm. (⅖—⅘ in.) long; flowers pink, 1.5 cm. long, 10—20 in stalked clusters from upper leaf bases; pods 2—3 cm. long, with 3—4, nearly circular segments. This was first described from our area but had been overlooked until 1918. It proves to be quite frequent from Missouri River westward (also Wells County, Stevens 696), growing on hillsides, especially on the rounded crest. June. The general appearance of the plant is that of a moderately large *Astragalus* but flowers are distinctly more of a pink color.

Desmodium TICK-TREFOIL

Our species perennial herbs; leaflets 3; flowers in large, dense or loose, branching, terminal clusters; pod flat, several seeded, constricted between seeds and breaking apart when ripe, each segment becoming a "stick-tight" by many, short, hooked hairs (fig. 228). The name *Meibomia* has also been used for the genus.

Key to Species

Leaves close together, most of stem bare, slender.
                                    58. *Desmodium acuminatum*
Leaves scattered along the stout stem.   59. *Desmodium canadense*

58. *Desmodium acuminatum* (Michx.) DC. Stem 1—3 dm. (4—12 in.) high to the leaves, flower stalk 2—4 dm. (8—16 in.) long with a few short branches at top; leaflets thin, rounded ovate, acuminate, 5—13 cm. (2—5 in.) long and on long stalks; flowers pink, 6—7 mm. (¼ in.) long; pods 2—3-seeded, the joints concave above, rounded below. This is a rare plant in North Dakota, extending from the eastern wooded area; recorded from Cass and Ransom Counties. July. Formerly known as *D. grandiflorum* Walt.

59. *Desmodium canadense* (L.) DC. Stem 5—10 dm. high, leafy, branched only at top; leaflets oblong or lanceolate, 3—8 cm. long; flowers pink, 10 mm. long, in a dense, branched, terminal cluster 1—3 dm. long; pod 3—5-seeded, 2—3 cm. long, segments oblong (fig. 228). Occasional in grassland at edge of brush or woods, west to Missouri River. July.

60. *Apios americana* Med. GROUND-NUT. Twining vine from a large, tuberous root; leaflets 5—9, ovate to lanceolate, 3—10 cm. long; flowers brownish purple, 1 cm. long, in a thick cluster 3—8 cm. long from leaf base; keel slender, twisted; pod slender, 6—10 cm. long. I collected a single plant of this without flowers in a thicket in sandhills near Sheldon, Ransom County, in 1926. Previously known as *A. tuberosa* Moench.

61. *Amphicarpa bracteata* (L.) Fern. HOG PEANUT. (Fig. 239). Slender, twining vine, climbing to 1 m. (3 ft.) or more, often forming a dense cover on ground or over weeds and bushes; leaflets 3, ovate, 3—8 cm. (2—3 in.) long, very thin; flowers purplish white, 10—15 mm. (½ in.) long, about 5—10 in a slender cluster from leaf base; pods 2—4 cm. long, flattened, hairy; seeds short oblong, flattened, greenish brown, 5 mm. long. Late Aug. Common in woods west to Missouri River. Besides the usual flowers, the plant produces long, thread-like branches near the ground, from which flower buds enter the ground and produce 1-seeded pods with a seed somewhat larger than that from ordinary flowers and often irregular in shape. The plant has been variously listed as *A. comosa* and *A. monoica*. Metcalf (Journ. Wash. Acad. Sci. 10:197, 1920) reported *A. pitcheri* (*comosa*) from Cannon Ball. Two of our specimens from Ransom County (Bell 434, Bergman 1078) seem to be that form, distinguished by long brown hairs on stems. Fernald has lately considered that as only a variety of *bracteata*.

62. *Strophostyles leiosperma* (T. & G.) Piper. WILD BEAN. Annual vine; stems usually trailing; leaflets 3, lanceolate to oblong or ovate, 1.5—5 cm. long; flowers pale purple, 5—8 mm. long, in small, rounded clusters on long stalks from leaf bases; pods cylindrical, 2—4 cm. long; seeds oblong, 4 mm. long. Dry, open ground. Aug. Records from Sargent, Grant and Adams Counties. Formerly called *S. pauciflora* Benth.

Vicia and **Lathyrus**   VETCH and VETCHLING

Vines, climbing by tendrils, trailing, or short and somewhat erect; leaves pinnately compound; flowers medium to large, showy; pods several seeded; seeds rounded. The following key includes species of both genera since they are not separated readily. Native perennials except No. 63.

Key to Species

Flowers purple.
  Leaflets 4—8, linear to lanceolate, 2.5—6 cm. (1—2½ in.) long.
                      66. *Lathyrus palustris*
  Leaflets 8—14, smaller or wider.
    Leaflets 2—8 mm. wide; slender plants.
      Flowers 1—2 at leaf bases; annual weed.   63. *Vicia angustifolia*
      Flowers 3—15 in a long stalked cluster.
        Flowers 3—9, reddish purple.   64. *Vicia americana*
        Flowers 1—4, bluish purple.   65. *Vicia sparsifolia*
    Leaflets 10—15 mm. wide, thick; coarse plant. 67. *Lathyrus venosus*
Flowers yellowish white, woodland plant.   68. *Lathyrus ochroleucus*

63. *Vicia angustifolia* (L.) Reichard. NARROW-LEAVED VETCH. (Fig. 229). Introduced annual; stems 3—6 dm. (1—2 ft.) long, climbing or trailing; leaflets 6—10, oblong to linear, 1—2.5 cm. (½—1 in.) long; flowers purple, 1—1.5 cm. long, usually 2 at leaf base; pods 4—5 cm. (2 in.) long; seeds spherical to slightly oblong, black or gray with black spots, 3 mm. (⅛ in.) wide. Occasional to frequent in grain fields. July.

64. *Vicia americana* Muhl. WILD VETCH. Stems slender, usually in tangled masses, spreading or climbing to 8 or 10 dm.; leaflets 8—12, oblong, 1—3 cm. long; flowers reddish purple, 1.5—2 cm. long, 3—9 close together in a one-sided cluster on a stalk 5—10 cm. long from leaf base; pods somewhat flattened, 2.5—3 cm. long; seeds short oblong, slightly flattened, greenish brown, 3—3.5 mm. long. Common in eastern part of State, especially in open or brushy areas. June-Aug.

65. *Vicia sparsifolia* Nutt. PRAIRIE VETCH. Stems slender, usually little branched, spreading to nearly erect, 1—4 dm. high; leaflets 8—12, linear; flowers bluish purple or lavender, 1.5—2 cm. long, 1—4 in a spreading cluster; pods flattened, 2—3 cm. long; seeds nearly spherical, greenish, 3 mm. long. This is the common prairie vetch and that name is here proposed for it. In the Red River Valley it occurs on railroad grades, no doubt brought with gravel. Bergman decided not to recognize this form but it seems to me quite distinct. Kearney and Peebles (Fl. Pl. and Ferns of Ariz.) recently have followed a common usage of reducing it to *V. americana*, var. *linearis* (Nutt.) Wats. They state that the typical form, this variety and var. *truncata* (Brewer.) Wats. with leaflets truncate or emarginate at tip, intergrade freely. Most of our specimens have leaflets slightly oblanceolate, rounded or truncate at tip with a sharp point. On one specimen from Washburn, McLean County, many of the broader leaflets are truncate and 3-toothed. A number of specimens have leaflets acute or somewhat rounded at tip.

66. *Lathyrus palustris* L. MARSH VETCHLING. Stems slender, 3—10 dm. long; leaflets 4—8, linear to lanceolate, 2.5—6 cm. long; flowers purple, 1.2—1.5 cm. long, in a loose, stalked cluster of 2—6 from leaf base; pods cylindrical, 4—5 cm. long; seeds nearly spherical, brown, 3—4 mm. long. In low grassland or brush. We have specimens from only Cass, Barnes, Pembina and Rolette Counties, but it probably is more common than these would indicate. July, Aug. The long leaflets will usually distinguish this from *Vicia americana*, which often grows with it.

67. *Lathyrus venosus* Muhl. BUSHY VETCH. Very coarse; stems often little branched and nearly erect, 3—10 dm. (1—3 ft.) high, larger plants clinging to other plants; leaflets 8—14, broadly oblong or elliptic, thick, 1—3 cm. (2/5—1⅕ in.) long; tendrils poorly developed; flowers purple, 1.2—1.5 cm. long, 10—15 in a dense, stalked cluster from leaf base; pods cylindrical, 4—5 cm. long; seeds nearly spherical, greenish brown, 3—5 mm. long. July. A very showy species, especially in brushy areas (fig. 13). In grass away from brush, the stems often are 2—4 dm. high and do not bloom.

68. *Lathyrus ochroleucus* Hook. YELLOW VETCHLING. Stems slender, weak but leafy, trailing or climbing, 4—10 dm. long; leaflets 6—8, ovate, broadly oblong or elliptic, thin, pale green, 1.5—2.5 cm. long; flowers yellowish white, 1.5 cm. long, 5—10 in a loose, stalked cluster from leaf base; pod cylindrical, 4 cm. long. June. Local in woods, west to Barnes County; common in Turtle Mts. (fig. 13).

GERANIUM FAMILY Geraniaceae

Herbs with opposite, usually rounded, palmately lobed leaves; flowers usually small, pink or purplish; sepals 5; petals 5; stamens usually 10; pistil elongated, consisting of 5 parts, each 1-seeded, splitting apart from base upward at maturity (fig. 114). The commonly cultivated geraniums belong to the genus *Pelargonium*.

Key to Species

Leaves palmately lobed; pistil tips not twisting.
  Petals 15—20 mm. (¾ in.) long; perennial plant in woods.
                            1. *Geranium maculatum*
  Petals 2—7 mm. long; annual or biennial.
    Sepals with bristle tips 1—2 mm. long; seeds netted.

Flowers in dense clusters of 5—10.   2. *Geranium carolinianum*
   Flowers usually in pairs.   3. *Geranium bicknellii*
Sepals without bristle tips; seeds smooth.   4. *Geranium pusillum*
Leaves pinnately divided; pistil tips twisting when ripe.
                                          5. *Erodium cicutarium*

1. *Geranium maculatum* L. WILD GERÁNIUM. CRANESBILL. Perennial;
   stems 2—5 dm. (8—20 in.) high; leaves rounded in general outline,
   5—8 cm. (2—3 in.) wide, deeply cut into 5 lobes which are 1—2 cm.
   wide and more or less toothed or lobed along edges; flowers 1—several
   on upper branch tips, light purple, 2—3 cm. wide. June, July. This still
   grows in aspen woods on the Minnesota side of the river at Fargo and
   is common a little farther east. We have no definite North Dakota
   record, though it probably occurred formerly in a few places on this
   side of the river.

2. *Geranium carolinianum* L. CAROLINA CRANESBILL. (Fig. 240). Annual;
   stem erect, or widely branched and spreading, 1—3 dm. high; leaves
   rounded, 2—5 cm. wide, divided into narrow lobes; flowers pale purplish or pink, 5 mm. wide, in dense, rather flat clusters 2—5 cm. wide;
   seeds oblong, 2 mm. long, netted with fine gray lines. Occasional or locally common in open ground or wooded areas. Late June. We have it
   from 6 scattered counties.

3. *Geranium bicknellii* Britton. BICKNELL'S CRANESBILL. Very similar to
   last but more slender; flowers usually in pairs, style tips over 4 mm.
   long (3 mm. or less in *carolinianum*). This seems confined to wooded
   areas. We have it from Turtle Mts., Killdeer Mts. and McKenzie
   County.

4. *Geranium pusillum* Burm. SMALL CRANESBILL. Annual; stems weak,
   prostrate or nearly, 1—3 dm. long; leaves rounded, 1—5 cm. wide,
   deeply 5—7-lobed, the lobes toothed or lobed at tip; flowers purplish,
   5 mm. wide, usually in pairs, fruiting branches becoming 1 dm. or more
   long. Specimens from Cass, Grand Forks, Pembina and LaMoure
   Counties. Introduced in lawn grass seed, not established.

5. *Erodium cicutarium* (L.) L'Her. STORKSBILL. ALFILERIA. Introduced
   annual; stems weak, spreading, 1—3 dm. long; leaves 3—7 cm. long,
   ovate or oblong in outline, pinnately divided into 5—10 oblong segments
   which are again divided into fine segments; flowers purple, 8—10 mm.
   wide, several on a stalk 5—10 cm. long from leaf base; pistil tips 3—4
   cm. long. Has been found at Fargo and Valley City.

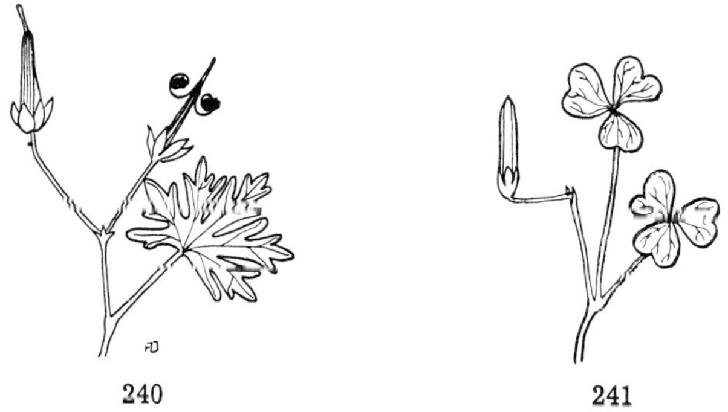

240   241

240. Cranesbill (*Geranium carolinianum*).

241. Wood Sorrel (*Oxalis stricta*).

WOOD SORREL FAMILY Oxalidaceae

Low herbs with alternate, palmately compound leaves; leaflets
3, heart-shaped, broadest at tip, narrowed to base (fig. 113); flowers
several in a cluster on a stalk from leaf base; sepals 5; petals 5,
wedge shaped, slightly united at base; stamens 10; pistil 1, but styles
5 and ovary 5-chambered; fruit a many seeded pod, splitting lengthwise, the seeds thrown out violently. The plants have a sharp taste
due to oxalic acid.

Key to Species

Plants from brown, scaly bulbs; flowers violet to white.
1. Oxalis violacea

Leafy stemmed plants; flowers yellow.
  Stems and pod densely hairy, stems green.   2. *Oxalis stricta*
  Stems and pods loosely hairy, stems red below.   3. *Oxalis europaea*

1. *Oxalis violacea* L. PINK WOOD SORREL. Perennial from a deep, brown,
    scaly bulb; leaf and flower stalks 1—2 dm. (4—8 in.) high; leaflets
    wider than long, 1—2.5 cm. (⅖—1 in.) wide, smooth, thick; flower
    violet to white, 1 cm. wide, funnel shaped; pods oblong, smooth, 4—5
    mm. (⅕ in.) long. Common on prairies or in fields, Richland to La-
    Moure and Grand Forks Counties. May, June.

2. *Oxalis stricta* L. UPRIGHT YELLOW WOOD SORREL. (Fig. 241). Annual:
    stem erect or spreading, 5—20 cm. high; stems and leaves covered with
    short, dense hairs; leaflets 10—20 mm. wide; flowers yellow, 10—15 mm.
    wide, 2—10 on a stalk from leaf base, individual flower stalks bent
    downward; pods columnar, 5-angled, 15—20 mm. long; seeds red, ovate.
    flat, cross-ridged. Common, woods and fields. June-Sept. In woods
    the plants may be slender, 2—4 dm. high. In fields they are often
    tufted, spreading, mat-like. A similar plant with a creeping stem,
    growing as a weed in greenhouses, is O. *repens* Thunb.

3. *Oxalis europaea* Jordan. LADY'S SORREL. Perennial by short rhizomes:
    stems erect, 1—4 dm. high, lower stems slightly fleshy, smooth and
    red; leaflets 10—15 mm. wide, only slightly 'hairy; flowers yellow, 10—
    15 mm. wide, usually 2 or 3 together, individual flower stalks not bent
    downward; pods columnar to conical, 10—15 mm. long, with few if
    any hairs. Common in open ground or in partial shade. June-Sept.
    We have more specimens of this than of the preceding form. It has
    been called O. *cymosa* Small, which many recent authors regard as only
    a variety of O. *europaea.*

FLAX FAMILY Linaceae

Annual or perennial herbs with slender stems and narrow, entire, alternate leaves; flowers blue or yellow (some cultivated forms
white or red), terminating the upper branches; sepals 5; petals 5,
wedge shaped, slightly united at base; stamens 10; pistil with 1
ovary and 5 slender styles; fruit a rounded, 10-seeded pod, splitting
from the top into 5 or 10 parts. Flowers open in morning and
corollas usually fall off by noon in bright weather.

Key to Species

Flowers blue or white.
  Cultivated annual; leaves elliptic or lanceolate, 3—8 mm. (⅛—⅓
    in.) wide.   1. *Linum usitatissimum*
  Native perennial; leaves linear, 1—3 mm. wide.   2. *Linum lewisii*
Flowers yellow; native annuals.
  Flowers 8—10 mm. wide; sepals persistent.   3. *Linum sulcatum*
  Flowers 15—25 mm. wide; sepals falling off as fruit matures.
4. Linum rigidum

1. *Linum usitatissimum* L. Common Flax. Annual; stems erect, much branched, 3—10 dm. (1—3 ft.) high; leaves smooth, 2—4 cm. (⅘—1⅗ in.) long; flowers blue or white, 2—3 cm. wide; pods rounded, short pointed, 6—8 mm. (¼ in.) wide. July-Sept. Volunteer flax plants are common along roadsides and especially about grain elevators. In recent years, white flowered and yellow seeded varieties have been grown to a considerable extent.

2. *Linum lewisii* Pursh. Lewis' Wild Flax. Perennial; stems 3—8 dm. high, very slender, branches often drooping; leaves 1—3 cm. long, 1—2 mm. wide; flowers blue, 2—3 cm. wide, hanging from spreading branches, forming 1-sided clusters 1—2 dm. long; pods rounded, not pointed, 5 mm. wide; seeds very thin, blackish, 3—3.5 mm. long. Common on prairie. July, Aug. This plant was named for Capt. Merriweather Lewis of the Lewis and Clark Expedition which first discovered it in Montana. It is similar to *L. perenne* and other European species which are frequently grown as ornamentals.

3. *Linum sulcatum* Riddell. Grooved Flax. Annual; stems slender, 2—4 dm. high, branches sometimes drooping; leaves linear or lanceolate, 1—2 cm. long; flowers yellow, 8—10 mm. wide; pods 3—4 mm. wide; seeds light brown, 2 mm. long. Lower prairie, eastern part of State to Oliver and Bottineau Counties. July.

4. *Linum rigidum* Pursh. Stiffstem Flax. Annual, stems rather stiff, sometimes compactly branched, 1—4 dm. high; leaves linear or lanceolate, 1—3 cm. long; flowers golden yellow, 15—25 mm. wide; pods 4—5 mm. wide; seeds light brown, 3 mm. long. Higher, especially sandy, prairie, common. July. The flowers are quite showy in early forenoon. *L. compactum* A. Nels. was described as apparently perennial from rhizomes. Our low, densely branched plants seem only dwarf specimens of *L. rigidum.* Rydberg (Fl. Pl. and Pr.) describes *L. rigidum* as perennial, but that is surely not true of our plants.

RUE FAMILY   **Rutaceae**

Aromatic trees or shrubs; leaves usually compound; sepals and petals 3—5, often 4; stamens 3—5; pistils 1—5, separate or united; fruit various. The citrus fruits belong to this family. We have only one species.

1. *Zanthoxylum americanum* Mill. Prickly Ash. (Fig. 55). Stout, widely branched, prickly shrub 1—3 m. (3—10 ft.) high; leaves alternate, pinnately compound; leaflets 3—11, oblong or ovate, 1.5—5 cm. (½—2 in.) long; flowers yellow, 4—5 mm. (⅕ in.) wide, in dense clusters 1 cm. wide at old leaf base; petals 4—5, sepals none; fruit a rather fleshy, 1-seeded pod, 4—6 mm. long, splitting lengthwise; seeds smooth, black. Blooms early May, fruit ripe July. Woods; Richland, Cass and Grand Forks Counties. The leaves and fruits are aromatic. The flowers are dioecious, one plant bearing only pistillate, another only staminate flowers.

CALTROP FAMILY   **Zygophyllaceae**

Herbs, shrubs or trees with compound leaves; flowers small, solitary at leaf base; sepals and petals 5; stamens 10; pistil 1, ovary 2—5-celled and angled.

1. *Tribulus terrestris* L. Puncture Vine. Annual, stems 2—8 dm. (8—32 in.) long, branched, creeping, forming a mat; leaves opposite, pinnately compound; leaflets 10—14, oblong, hairy, 3—10 mm. (⅛—⅖ in.) long; flowers yellow, 8—10 mm. wide, a single one from a leaf base; fruit splitting into 3—5, usually 1 seeded segments, each with 2 stout spines 5—10 mm. long and some smaller ones (fig. 112). A specimen was received from Mrs. N. V. Jaccard of Havana, Sargent County, in 1934. We are not certain that it grew there, but the weed has been working

northward in recent years. A second record is from Hettinger, Adams County, in 1947. One plant was found at the edge of a grain field where the previous crop had been harvested by a machine from Oklahoma or Texas. The spines of the fruits stick into auto tires but do not actually puncture them. Caltrop was a pronged iron used by the Romans to stop cavalry. The fruit of this plant resembles it in shape.

MILKWORT FAMILY Polygalaceae

Our species are small plants with narrow leaves and small, white or greenish flowers in dense clusters at stem tips; flowers somewhat irregular in shape; petals 5 or 3, the lower one hollowed, pointed or crested, the others sometimes partly united; stamens 8, filaments usually united at base; pistil 1, ovary usually 2-celled; fruit a rounded pod, splitting lengthwise; seeds obovoid with 2 small, whitish appendages at smaller end.

Key to Species

Annual; leaves in whorls of 4 or 5; flowers not showy.
1. *Polygala verticillata*
Perennial; leaves alternate; flowers showy.
  Leaves linear; plants of hill tops.
2. *Polygala alba*
  Leaves lanceolate; plants of low prairie.
3. *Polygala senega*

1. *Polygala verticillata* L. WHORLED MILKWORT. Slender annual; stems widely branched, 1—3 dm. (4—12 in.) high; leaves 1—3 cm. (⅖—1⅕ in.) long, 1—3 mm. wide (fig. 87); flowers greenish, 3—4 mm. (⅛ in.) wide, in clusters 0.5—1 cm. long. Prairie, chiefly western. July-Sept. We have two records for Ransom County and have it from Clay County, Minnesota. It seems not common but is inconspicuous and easily overlooked.

2. *Polygala alba* Nutt. WHITE MILKWORT. Perennial with many, slender, erect stems 1—2 dm. high from a thick root; leaves alternate, few on upper part of stems, 1—2.5 cm. long, 1 mm. wide; flowers white, 4—5 mm. wide, in dense, finger-like clusters 2—10 cm. long. Common on dry hills. June, July. This is quite showy. Chiefly western half of State; Sargent County most eastern record.

3. *Polygala senega* L. SENEGA SNAKEROOT. Perennial; stems usually several, spreading at base, 1—2 dm. long; leaves alternate, lanceolate, 1—3 cm. long; flowers greenish white, 4—5 mm. wide, in oblong clusters 1—4 cm. long. Low, moist prairie, Ransom and Morton Counties and Cavalier to McHenry Counties. June. The root is used medicinally and has been in considerable demand.

SPURGE FAMILY Euphorbiaceae

Our species are herbs with opposite or alternate, simple leaves and milky sap; flowers small but clustered and often surrounded by colored leaves (bracts); stamen 1, several attached to a small cup, which bears several, yellowish glands and sometimes, white, petal-like appendages; pistil 1 in each cup, the 3-celled ovary projecting; pod 3-seeded (fig. 82), splitting lengthwise and throwing the rounded or angular seeds.

This is a large family of chiefly tropical plants. Poinsettia is a shrubby species (good to study flower structure). Castor-oil plant is grown here to some extent as an ornamental. Some tropical kinds have stinging hairs. Ours all belong to the genus *Euphorbia* which some authors separate into several genera.

Key To Species

Leaves alternate; flowers in terminal clusters with prominent bracts.
  Bracts of flower clusters white or white edged.   1. *Euphorbia marginata*
  Bracts of flower clusters green or yellowish.
    Plants annual; leaves oblong, ovate or spatulate.
      Leaves spatulate, toothed; locally common westward.
                             2. *Euphorbia dictyosperma*
      Leaves oblong to ovate, not toothed; rare plant.
                             3. *Euphorbia peplus*
    Plants perennial; leaves linear to lanceolate.
      Stems 4—8 dm. (16—32 in.) high; leaves 3—10 mm.
        (⅛—⅖ in.) wide.   4. *Euphorbia esula*
      Stems 1—3 dm. high; leaves 1—3 mm. wide, crowded.
                           5. *Euphorbia cyparissias*
Leaves opposite; flowers few in leaf axils; annual plants.
  Leaves without teeth.
    Stems mostly creeping on the ground; leaves rounded or oblong.
      Leaves rounded, 3—5 mm. (⅙ in.) long; flowers greenish.
                         10. *Euphorbia serpens*
      Leaves oblong or ovate, 4—12 mm. long; flowers white.
                         6. *Euphorbia geyeri*
    Stems erect; leaves linear or lanceolate, 1—2.5 cm. long.
                         7. *Euphorbia petaloidea*
  Leaves with small teeth, at least on apical part.
    Stems erect or mostly so.
      Leaves broadly, obliquely ovate.   8. *Euphorbia maculata*
      Leaves narrowly oblong or spatulate.   11. *Euphorbia serpyllifolia*
    Stems creeping, forming a mat.
      Stems densely covered with fine hairs.   9. *Euphorbia supina*
      Stems not hairy.
        Leaves oblong, ovate or obovate, with a purple spot.
                     11. *Euphorbia serpyllifolia*

       Leaves oblong to linear, not spotted. 12. *Euphorbia glyptosperma*

1. *Euphorbia marginata* Pursh. SNOW-ON-THE-MOUNTAIN. Stout annual;
stem erect, 3—10 dm. (1—3 ft.) high, widely forking above, soft hairy;
leaves oblong or ovate, 5—9 cm. (2—4 in.) long, small upper ones in
flower clusters broadly white edged or all white; pods 6 mm. (¼ in.)
wide, soft hairy; seeds 3—4 mm. long, rounded, gray, rough. Records
from Morton and Dunn Counties, perhaps all of plants escaped from
cultivation. It is often grown as an ornamental and is a common native weed from Nebraska southward. Aug., Sept.

2. *Euphorbia dictyosperma* Fisch. & Mey. Annual; stem smooth, erect,
3—8 dm. high, branched at top; leaves oblong or spatulate, 1—3 cm.
long; flowers greenish; pods 3 mm. wide, rough; seeds rounded, 1.5 mm.
long, brown and netted. Prairie, frequent in western part of State.
July. Collected near Fargo in 1914 but not observed recently. The
name *E. arkansana* Engelm. & Gray has been applied to this plant, but
conservative authors have not separated it from the European species.

3. *Euphorbia peplus* L. PETTY SPURGE. Annual; stem erect, 1—3 dm. high;
leaves ovate to obovate, 1—4 cm. long, not toothed; flowers greenish;
pods 1—3 mm. wide; seeds oblong, white, coarsely pitted, 1.5 mm. long.
A specimen was collected in a flower bed at Fargo in 1911. It is reported as widely distributed in America, introduced from Europe.

4. *Euphorbia esula* L. LEAFY SPURGE. Deep rooted perennial; stems
erect, 4—8 dm. (16—32 in.) high, smooth, hard; leaves linear to oblong
or lanceolate, 2—6 cm. (1—3 in.) long; flowering branches often forming an umbel at tip of stem, later branching from farther down; bracts
just below flowers rounded, 6—10 mm. (⅙ in.) long; flowers yellowish
green; pods 3—4 mm. long, a little longer than wide; seeds oblong, gray,
2—2.5 mm. long. Common in fields and along roadsides, a very persistent weed. Blooms mostly in June.

242   243

242. Thyme-leaved Spurge (*Euphorbia serpyllifolia*).
243. Mallow (*Malva rotundifolia*)

5. *Euphorbia cyparissias* L. CYPRESS SPURGE. GRAVEYARD MOSS. Perennial; stems often densely clustered, erect or spreading, 1—3 dm. high; leaves numerous, linear, 1.5—3 cm. long; flowers similar to preceding. Frequently planted as an ornamental, not known to have spread and become established as a wild plant. One specimen from Kidder County in 1938.

6. *Euphorbia geyeri* Engelm. GEYER'S SPURGE. Annual; stems spreading on ground, 1—3 dm. long; leaves oblong to ovate, 4—12 mm. long; flowers solitary at leaf bases, bearing small, white, petal-like bodies; pods 2—3 mm. long; seeds rounded, gray, 1.5 mm. long. Sandy soil, Richland, Cass, Billings and McHenry Counties.

7. *Euphorbia petaloidea* Engelm. WHITE-FLOWERED SPURGE. Annual; stem slender, erect, widely forking, 1—6 dm. high; leaves linear to oblong, 1—2.5 cm. long; flowers solitary at leaf bases with 4—5 white, petal-like bodies 2—3 mm. long; pods 2 mm. long; seeds rounded, gray, 2 mm. long. Sandy soil. We have it from Morton, Slope and Billings Counties only. This plant has more showy flowers than other species of this sort. Wheeler (Rhodora 43:134, 1941) uses the name *E. missurica* Raf.

8. *Euphorbia maculata* L. NODDING SPURGE. Annual; stems erect, branches curving or nodding, 2—6 dm. high; leaves oblong or ovate, often curved, base oblique, reddish blotched in the middle, finely toothed; flowers in leaf axils but mostly crowded toward ends of branches, bearing small, white, petal-like bodies; pods 1.5 mm. long, seeds ovoid-oblong, 1.25 mm. long, somewhat 4-angled, brownish. One plant was found in 1929 along a railroad near Fargo. It is a common plant in central U. S. This was formerly called *E. nutans* Lag. or *E. preslii* Guss. and the name *E. maculata* applied to the next species. Both could equally well be called "spotted".

9. *Euphorbia supina* Raf. SPOTTED SPURGE. Annual; stems very hairy, often red, spreading in a dense mat 1—5 dm. (4—20 in.) wide; leaves oblong, 4—15 mm. (⅙—⅗ in.) long, usually with a reddish blotch in the middle; pods 1—2 mm. long; seeds 4-angled, gray, 0.75 mm. long. Records from Cass, Morton and McKenzie Counties. At Fargo it has grown for many years in the railroad yards. It is a very common plant farther south. Formerly called *E. maculata* L. by incorrect interpretation of the original description.

10. *Euphorbia serpens* HBK. ROUND-LEAVED SPURGE. Annual; stems forming a dense, green mat 1—4 dm. wide; leaves rounded, 2—5 mm. long; flowers very small, greenish; pods 1 mm. long; seeds gray, 4-angled. We have this from Barnes, Sioux, Morton and Billings Counties. It is another species which is common in dry soil farther south.

11. *Euphorbia serpyllifolia* Pers. THYME-LEAVED SPURGE. (Fig. 242). Annual; stems usually quite red, forming a mat 1—6 dm. wide, or small plants nearly upright; leaves oblong, 3—12 mm. long, often with a central reddish streak; flowers very small, greenish or whitish; pods 2 mm. long, seeds gray, 4-angled, 1 mm. long. Very common in fields, prairies, roadsides. June-Sept.

12. *Euphorbia glyptosperma* Engelm. RIDGE-SEEDED SPURGE. Annual; stems mat-like or somewhat erect, 1—3 dm. long; leaves oblong or linear-oblong, 2—10 mm. long; flowers very small, pinkish or whitish; pods 1.5 mm. long; seeds gray, hardly 1 mm. long, somewhat 4-angled. Common on prairies and in dry soils, chiefly westward. July-Sept.

WATER STARWORT FAMILY   Callitrichaceae

Small, aquatic plants with opposite leaves and very small, greenish flowers in leaf axils; sepals and petals 0; stamen 1; pistil 1, ovary 4-celled, splitting at maturity into 4, 1-seeded parts.

Key to Species

Upper, floating leaves obovate or spatulate, 3-nerved.
                                      1. *Callitriche palustris*
All leaves linear, 1-nerved.   2. *Callitriche hermaphroditica*

1. *Callitriche palustris* L. Perennial; stems 2—30 cm. (1—12 in.) long, the tips usually floating; submerged leaves linear, 1—1.5 cm. long (fig. 68); floating leaves petioled, 3-nerved, blades 3—10 mm. (⅛—⅖ in.) long; flowers with a pair of bracts 1 mm. long; fruit oval, 1.5 mm. long. In ponds or pools; Cass, LaMoure, Barnes and Stutsman Counties.

2. *Callitriche hermaphroditica* L. Perennial; stems usually all submerged, 1—4 dm. (4—10 in.) long; leaves all linear, slightly notched at tip, 0.5—2 cm. long; no bracts below flowers. Stutsman, LaMoure, Cavalier and Ward Counties. This has usually been called *C. autumnalis* L. These

plants are probably more common than our collections would indicate, but small water plants are easily overlooked or not distinguished from each other. Also, conditions favorable for their growth are neither frequent nor dependable in North Dakota.

Rydberg (Fl. Pl. and Pr.) credits *Floerkea proserpinacoides* Willd., FALSE MERMAID, to North Dakota. It is somewhat similar to *Callitriche* in appearance, but is annual and grows in wet grassland. The leaves are pinnate, of about 5 leaflets; flowers 3—5 mm. wide, peduncled in leaf axils; sepals 3; petals 3; stamens 6; ovaries 3—5, each 1-seeded in fruit. We have not yet found this plant.

SUMAC FAMILY Anacardiaceae

Our species shrubs with alternate, compound leaves; flowers small, white or yellow; sepals, petals and stamens usually 5 each; pistil 1; fruit a hard, stone-like seed surrounded by thin flesh, persisting all winter.

Key to Species

Leaflets 5—31; shrub 1—3 m. (3—10 ft.) high.   1. *Rhus glabra*
Leaflets 3; shrubs 0.3—2 m. high.
   Leaflets 1—4 cm. (⅖—1⅗ in.) long, hairy; fruit hairy, red.
                                       2. *Rhus trilobata*
   Leaflets 3—15 cm. long, smooth; fruit smooth, white.   3. *Rhus radicans*

1. *Rhus glabra* L. SMOOTH SUMAC. (Fig. 244). Widely forking shrub, 1—3 m. (3—10 ft.) high; leaflets ovate to lanceolate, 3—8 cm. (1—3 in.) long, sharply toothed; flowers yellow, 3 mm. (⅛ in.) wide, in a thick, ovate, terminal cluster 1—3 dm. long; fruit red, 3—5 mm. long, densely soft hairy. Infrequent or locally abundant; Richland to Ramsey and Pembina Counties. Blooms June; fruit ripe Sept.

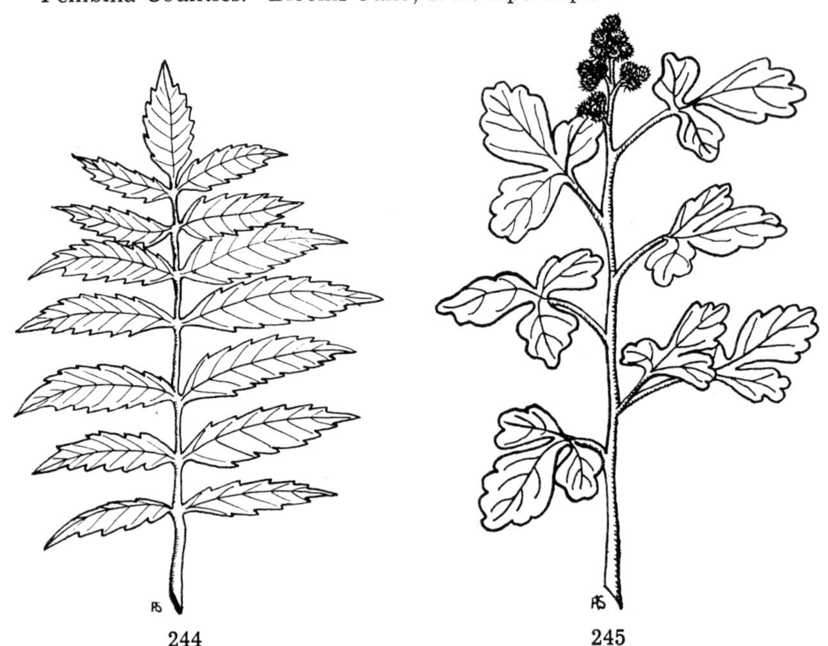

244   245

244. Smooth Sumac (*Rhus glabra*).

245. Ill-scented Sumac (*Rhus trilobata*).

2. *Rhus trilobata* Nutt. Ill-scented Sumac. Skunk Bush. (Fig. 245).
Much branched shrub, 1—2 m. high; leaflets 3, obovate, smooth above,
hairy below, 3-lobed or smaller ones entire, 1—3 cm. long; flowers
yellow, 2—3 mm. long, in dense, rounded clusters 1—2 cm. long; fruit
red, 5 mm. long, densely soft hairy. Common on hillsides west of
Missouri River; records also from Emmons and LaMoure Counties.
Blooms late May.

3. *Rhus radicans* L. Poison Ivy. (Fig. 246). Shrub, in North Dakota
usually 3—8 dm. (1—2½ ft.) high; leaflets ovate, 5—15 cm. long,
smooth, often toothed; flowers white, 3 mm. long, in loose, drooping
clusters 5—10 cm. long; fruit rounded, 5 mm. long, becoming white,
shiny. A common and troublesome plant, growing in all sorts of places
all over State. Usually in woods or brush, sometimes in the open, especially near brushy areas; usually in moist places, but also on exposed
buttes (probably moist pockets). Late June.

Poison ivy is greatly variable in habit of growth. We do not have the
high climbing vines nor tall shrubs. Our plants (var. *rydbergii* (Small)
Rehder) usually have an upright, scarcely branched stem, 1—4 dm.
high, the long petioles of the leaves making a considerable part of the
total height of the plant. The name "Poison Oak" refers to the same
plant. Many authors place the smooth fruited, poisonous sumacs in a
separate genus, *Toxicodendron*.

STAFF-TREE FAMILY Celastraceae

Shrubs, trees or vines with simple leaves and small flowers;
sepals, petals and stamens 4 or 5; pistil 1, ovary 2—5-celled; fruit a
berry or pod splitting open, showing bright colored interior. The
flowers are peculiar in having a fleshy disc below the ovary.

Key to Species

Woody vine, 2—10 m. (7—35 ft.) high; flowers greenish white.
1. *Celastrus scandens*
Shrub, 1—2 m. high; flowers dark purple.
2. *Evonymous atropurpureus*

1. *Celastrus scandens* L. Climbing Bittersweet. Woody vine, 2—10 m.
(7—35 ft.) high, twining around trees or bushes, stem becoming 1—2
cm. (⅖—⅘ in.) thick; leaves alternate, oblong, finely toothed, 3—10 cm
long, rather thick and smooth; flowers greenish white, 3—4 mm. (⅛
in.) wide, in cluster 2—5 cm. long at end of branch; fruit rounded, 8—10
mm. wide, becoming orange colored and splitting in 4 parts, showing
seeds covered by bright red pulp. Frequent in woods or thickets.
Blooms June; fruit ripe Sept.

2. *Evonymous atropurpureus* Jacq. Wahoo. Shrub 1—2 m. high with
green bark; leaves alternate, elliptic to obovate, 5—15 cm. long; flowers
dark purple, 5—8 mm. wide, in much branched cluster 3—8 cm. wide
at leaf base; fruit 1—2 cm. wide, 4-lobed, splitting to show seeds covered with bright red pulp (fig. 61). Known only from woods of Sheyenne River near Ransom-Richland county line, where it is fairly common. Blooms late June; fruit ripe Sept. Several species of Europe and
Asia are grown as ornamentals.

MAPLE FAMILY Aceraceae

Trees with opposite, simple or compound leaves; flowers small,
clustered; sepals usually 4—5; petals 4—5, or 0 (in ours), stamens
4—9; pistil 1, ovary 2-lobed; fruit splitting in 2 parts, each with a
long wing (fig. 47).

Key to Species

Leaves compound; flowers dioecious.   1. *Acer negundo*
Leaves simple; flowers perfect or monoecious.
  Leaves ovate, sharply and deeply toothed and lobed.
           2. *Acer saccharinum*
  Leaves rounded, with 3—5, wide, short lobes.   3. *Acer saccharum*

246   247

246.   Poison Ivy (*Rhus radicans*).
247.   Red Mallow (*Sphaeralcea coccinea*).

1. *Acer negundo* L. BOX-ELDER. Tree 5—10 m. (17—35 ft.) high with gray, narrowly ridged bark; young branches green or purple; leaves pinnate, of 3 or 5 leaflets; leaflets ovate, 5—15 cm. (2—6 in.) long, often with 1 or 2 short lobes; flowers from lateral buds, filaments of stamens becoming 2—3 cm. long; body of fruit narrowly oblong, 10—15 mm. long, wing 15—20 mm. Common along streams and widely planted Blooms late April, fruit falling in winter or spring.

2. *Acer saccharinum* L. SILVER MAPLE. SOFT MAPLE. Tree 5—20 m. high with flaky bark; young branches red; leaves ovate, 5—10 cm. long, deeply and sharply palmately lobed and cut, silvery below; flowers on short branches, filaments of stamens becoming 1—2 cm. long; body of fruit broadly oblong, 10 mm. long, wing 3—5 cm. Blooms early Apr., fruit ripe June, falling at once. This is not native and though often planted is little if any established as a wild plant.

3. *Acer saccharum* Marsh. SUGAR MAPLE. HARD MAPLE. Tree 5—20 m. high with dark bark, becoming rough in age; leaves 8—15 cm. long and usually wider, with 3—5 wide lobes, no small teeth; flowers monoecious, staminate on stalks 5—10 cm. long; body of fruit oblong, 5—7 mm. long, wing 2—3 cm. Blooms mid-May. We have one specimen from Sargent County in 1891. Mr. C. B. Waldron told me they saw two trees at Skunk Lake near Rutland. The locality has not been revisited. This tree is occasionally planted.

Amur Maple (*Acer ginnala* Maxim.) has been planted in recent years. It is hardy, seeds freely and could be expected to establish itself. It is a small, spreading tree with gray bark, resembling a box-elder. Leaves small, ovate, sharply cut or 3-lobed, very bright red in fall; flowers with white petals 3—5 mm. long in June; fruit similar to sugar maple.

BALSAM FAMILY Balsaminaceae

Herbs with weak, succulent stems and alternate, simple leaves; flowers showy, irregular in shape; sepals usually 3, the lower with a sac or spur; petals 5 or appearing as 3 by union of 2 pairs; stamens 5; pistil 1; fruit a pod, splitting lengthwise and throwing the few, large seeds. The garden Balsam (*Impatiens balsamina*) has this sort of fruit.

Key to Species

Corolla pale yellow, scarcely spotted.   1. *Impatiens pallida*
Corolla orange yellow, spotted with purple.   2. *Impatiens biflora*

1. *Impatiens pallida* Nutt. Pale Touch-me-not. Annual, stem erect, branched above, 1—1.5 m. (3—5 ft.) high, weak, watery; leaves thin, ovate, 3—10 cm. (1—4 in.) long, with rounded teeth, petioles long; flowers 1 or 2 from upper leaf bases, 2 cm. long, hanging horizontally (fig. 99); pod 2—3 cm. long, 2-seeded. Three specimens from Cass County seem to be this species but one other and one from Pembina County which had been placed here apparently represent a paler form of the next species.

2. *Impatiens biflora* Walt. Spotted Touch-me-not. Similar to last but flowers purple spotted. Frequent in woods, especially in shaded, boggy places. Aug. The flowers are much visited by hummingbirds.

BUCKTHORN FAMILY Rhamnaceae

Shrubs or trees with opposite, simple leaves and small, greenish flowers; sepals, petals and stamens 4 or 5 each; pistil 1, 2—3-celled; fruit a berry (in ours) or capsule with a few, large, bony seeds.

Key to Species

Thorny shrub 1—5 m. (3—17 ft.) high, escaped from cultivation.
    1. *Rhamnus cathartica*
Shrub, not thorny, 3—10 dm. (1—3 ft.) high, native in swamps.
    2. *Rhamnus alnifolia*

1. *Rhamnus cathartica* L. Common Buckthorn. Stout, thorny shrub, 1—5 m. (3—17 ft.) high; leaves 3—6 cm. (1—2 in.) long, oblong, smooth, with fine, but hardly sharp teeth; flowers dioecious, greenish yellow, 4—5 mm. (⅕ in.) wide, closely clustered at leaf base; berry black, rounded, 6—8 mm. wide. Blooms late in May, fruit ripe Sept., persisting. Frequently planted for hedges but not recommended because it is an alternate host for oat rust. It makes an excellent hedge and holds its leaves until late October. The berries are bitter but are eaten by birds. Plants are well established in woods and elsewhere at Fargo and probably at other places, though we have collected it only in Cavalier County.

2. *Rhamnus alnifolia* L'Her Alder Buckthorn. Low, often trailing shrub, 3—10 dm. high; leaves 3—10 cm. long, elliptic or ovate, finely toothed; flowers 1—3 at leaf base; berry black, 7—8 mm. wide. Rare in shaded, very wet ground; Ransom, Pembina and Rolette Counties (fig. 15).

GRAPE FAMILY Vitaceae

Woody vines, climbing by tendrils or otherwise, the tendril opposite a leaf; flowers small, greenish, in a cluster opposite a leaf; sepals, petals and stamens 4 or 5 each; pistil 1, ovary 2—6-celled; fruit a rounded berry with 2—8 large seeds. Flowers sometimes perfect, sometimes dioecious. Petals fall off when flower opens.

Key to Species

Leaves simple, rounded, usually deeply lobed.   1. *Vitis vulpina*
Leaves palmately compound of 5 leaflets.   2. *Parthenocissus inserta*

1. *Vitis vulpina* L. Wild Grape. Woody vine, climing to 10 m. (35 ft.) by stout tendrils; bark dark reddish brown, splitting in narrow shreds; leaves rounded, 5—20 cm. (2—8 in.) wide, sometimes 3-lobed, sharply, coarsely toothed (fig. 46); flowers 5—6 mm. (¼ in.) wide in branching clusters 2—5 cm. long opposite a leaf; berry blue black, 8—10 mm. wide. Frequent in woods along streams. Blooms June; fruit ripe Aug. *V. riparia* Michx. is now regarded as a separate species, but we seem not to have it.

2. *Parthenocissus inserta* (Kerner) Fritsch. Virginia Creeper. Woodbine. Woody vine, climbing to 10 m.; tendrils small, weak; bark gray, not shreddy, rather spongy; leaflets 5, lanceolate or elliptic, toothed, 5—20 cm. long; flowers greenish, 4—5 mm. wide; berry blue, 5—7 mm. wide; fruiting cluster 1—2 dm. wide, branches spreading, stout, red. Frequent in woods. Blooms June; fruit ripe Aug. Formerly known as *P. vitacea* (Knerr.) Hitchc. The eastern *P. quinquefolia* forms adhesive discs at the ends of the tendrils, which our plant does not.

LINDEN FAMILY Tiliaceae

Trees or shrubs with simple, usually alternate leaves; sepals and petals 4—5 each; stamens many; pistil 1, ovary 5-celled; fruit in ours nut-like. We have only the following.

1. *Tilia americana* L. Linden. Basswcod. Tree 5—20 m. (17—70 ft.) high, with very soft wood and tough inner bark; leaves rounded, 5—20 cm. (2—8 in.) long, thin, sharply toothed; flowers cream colored, 1—1.5 cm. wide, 15—50 in a compact cluster on a slender stalk attached to a "wing" which becomes 5—15 cm. long, 1 cm. wide; fruit rounded, hard, hairy, 8—10 mm. (⅜ in.) wide. Locally common on low ground near streams, west to Benson County. July.

MALLOW FAMILY Malvaceae

Herbs with alternate, palmately ribbed, usually lobed leaves; flowers cup or funnel shaped, often large; sepals 5, petals 5, slightly united at base; stamens many, filaments united into a tube around style (fig. 108); pistil 1, stigmas 5 or 10, ovary 5-celled; fruit a pod, or splitting into 5—many parts which are one seeded or split open.

This family is easily recognized by the tube of stamens. In addition to sepals, there often are 3—10 small bracts just below the calyx. Hollyhock and several other ornamentals belong to this family.

Key to Species or Genera

Flowers at leaf base.
  Flowers lavender, white or purple; fruit splitting into segments.
                                                      1-5. *Malva*
  Flowers light yellow with purple center; fruit a pod.
                                                      6. *Hibiscus trionum*
Flowers in terminal clusters.
  Flowers orange to pink; low plant, 1—3 dm. (3—12 in.) high.
                                                      7. *Sphaeralcea coccinea*
  Flowers deep yellow; stout plant, 1—2 m. (3—7 ft.) high.
                                                      8. *Abutilon theophrasti*

Malva MALLOW

Leaves rounded, slightly or deeply lobed; flowers 1 or more at leaf base; bracts 2—3; fruit wider than long, splitting into 10—13 wedge-shaped sections. All introduced plants.

Key to Species

Corolla dark purple, 2—5 cm. (1—2 in.) wide; plant erect, 3—10 dm.
  (1—3 ft.) high.   5. *Malva sylvestris*
Corolla white to blue, 8—15 mm. (½ in.) wide.
  Plant erect, 1—2 m. (3—7 ft.) high; leaves much curled.
                                                      3. *Malva crispa*
  Plant 1—4 dm. high, stems often trailing; leaves not much curled.
    Flowers 15—20 mm. wide; fruit segments rounded on back.
                                                      2. *Malva neglecta*
    Flowers 8—12 mm. wide; fruit segments ridged on back.
      Sepals broad, almost rounded, forming an open cup.
                                                      4. *Malva parviflora*
      Sepals narrowly triangular, curled, closely enclosing fruit.
                                                      1. *Malva rotundifolia*

1. *Malva rotundifolia* L. SMALL MALLOW. (Fig. 243). Annual; stems trailing, 3—6 dm. (1—2 ft.) long, central stalk often erect; leaves rounded, often wider than long, 3—10 cm. (1—4 in.) long, with 5—7 broad, rounded lobes; flowers pale blue or nearly white, 8—12 mm. (⅓—½ in.) wide, 5—15 at a leaf base; sepals triangular, curled; fruit segments usually 10, with irregular, low ridges on back. A very common yard and garden weed. June-Oct. Grows vigorously after heavy frosts. This was formerly called *M. borealis* Wallm., or *M. pusilla* With.

2. *Malva neglecta* Wallr. COMMON MALLOW. Annual; stems trailing, 3—6 dm. long; leaves rounded, 5—10 cm. long; flowers pale blue, 15—20 mm. wide; fruit segments usually 13, rounded on back, not ridged. This has been collected in Cass, Barnes, Stutsman, Sargent, Dickey and Walsh Counties. It is the common species in eastern U. S. but rare here.

3. *Malva crispa* L. CURLED MALLOW. Annual; stem stout, erect, 1—2 m. high; leaves rounded, 5—10 cm. wide, much crisped, lobed about ⅓ of length; flowers purplish, 1 cm. wide, closely crowded at upper leaf bases (not stalked as in Nos. 1 and 2); calyx lobes triangular, larger than in No. 1; fruit segments rounded, not prominently ridged. Specimens from Grand Forks and Kidder Counties have been referred to this species. It appears in gardens, apparently introduced by accident. We do not know if it re-seeds itself.

4. *Malva parviflora* L. SMALL-FRUITED MALLOW. Annual; stem erect or spreading, 3—6 dm. high; leaves rounded, usually 5-lobed, 3—7 cm. wide; flowers pale blue, 8—10 mm. wide; sepals broad, forming a loose cup around fruit, fruit segments angled and strongly cross-ridged. Collected by Lunell in his yard at Leeds, Benson County, in 1914. It persisted there, at least for some years. A specimen from a farm garden in Barnes County (Stevens in 1944) seems to be this species but grew 1—1.5 m. high. Another from Traill County (Stevens 893) is typical.

5. *Malva sylvestris* L. HIGH MALLOW. Biennial; stem stout, erect, 3—10 dm. high; leaves rounded, 4—10 cm. long, with 3—7 short, rounded lobes; flowers purple, 2—4 cm. wide. Collected in Cass, LaMoure and Stutsman Counties, probably a garden escape.

6. *Hibiscus trionum* L. FLOWER-OF-AN-HOUR. Annual; stem widely branched, spreading at base, 3—6 dm. (1—2 ft.) high; leaves 3—5 cm. (1—2 in.) long, deeply divided into 3—7 obovate lobes which are pinnately lobed and toothed; flowers at leaf base, 3—4 cm. wide, light yellow with dark purple center and one edge of petals purple; 10—15 slender bracts below calyx; fruit a rounded pod, 2—3 cm. long, splitting lengthwise; seeds black, 2 mm. long, rounded triangular. Fields or more often gardens, locally abundant. Cass and Dickey Counties. This is quite troublesome as an annual weed but is a showy flower. The name refers to the fact that the flowers are open on bright days about 8—11 a. m. The pods are enclosed by the enlarging calyx which is thin, beautifully veined and hairy—a striking object, especially when covered with dew.

7. *Sphaeralcea coccinea* (Nutt.) Rydb. RED MALLOW. (Fig. 247). Perennial, spreading by roots; stems erect or spreading, 1—2 dm. high; leaves rounded, gray hairy, 2—6 cm. wide, deeply 3—5-lobed, the lobes usually further lobed or divided into narrow segments; flowers deep orange to pinkish, 1—1.5 cm. wide, saucer shaped, grouped in a dense cluster 2—5 cm. long at end of stem; fruit segments 5, densely gray hairy. Prairie, very common, especially on dry, poor soil. Late June, July. Formerly called *Malvastrum coccineum* (Nutt.) A. Gray. One of our most characteristic prairie flowers. Often called "wild geranium", suggested by both leaf and flower. The leaf hairs are interesting to examine with a magnifying glass. Each hair has 5 radiating branches.

8. *Abutilon theophrasti* Medic. VELVET LEAF. Very stout annual; stem erect, 3—20 dm. high; leaves 1—2 dm. long, rounded, with a slender tip, very soft hairy; flowers deep yellow, 2 cm. wide, at leaf bases; fruit of about 10 flattened, oblong pods forming a rounded, flat topped group, 2 cm. wide; seeds black, flattened, rounded triangular, 3—4 mm. long. A common weed from southern Iowa southward. Specimens from Richland, Traill and Grand Forks Counties, probably from seeds in forage crop seed and not established. Well developed pods were received from Divide County in 1948.

ST. JOHN'S-WORT FAMILY Hypericaceae

Mostly herbs with opposite, entire, dotted leaves; flowers yellow or pinkish, clustered at top of stem; sepals and petals 4—5; stamens 4—many, usually united at base in several groups (fig. 109); pistil 1, ovary 3—7-celled; fruit a pod with many small seeds, splitting lengthwise. This family is well represented farther south, east and on the Pacific Coast, but we have only two, rare, introduced species.

Key to Species

Stamens many.　　　　　　　　　　　1. *Hypericum perforatum*
Stamens 5—10 in 3 groups.　　　　　　2. *Hypericum majus*

1. *Hypericum perforatum* L. COMMON ST. JOHN'S-WORT. Perennial; stem erect, 3—8 dm. (1—2½ ft.) high; leaves linear to oblong, 1—3 cm. (½—1 in.) long; flowers in a flat cluster 5—15 cm. wide, petals narrow, spreading; pods 5—6 mm. (¼ in.) long. Collected at Fargo in 1911 by Clarence Waldron. It is regarded as a bad weed in many states.

2. *Hypericum majus* (A. Gray) Britton. GREATER ST. JOHN'S-WORT. Annual (?); stem erect, slender, 2—6 dm. high; leaves lanceolate or oblong-lanceolate, 2—6 cm. long; flowers yellowish, 4—5 mm. wide; pods 3—4 mm. long. Collected in Richland and Benson Counties.

WATERWORT FAMILY   Elatinaceae

Small rather fleshy plants growing on mud; leaves opposite, flowers very small. We have only the following.

1. *Elatine triandra* Schkuhr. MUD-PURSLANE. Annual; stems creeping on mud, 3—10 cm. (1—4 in.) long, often forming a mat; leaves oblanceolate or oblong, 5—10 mm. (⅕—⅖ in.) long; flowers at leaf base; sepals, petals and stamens 3 each or stamens 6; pods rounded, 1 mm. long, seeds numerous. Cass and Morton Counties.

ROCK-ROSE FAMILY   Cistaceae

Herbs or shrubs with opposite or alternate, small leaves; sepals 3—5; petals 3—5, quickly falling or entirely absent; stamens 8; pistil 1, ovary usually 1-celled; fruit a capsule with small seeds. Our species all rare.

Key to Species

Creeping, woody plant with scale-like leaves.   1. *Hudsonia tomentosa*
Erect plants, scarcely woody; leaves small but not scale-like.
  Petals 5, yellow, at least on upper flowers. 2. *Helianthemum bicknellii*
  Petals 3, greenish, persistent.   3. *Lechea stricta*

1. *Hudsonia tomentosa* Nutt. BEACH HEATHER. Creeping shrub, branches upright, 1—2 dm. (4—8 in.) high; leaves numerous, scale-like, 2 mm. ($\frac{1}{12}$ in.) long, finely hairy; flowers yellow, 5—8 mm. wide, along upper part of branches; pod 1-seeded. Known only from one locality in sand dune area, Ransom County.

2. *Helianthemum bicknellii* Fernald. FROSTWEED. Perennial; stem slender, erect, 1—6 dm. high; leaves alternate, narrowly oblong or lanceolate, 1—3 cm. long, rather thick, densely covered with very fine, star shaped hairs; upper flowers yellow, 1.5—3 cm. wide; lower flowers without petals, clustered at leaf bases; pods rounded, 3—4 mm. long. Richland and Pembina Counties. Formerly referred to *H. majus* (L.) BSP.

3. *Lechea stricta* Leggett. PINWEED. Perennial; stem erect, 1—3 dm. high, densely branched above; leaves alternate, crowded, 1—2.5 cm. long, linear, with coarse, straight hairs; flowers greenish, 1—2 mm. wide, in dense, slender, terminal clusters; pod rounded, 1—2 mm. long. Richland and Bowman Counties. In a sandy, eroded field in the former, I found it in quantity in one locality in 1936 (No. 228).

VIOLET FAMILY   Violaceae

Low herbs, often "stemless", each leaf and flower coming from a perennial, underground stem; leaves alternate, usually ovate; flowers slightly irregular, lower petal longest and having a sac or spur at base; sepals, petals and stamens 5 each, filaments of stamens very short; pistil 1; fruit a rounded or oblong 1-celled pod, splitting lengthwise, the ovoid, smooth seeds attached to 3 places on outer wall. These are well known spring flowers but species are often difficult to determine. Some of them seem to hybridize freely and plants vary much in characters. The early flowers often do not produce seeds, late flowers near or under the ground, producing seeds but not bearing petals.

Key to Species

Plants without leafy stems; flowers blue or purple.
  Leaves ovate, not divided.
    Plants without hairs.   1.   *Viola papilionacea*
    Plants quite hairy.   2.   *Viola sororia*
  Leaves divided into narrow segments.   3.   *Viola pedatifida*
Plants with leafy stems, these sometimes short.
  Flowers yellow.
    Leaves lanceolate to ovate or oblong; western prairie.
         4.   *Viola nuttallii*
    Leaves broadly ovate; eastern woods.   5.   *Viola eriocarpa*
  Flowers blue, purple or whitish.
    Flowers white, purplish on outer side; leaves pointed.
         6.   *Viola rugulosa*
    Flowers blue or purple; leaves rounded.
      Leaves 5—10 mm. (⅕—⅖ in.) long, hairy; western, open ground.
         7.   *Viola adunca*
      Leaves 10—15 mm. long, smooth; rare in woods.   8.   *Viola conspersa*

1. *Viola papilionacea* Pursh. COMMON BLUE VIOLET. Perennial from a short, underground stem; leaf blades smooth, ovate or cordate, 1—5 cm. (⅖—2 in.) long, petioles 2—20 cm. long; flowers blue or dark purple, 1—2 cm. wide. Common in woods and low grassland. Early May, June. Size and habit vary with conditions; spreading, short petioles in open places, upright tall ones in tall grass. Other species may be recognizable, but this is our most difficult group.

2. *Viola sororia* Willd. HAIRY BLUE VIOLET. Differs from the last in hairy leaf stalks and lower surfaces of leaves. It seems to be frequent, chiefly in dry, open woods early in the season. One would be inclined to suspect this to be a response to conditions, but the species is considered distinct by most botanists. A careful study of individual plants throughout the season would be a worthwhile project.

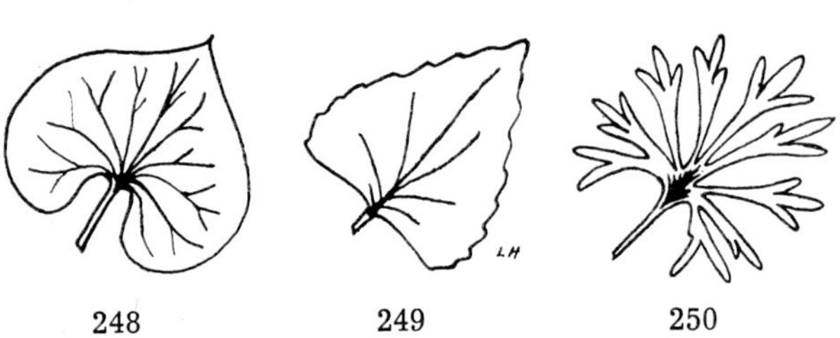

248   249   250

3. *Viola pedatifida* G. Don. PRAIRIE VIOLET. (Fig. 250). Perennial from a short, upright, underground stem; leaves 1—6 cm. long, deeply divided into narrow (occasionally wide) lobes; flowers blue or dark purple, 1—2.5 cm. wide. Common on prairie (fig. 12). May. Flowers larger than those of *V. papilionacea*.

4. *Viola nuttallii* Pursh. NUTTALL'S VIOLET. (Fig. 251). Underground stem short, upright; several slender stems 2—10 cm. long extending above ground; leaves broadly oblong or ovate to lanceolate, 1—5 cm. long; flowers 8—12 mm. wide, deep yellow with purple lines inside. Frequent on hills in central and common on prairie in western part of State. We have one record from Arvilla, Grand Forks County, but Ransom, Barnes and Ramsey form the general eastern limit (fig. 12). A very attractive little plant, named for Thomas Nuttall who collected the first specimens in 1811.

5. *Viola eriocarpa* Schwein. YELLOW WOOD VIOLET. (Fig. 249). Perennial from a short, thick, underground stem; leafy stems erect or spreading, 1—3 dm. long, leaves broadly ovate, 2—6 cm. long, with rounded teeth, more or less hairy; flowers yellow, 1—1.5 cm. wide, each from a leaf base. Frequent in woods, eastern part of State. This name seems

to have preference over *V. scabriuscula* Schwein. Whether or not our plants are distinct from the eastern *V. pubescens* Ait. seems a problem. Fernald (Rhodora 43:616) has used *V. pensilvanica* Michx. in place of *eriocarpa*.

6. *Viola rugulosa* Greene. PINK WOOD VIOLET. (Fig. 248). Perennial from slender rhizomes; leafy stems 1—3 dm. (4—12 in.) high; leaves rounded, 2—10 cm. (1—4 in.) wide, usually wider than long but with a slender tip, finely toothed, hairy below; flowers from leaf bases, 10—15 mm. (½ in.) wide, mainly white but pink or purple outside and yellowish at inner base. Woods and thickets, all over the State (fig. 12). Begins to bloom about May 1, our earliest violet. The plants spread freely by rhizomes.

251   Nuttall's Violet (*Viola nuttallii*).

7. *Viola adunca* Smith. SMALL BLUE VIOLET. Leafy stems 5—20 cm. long, often prostrate; leaves rounded or ovate, 5—25 mm. long, with small, rounded teeth; flowers 8—10 mm. wide, blue. Dry woods or open ground; frequent over most of State (fig. 12). These might be mistaken for small plants of *V. papilionacea*, but careful examination will show leafy stems. It is very similar to *V. arenaria* DC. which is widely distributed.

8. *Viola conspersa* Reichenb. DOG VIOLET. Leafy stems 5—20 cm. long, usually spreading; leaves rounded, 1—3 cm. long, with small, rounded teeth; flowers pale blue or purple, 8—15 mm. wide. Quite local, usually in aspen woods. Records for Richland and Dunn Counties. A specimen was secured at Fargo about 1912 but apparently not preserved, though we have some from the Minnesota side of the river. The more rounded, not pointed leaves, will usually distinguish it from *V. papilionacea*, though the slender spur of the lower petal (6 mm.) is more characteristic.

LOASA FAMILY Loasaceae

Coarse, very rough herbs, mostly tropical; leaves alternate, simple; petals 5 or 10, attached on top of ovary (fig. 100); stamens many; styles 3, more or less united; fruit a pod, opening at top. Some tropical species have stinging hairs. The leaves of ours have short, stout hairs bearing rings of barbs.

Key to Species

Stout plant, 4—10 dm. (16—40 in.) high; flowers 8—12 cm. (3—5 in.) wide.   1. *Mentzelia decapetala*
Small plant, 3 dm. or less high; flowers 5 mm. (⅕ in.) wide.
                                                                    2. *Mentzelia disperma*

1. *Mentzelia decapetala* (Pursh) Urban & Gilg. EVENING STAR. "SCORIA LILY." Biennial; stem stout, 4—10 dm. (16—40 in.) high; leaves lanceolate, 5—15 cm. (2—6 in.) long, deeply pinnately toothed or lobed; flowers cream colored, 8—12 cm. (3—5 in.) wide, at ends of branches, petals 10, oblanceolate; pod oblong, 3—5 cm. long; seeds many, flattened. On buttes in burned, crumbled clay ("scoria"), or on clay slopes, Missouri River westward. There is a large colony on one hill about 10 miles north of Lisbon, Ransom County. Aug. Flowers open in evening and resemble a cactus flower.

2. *Mentzelia disperma* S. Wats. Annual (?); stems slender, often widely branched, 1—3 dm. high; leaves linear or oblong, 2—5 cm. long; flowers yellow, 5 mm. wide, petals 5, rounded; pods slender, 2 cm. long. July. First found in 1935 (Stevens 138) on a small butte about 5 miles east of Glen Ullin. Very few plants have been seen there on later visits. In 1946, I found it abundant on a hillside at the Logging Camp Ranch, Slope County. The flowers seemed open only about 3—4 p. m.

CACTUS FAMILY Cactaceae

Fleshy stemmed plants without true leaves, with tufts of stout spines and very small spines (glochids); flowers showy with numerous petals and similar sepals borne on top of ovary. Fruit fleshy, many seeded. About 1,000 species of cactus are found in warm, dry parts of North and South America, but few occur as far north as the Dakotas.

Key to Species

Stems rounded, covered with spine-tipped tubercles; flowers arising between tubercles; fruit a smooth berry wedged between tubercles.
    Flowers red, fruit green, becoming brown.   1. *Mamillaria vivipara*
    Flowers yellow, mature fruit bright red.   2. *Mamillaria missouriensis*
Stems jointed, the segments 3—10 cm. (1—4 in.) long, more or less flattened; fruits on upper ends of stems.
    Segments much flattened, 5—10 cm. long.   3. *Opuntia polycantha*
    Segments little flattened, 2.5—5 cm. long.   4. *Opuntia fragilis*

1. *Mamillaria vivipara* (Nutt.) Haw. BALL CACTUS. Stems 5—8 cm. (2—3 in.) in diameter, usually branched to form a group of 3—12 such balls; flowers red, 5 cm. wide; fruit brownish, oblong, 10—15 mm. (½ in.) long. Frequent on prairie, especially on stony soil, over most of State except Red River Valley. Flowers June, July; fruit ripe in fall.

2. *Mamillaria missouriensis* Sweet. Similar to No. 1 except flowers yellowish, fruit bright red, maturing the next spring. Local in western part of State; Slope and McKenzie Counties.

252. Prickly Pear (*Opuntia polycantha*).

3. *Opuntia polycantha* Haw. PRICKLY PEAR. (Fig. 252.) Stem segments flattened, 5—10 cm. long and nearly as wide. Flowering about July 1, the flowers 5—8 cm. (2—3 in.) wide, bright yellow, fading into bronze tints. Fruits 1—2 cm. long. Quite common, Missouri River westward, on hillsides, buttes and clay flats.

4. *Opuntia fragilis* (Nutt.) Haw. BRITTLE PRICKLY PEAR. An inconspicuous plant, often hidden in the grass and rarely flowering. Stem segments little flattened, breaking apart readily. It seems most characteristic on strongly saline clay flats, but grows also on buttes and stony knolls. It thrives on quartzite outcrops in southeastern S. D.

OLEASTER FAMILY Elaeagnaceae

Shrubs or trees; leaves and young stems silvery with scale-like hairs; flowers small, yellow; petals 4, stamens 4, attached on top of ovary; pistil 1; fruit more or less fleshy, 1-seeded.

Key to Species

Leaves opposite; flowers dioecious; fruit bright red.
  Plants thorny; leaves gray on both sides.   1. *Shepherdia argentea*
  Plant not thorny; leaves green above.   2. *Shepherdia canadensis*
Leaves alternate; flowers perfect; fruit silvery.   3. *Elaeagnus argentea*

1. *Shepherdia argentea* Nutt. BUFFALO BERRY. Thorny, much branched shrub, 2—5 m. (7—17 ft.) high; leaves opposite, narrowly oblong, 2—5 cm. (1—2 in.) long, gray on both sides; staminate flowers yellow, 3 mm. (⅛ in.) wide in clusters at leaf base; pistillate flowers 1 mm. wide; fruit bright red, 3 mm. wide. Flowers of staminate trees are quite showy at end of April. Pistillate flowers are inconspicuous. The fruit, ripe in September, is highly esteemed for jelly making. Occurs over most of State, but is most abundant in western part, along streams, coulees and hillsides, often forming thickets. A colony in northeastern Richland County, near Kindred, is the easternmost locality known to me, but it has been found in western Minnesota.

2. *Shepherdia canadensis* (L.) Nutt. RABBITBERRY. Bushy plant, not over 1 m. high, known only in Pembina, Turtle and Killdeer Mountain areas. The flowers are less showy than in No. 1.

3. *Elaeagnus argentea* Pursh. SILVERBERRY. WILD OLIVE. Slender shrub 1—2 m. high; leaves alternate, oblong or ovate, 2—6 cm. long; flowers yellow, 6—8 mm. wide, very fragrant; fruit oblong, 1 cm. long. June. Russian Olive (*E. angustifolia* L.) has been planted commonly but seems not to have become established. It is similar to silverberry but becomes a tree.

LOOSESTRIFE FAMILY Lythraceae

Herbs with small, pink or purple flowers at leaf bases; petals 4—6, attached to calyx tube; stamens 4—8; pistil 1; fruit a capsule with many, small seeds.

Key to Species

Flowers densely clustered at leaf bases; petals 4; capsule rounded,
   3 mm. (⅛ in.) wide, thin walled.   1. *Ammannia coccinea*
Flowers solitary at upper leaf bases; petals 5; capsule narrow,
   5—8 mm. long, strongly ribbed.   2. *Lythrum alatum*

1. *Ammannia coccinea* Rottb. AMMANNIA. Annual; stems stiff, widely branched below at right angles; leaves opposite, narrow, 2—5 cm. (1—2 in.) long, smooth; flowers 5 mm. (⅕ in.) wide, densely clustered at leaf bases; petals purple, falling very quickly; pod rounded, 3 mm. long. July, Aug. First found at Fargo in 1923 on a low place near railroad. Later found quite common in Cass County along ditches and in low places in fields.

2. *Lythrum alatum* Pursh. LOOSESTRIFE. Perennial, stems slender, little branched, 3—6 dm. (1—2 ft.) high, closely set with alternate, lanceolate or ovate-lanceolate leaves 2—5 cm. long; flowers dark purple, 1 cm. wide. We have several collections from Richland County only (fig. 19). It is quite a common plant along streams or in other low places farther south and east.

EVENING PRIMROSE FAMILY Onagraceae

Perennials or annuals with usually showy, white, yellow or purple flowers; petals 4, (2 in *Circaea*) attached to a short or long, slender tube which comes from top of ovary (fig. 104); stamens usually 8, pollen usually held together by cobweb-like threads; stigma usually 4-lobed; fruit a many seeded capsule (1-seeded in *Gaura* and *Circaea*). Flowers of most species open in late afternoon or at dusk; white flowers usually become pink when they fade next morning.

Key to Species or Genera

Petals 2, deeply 2-lobed; stamens 2; flowers 3 mm. (⅛ in.) wide;
   fruit a small, sticky bur.
   Plants 2—4 dm. (8—16 in.) high in woods; leaves rounded at base.
                      15. *Circaea quadrisulcata*
   Plants 1—2 dm. high in wet, shady places; leaves more heart-
     shaped.   16. *Circaea alpina*
Petals 4, stamens 8; flowers often larger; fruit not sticky.
   Slender plants 1—2 dm. high; sepals upright at flowering time.
                      5. *Boisduvalia glabella*
  Much larger or stemless; sepals bent back at flowering time.
    Seeds with a tuft of fine hairs at tip.   1-4. *Epilobium*
    Seeds without a tuft of hairs.

Flowers yellow.   6-13. *Oenothera*
Flowers white or pink (especially when old).
   Flowers 1 cm. wide; fruit 1-seeded, nut-like. 14. *Gaura coccinea*
   Flowers 4—8 cm. wide; fruit a many-seeded capsule
                                     6-13. *Oenothera*

Epilobium   WILLOW-HERB

Annuals or perennials, with many, narrow, alternate leaves; flowers white to purple in branching clusters; pods long and slender, many seeded; seeds with a tuft of fine hairs at tip.

Key to Species

Flowers purple, 1—3 cm. (⅖—1⅕ in.) wide; plants showy, 1 m. (3 ft.)
  high.   1. *Epilobium angustifolium*
Flowers white or pinkish, 5—8 mm. (¼ in.) wide; plants 3—8
  dm. (12—32 in.) high.
   Leaves 1—2 cm. wide; plant stout and leafy. 4. *Epilobium glandulosum*
   Leaves 5—10 mm. wide; plants rather slender.
     Annual; stigma 4-cleft.   2. *Epilobium paniculatum*
     Perennial; stigma entire or nearly.   3. *Epilobium leptophyllum*

1. *Epilobium angustifolium* L. FIREWEED. Leafy perennial; stems 1—2 m. (3—7 ft.) high; leaves lanceolate, 5—15 cm. (2—6 in.) long; flowers purple or rose colored, 1—2.5 cm. wide, in showy, terminal clusters; pods 5—10 cm. long. Wet, especially wooded places. July, Aug. We have specimens from most of the eastern and northern counties but it is relatively rare and local in North Dakota. In burned lands of northern Minnesota one may see acres of it—a very showy display.

2. *Epilobium paniculatum* Nutt. WILLOW-HERB. Slender leaved annual, 3—6 dm. high; flowers 8 mm. wide; pods 2—3 cm. long. Uncommon in boggy places, chiefly northern parts of State.

3. *Epilobium leptophyllum* Raf. Very similar to last but perennial; pods 4—5 cm. long. Richland, Ransom, Benson and Pembina Counties. Also known as *E. densum* Raf.

4. *Epilobium glandulosum*, var. *adenocaulon* (Hausskn.) Fern. Perennial, 3—9 dm. high, with many leaves 1—2 cm. wide; pods 4—6 cm. long. Our commonest willow-herb, growing along edges of ponds, river banks, etc. July, Aug.

5. *Boisduvalia glabella* (Nutt.) Walp. Low, branching annual, 1—3 dm. high; leaves lanceolate, 1—1.5 cm long, finely toothed; flowers at leaf bases, purple, 3—5 mm. wide; pod 6—8 mm. long. July. A small plant, somewhat like *Gaura*, with very inconspicuous flowers. I collected this in a dried up water hole in a prairie pasture near New England, Hettinger County, in 1935. It was identified by Dr. P. A. Munz. The following year Mrs. Anna Meissner of Mott, found specimens in the eastern part of the same county. The general range of the species is Calif. to B. C. and Nev.

Oenothera   EVENING PRIMROSE

Usually stout biennials or perennials with simple, alternate leaves; flowers showy, yellow or white; petals 4, sepals 4; stamens 8; pods many seeded.

Key to Species

Flowers yellow.
  Stemless plant, similar to a dandelion in habit.   12. *Oenothera flava*
  Plants with leafy stems 2—10 dm. (8—40 in.) high.
    Prairie perennial; flowers open in daytime. 13. *Oenothera serrulata*
    Weedy biennials; flowers open at night.

Low, spreading plants; flowers at leaf bases. 7. *Oenothera laciniata*
Upright plants 1—1.5 m. (3—5 ft.) high; flowers in long spikes.
Petals 5 mm. (⅕ in.) wide; rare plant. 8. *Oenothera rhombipetala*
Petals 1—2 cm. wide; common weed.   6. *Oenothera strigosa*
Flowers white, fading to pink.
Low, short stemmed plant on barren clay soils. 11. *Oenothera caespitosa*
Plants with tall stems, usually in sandy soils.
Annual; stems hairy, 2—4 dm. high.   9. *Oenothera albicaulis*
Perennial; stems smooth, 4—10 dm. high.   10. *Oenothera nuttallii*

6. *Oenothera strigosa* (Rydb.) Mack. & Bush. COMMON EVENING PRIMROSE. (Fig. 255). Biennial; stems stout, quite woody, often widely branched at base, 1—1.5 m. (3—5 ft.) high; leaves lanceolate, 2.5—10 cm. (1—4 in.) long; flowers yellow, 2.5—4 cm. wide, each one at base of a small leaf, forming a long cluster; pods 2.5—3.5 cm. long. Roadsides and old fields. Common. July-Sept. Dr. R. R. Gates described *O. rubricapitata* and *O. albinervis* from North Dakota material (Phil. Trans. Roy. Soc. Lond. Sec. B. 226:239-355, 1936.) It is a puzzling group. The name *O. biennis* is often used to cover the common forms.

7. *Oenothera laciniata* Hill. Annual; stems spreading, 1—5 dm. high; leaves oblong or lanceolate, 2.5—5 cm. long, toothed or pinnately lobed; flowers 1—2 cm. wide at leaf bases. Grant and Billings Counties are our only records.

8. *Oenothera rhombipetala* Nutt. Similar to No. 6, but has narrow petals. One specimen from Richland County.

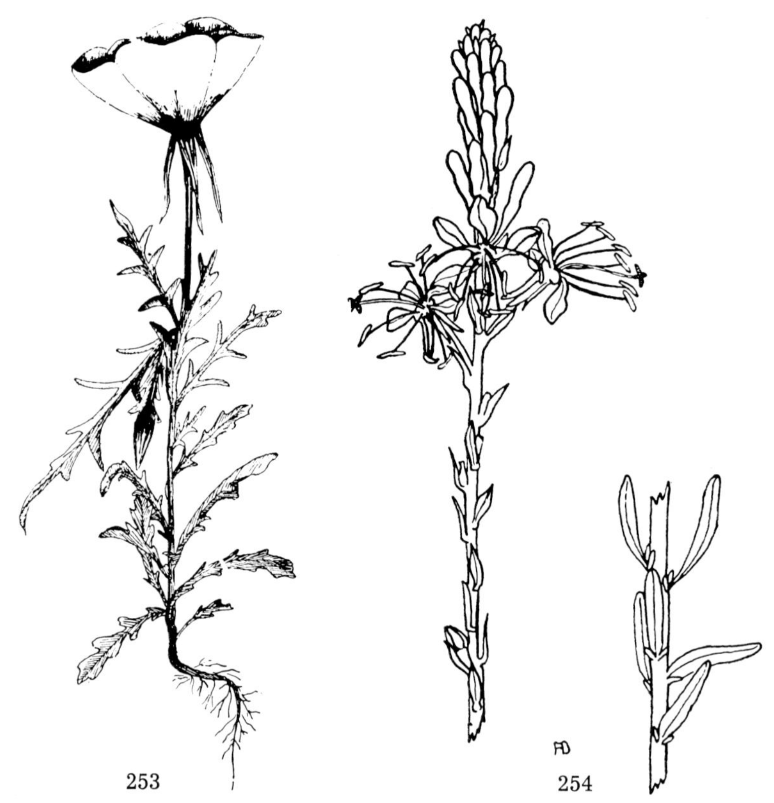

253. Evening Primrose (*Oenothera albicaulis*).
254. Gaura (*Gaura coccinea*).

9. *Oenothera albicaulis* Pursh. (Fig. 253). Annual, finely hairy; stems branching, 1—3 dm. high; leaves lanceolate, 2.5—10 cm. long, deeply pinnately lobed; flowers white, 3—7 cm. wide. Frequent in sandy soil west of Missouri River. July.

10. *Oenothera nuttallii* Sweet. WHITE-STEMMED EVENING PRIMROSE. Perennial, spreading by roots; stems 6—12 dm. (2—4 ft.) high; leaves linear to oblong, 2.5—8 cm. (1—3 in.) long, entire or toothed; flowers white, 2.5—6 cm. wide; pods slender, 2—3 cm. long. Frequent in very sandy soil. July, Aug. The stems are white and smooth as if enameled and often rather bare of leaves. The flowers turn pink on fading. Formerly referred to *O. pallida* Lindl.

11. *Oenothera caespitosa* Nutt. BUTTE PRIMROSE. "GUMBO LILY." Perennial, stem very short; leaves clustered, 5—15 cm. long, oblanceolate, usually toothed; flowers white, 3.5—8 cm. wide; pods woody, rough, about 2.5 cm. long. Frequent on bare, clay buttes or flats west of Missouri River. A showy plant which is being grown as an ornamental. It is of course not a lily, but the name "gumbo lily" is commonly used in this area.

12. *Oenothera flava* (A. Nels.) Munz. Stemless perennial; leaves 5—30 cm. long, narrow, pinnately toothed or with narrow lobes; flowers yellow, 2.5—5 cm. wide. Clay flats; Cass, Towner, McLean, Bowman, Stark and Divide Counties. June. We have considered this rare, but it has been sent in as a weed two or three times.

255   256

255. Evening Primrose (*Oenothera strigosa*).

256. Enchanter's Nightshade (*Circaea quadrisulcata*).

13. *Oenothera serrulata* Nutt. Tooth-leaved Evening Primrose. Perennial; stems slender, erect, 1.5—5 dm. (6—20 in.) high; leaves linear—oblong, 2—5 cm. (1—2 in.) long, with fine teeth; flowers yellow, 2—3 cm. wide; capsules slender, 1.5—3 cm. long. Common on prairie. June-Aug. Flowers open in day time.

14. *Gaura coccinea* Pursh. Gaura. "Honeysuckle." "Waving Butterfly." (Fig. 254). Perennial, spreading by roots; stems spreading or erect, 2—3 dm. high; leaves linear, 1—3 cm. long, slightly toothed, usually gray; flowers white to scarlet, 1 cm. wide in terminal spikes; fruit rounded or oblong, 4-sided toward tip, 5 mm. long, 1-seeded. June, July. Common on prairie. Flowers open about 5 p. m., when the patches are quite showy. I had thought the flowers were white and faded into pink, but in 1946 I found many patches in an old field at Amidon, Slope County, where there were great differences in the color of the fresh flowers. Locally it is often called honeysuckle, which the flowers do resemble.

15. *Circaea quadrisulcata* (Maxim.) Franch. & Sav. Enchanter's Nightshade. (Fig. 256). Perennial; stems erect, 3—6 dm. high; leaves opposite, ovate, 5—10 cm. long, very thin; flowers white, 3 mm. wide, in slender clusters at top of plant; fruit rounded, 4 mm. long, covered with hooked hairs. Woods, eastern edge of State. July, Aug. Previously referred to *C. lutetiana* L., the European plant.

16. *Circaea alpina* L. Similar to last but small, fragile, 1—2 dm. high; leaves cordate, 2—5 cm. long. Shaded places around cold springs. Barnes, Benson, Pembina and Bottineau Counties (fig. 15).

WATER MILFOIL FAMILY Haloragidaceae

Water plants with very small or finely divided leaves and very small flowers; flower parts 2—4, stamens sometimes 8; fruit separating into 4, 1-seeded pieces.

Key to Species

Submerged leaves divided into narrow segments (double comb-like—
    fig. 66); stamens 4 or 8.
  Fruit segments smooth; stamens 8.   1. *Myriophyllum exalbescens*
  Fruit segments rough and with 2 ridges; stamens 4.
           2. *Myriophyllum heterophyllum*
All leaves entire, narrow, in whorls (fig. 68); stamen 1, no petals nor
  sepals.   3. *Hippuris vulgaris*

1. *Myriophyllum exalbescens* Fern. Water Milfoil. Stems floating in water, flowering tips projecting a little; leaves in a whorl of 4 at each node. A common plant growing with pondweeds. Previously referred to the European *M. spicatum* L.

2. *Myriophyllum heterophyllum* Michx. Recorded from Barnes and LaMoure Counties only.

3. *Hippuris vulgaris* L. Marestail. A perennial in wet places, usually seen as stout stems 3—6 dm. high, densely covered with narrow leaves, protruding above shallow water or from dried up pools. Six widely scattered records.

GINSENG FAMILY Araliaceae

Perennial herbs with divided leaves, small, clustered flowers and berry-like fruits.

1. *Aralia nudicaulis* L. Wild Sarsaparilla. Perennial from a short, thick rhizome; leaves twice compound, standing 3—6 dm. (1—2 ft.) high; leaflets oblong, 3—6 cm. long; flowers yellowish white, 3 mm. (⅛ in.) wide in umbels from rhizome; berries 5 mm. wide, dark purple. Local-

ly abundant in woods; Richland, Cass, Pembina, Ramsey, Benson, Bottineau and Dunn Counties. July. This plant is frequently mistaken for ginseng (*Panax quinquefolium*), which has a 5-foliate leaf more like that of a blackberry and does not occur in North Dakota.

CARROT FAMILY   Ammiaceae or **Umbelliferae**

Herbs with alternate, usually compound leaves and small white or yellow flowers in simple or compound umbels; petals and stamens 5 each, sepals minute or absent, all attached to top of ovary; styles 2; fruit dry, splitting into 2, 1-seeded halves (fig. 83). Shape and structure of fruit is an essential feature of the technical classification. Each half-fruit has usually 3 ribs besides one at each edge. The plants contain aromatic oils (as in dill, fennel and caraway) which give characteristic odors.

Key to Species

Plants 5—20 dm. (1½—7 ft.) high, leafy stemmed.
  Flowers white.
    Leaves large, 1—4 dm. wide, coarsely lobed.   20. *Heracleum lanatum*
    Leaves medium sized, variously compound.
      Leaves with 3—7 coarse lobes, palmate or ternate.
        Leaves with 5—7 palmate divisions (fig. 257); fruit covered
          with hooked bristles.   1. *Sanicula marylandica*
        Leaves ternate with 3—7 lobes; fruit smooth.
              2. *Cryptotaenia canadensis*
      Leaves with 7—many leaflets, usually pinnate, at least in part.
        Fruits 2—3 cm. (1 in.) long, black; leaves soft hairy.
             3. *Osmorhiza longistylis*
        Fruits 3—5 mm. (1/6 in.) long, green to brown; leaves usually
          smooth.
          Leaf segments 1—2 cm. wide, toothed; marsh plants.
            Leaves 1—3 times ternate or somewhat pinnate (fig. 258).
               11. *Cicuta maculata*
            Leaves once pinnate (fig. 259).
              Leaflets lanceolate to linear, 4—10 cm. long. 10. *Sium suave*
              Leaflets ovate or oblong, 1—3.5 cm. long.   9. *Berula pusilla*
        Leaf segments narrow and numerous.
          Marsh plant bearing bulblets at upper leaf bases.
              12. *Cicuta bulbifera*
          Introduced plants without bulblets.
            Fruit with bristly ribs.   4. *Daucus carota*
             Fruit smooth.
               Fruit 3—4 mm. long; leaves 3—5 cm. wide.
                 17. *Carum carvi*
               Fruit 2—3 mm. long; leaves 10—20 cm. wide.
                8. *Conium maculatum*
  Flowers yellow.
    Plants 1—2 m. (3—7 ft.) high; all leaves pinnate, 2—4 dm. long.
          19. *Pastinaca sativa*
    Plants 6—10 dm. (2—3 ft.) high; leaves ternate or little lobed.
      All leaves 2—3 ternate.   6. *Zizia aurea*
      Lower leaves cordate, not divided (fig 260).   5. *Zizia aptera*
Plants 1—3 dm. high; stemless or short-stemmed; leaves finely divided.
  Leaves finely hairy, fruit flat and thin.
    Flowers yellow.   16. *Lomatium foeniculaceum*
    Flowers white.
      Fruit circular or broadly oblong, 6—8 mm. long.
          14. *Lomatium orientale*

Fruit oblong, 10—12 mm. long.　　15. *Lomatium macrocarpum*
Leaves not hairy; fruit not flat and thin.
　Flowers yellow; fruit oblong, not winged.　7. *Musineon divaricatum*
　Flowers white; fruit with several, winged ribs.

13. *Cymopterus acaulis*

1. *Sanicula marylandica* L. BLACK SNAKEROOT. Perennial; stem erect, 6—10 dm. (2—3 ft.) high; leaves few, the basal on petioles 1—3 dm. long; leaflets oblanceolate, 5—15 cm. (2—6 in.) long, sharply toothed or somewhat lobed (fig. 257); flowers white, in few flowered, terminal clusters; fruit rounded, 5—7 mm. (¼ in.) long, covered with soft, hooked bristles. Common in woods. July, Aug.

2. *Cryptotaenia canadensis* (L.) DC. HONEWORT. Perennial; stem erect, 3—9 dm. high; leaflets 3, oblong-ovate, 3—10 cm. long, sharply toothed, often deeply divided into 2 or 3 segments; flowers white, in slender clusters at ends of branches; fruit linear-oblong, slightly flattened, 4—8 mm. long. Woods; southeast 4 counties, also Pembina and Morton. July.

3. *Osmorhiza longistylis* Torr. SWEET CICELY. Perennial from thick roots; stems spreading, 2—4 dm. high; leaves large, few, at first ternate, then pinnate; leaflets lanceolate, 2—5 cm. long, deeply toothed or lobed, thin, soft; flowers white, in clusters 3—5 cm. wide; fruit narrowly oblong-oblanceolate, not flattened, 2—3 cm. long, black at maturity, the styles remaining to form long, slender hooks. Frequent in woods. Blooms May; fruit ripe Aug. The roots are fragrant and edible. One specimen from Abercrombie, Richland County, is var. *villicaulis* Fern., stems covered with long, soft hairs.

4. *Daucus carota* L. WILD CARROT. Biennial; stem erect, 4—12 dm. high; leaves finely divided, 1—2 dm. long; flowers white or pinkish, in clusters 1—1.5 dm. wide; fruit oblong, slightly flattened, 2—4 mm. long, the ribs bristly. A rare weed here, not established; collected in Cass, Grand Forks, Ramsey and Benson Counties.

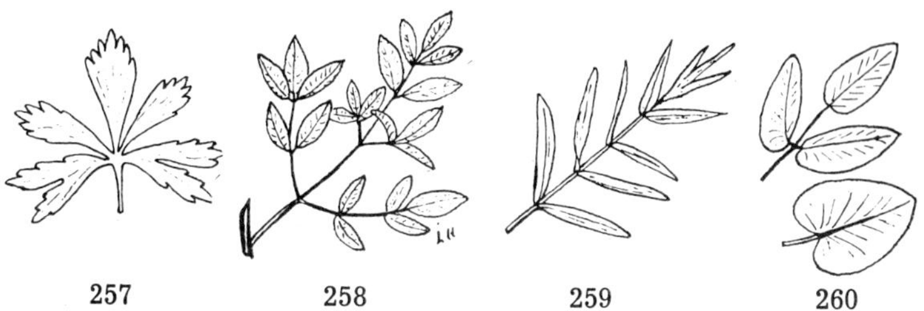

257　　　　　　258　　　　　　259　　　　　　260

5. *Zizia aptera* (A. Gray) Fern. MEADOW PARSNIP. Perennial; stem erect, 3—6 dm. (1—2 ft.) high; leaves smooth, thick, the lowest cordate, 4—10 cm. (1½—4 in.) long; the upper divided into 3 oblong or lanceolate leaflets which are toothed or lobed (fig. 260); flowers yellow, in clusters 3—5 cm. wide; fruit oblong, slightly flattened, ribs not prominent, 2—4 mm. (⅛ in.) long. Common in coulees and moist meadows. May, June. Formerly called *Z. cordata*, a name given earlier to a different plant.

6. *Zizia aurea* (L.) Koch. MEADOW PARSNIP. Perennial; stem erect, 4—8 dm. high; leaves all divided into lanceolate leaflets or lobes 2—5 cm. long; flowers and fruit as in *Z. aptera*. In or near woods, eastern half of State. June.

7. *Musineon divaricatum* (Pursh) Nutt. WILD PARSLEY. Perennial from a thick root; stem with spreading branches, 5—10 cm. long; leaves, smooth, thick, glossy, 2—6 cm. long, pinnately divided in narrow leaflets and lobes; flowers yellow, in clusters 3—6 cm. wide on a stalk

1—2 dm. high; fruit oblong, slightly flattened, smooth, 3—6 mm. long. Clay flats, especially below buttes, chiefly west of Missouri River; also collected at Valley City, Barnes County, and Kulm, LaMoure County. June.

8. *Conium maculatum* L. POISON HEMLOCK. Biennial; stem erect, 1—2 m. high; leaves 1—3 dm. long, soft, divided into narrow segments but main divisions widely spreading; flowers white, in clusters 1—2 dm. wide; fruit rounded, 2 mm. long, ribs prominent. Well established in a farmyard near Buffalo, Cass County, in 1924. This was used by the ancient Greeks to execute criminals and is said to have been given to Socrates.

9. *Berula pusilla* (Nutt.) Fern. CUT-LEAVED WATER PARSNIP. Perennial; stems weak, spreading or floating, 3—6 dm. long; leaves 1—3 dm. long, pinnately compound of 11—19 oblong leaflets, some leaves with pinnately divided or lobed leaflets; flowers white, in clusters 5—10 cm. wide; fruit rounded, 1—2 mm. long. In flowing springs in Richland-Ransom sandhill area and in one near Williston, Williams County. Formerly included in the European *B. erecta* (Huds.) Coville.

10. *Sium suave* Walt. WATER PARSNIP. Perennial from a hardened base and mass of slender roots; stem stout, erect, 1—2 m. (3—7 ft.) high; leaves 1—4 dm. (4—16 in.) long, pinnate with 7—17, lanceolate, finely toothed leaflets 4—10 cm. (1½—4 in.) long (fig. 259). flowers white, in clusters 1—2 dm. wide; fruit rounded, 3 mm. (⅛ in.) long, ribs prominent. Common in wet places. July, Aug. Formerly called *S. cicutaefolium* Gmel. In general appearance it resembles water hemlock but is easily distinguished by either roots or leaves. Lowest leaves, developing in water, are finely divided.

11. *Cicuta maculata* L. WATER HEMLOCK. Perennial from several thick roots 5—15 cm. long; stem erect, 1—2 m. high, usually purplish; leaves 1—3 dm. long, 2—4 times ternately (in 3 parts) divided into lanceolate leaflets 2—5 cm. long (fig. 258); flowers white, in clusters 1—2 dm. wide; fruit oblong, 4 mm. long, ribs yellowish, rounded. Common in wet ground, especially· northeast. July. Small leaves may have only 3 leaflets, large ones may be somewhat pinnate. The roots, especially, are very poisonous. The plant seems not to cause trouble in hay.

12. *Cicuta bulbifera* L. BULBOUS WATER HEMLOCK. Similar to last but leaflets linear; flowers often or usually replaced by bulblets. Specimens only from Richland and Pembina Counties.

13. *Cymopterus acaulis* (Pursh) Raf. WILD PARSLEY. Perennial; leaves and flower stalks from a thick root; leaves 5—15 cm. long, divided into linear segments; flower stalk 5—15 cm. high; flowers white, the cluster 2—5 cm. wide; fruit 6—8 mm. long, nearly as wide, all ribs extending into thin wings 1—2 mm. wide. Dry prairies and hills, chiefly west of Missouri river. Lunell found it in Benson County. Early May. In flower the plant closely resembles No. 14, but leaf segments are narrower and smoother, base of petiole slender, not winged.

14 *Lomatium orientale* Coult. & Rose. WILD PARSLEY. Perennial from a thick root, leaves 5—15 cm. long, divided into narrowly lanceolate segments, somewhat hairy, base of petiole broad and winged; flower stalks 5—20 cm. high; flowers white, in clusters 3—6 cm. wide; fruit circular to broadly oblong, 5 mm. long, flat, lateral ribs winged. Common on prairie. Late Apr.-May.

15. *Lomatium macrocarpum* (H. & A.) Coult. & Rose. Taller than last, leaves finely divided; fruit oblong, 10—15 mm. long. On clay flats, Walsh to Ward Counties.

16. *Lomatium foeniculaceum* (Nutt.) C. & R. YELLOW WILD PARSLEY. Perennial from thick root; leaves 5—15 cm. long, finely divided and somewhat hairy; flowers yellow, in clusters 5—10 cm. wide, fruit circular or broadly oblong, 7—10 mm. long, flat, lateral ribs winged. Prairie, chiefly west of Missouri River, also LaMoure, Barnes, Grand Forks Counties and probably local elsewhere. May.

17. *Carum carvi* L. CARAWAY. Biennial; stem erect, 3—6 dm. high; leaves 7—15 cm. long, finely divided into narrow segments; flowers white, in clusters 5—10 cm. wide; fruit oblong, 3—4 mm. long. An occasional garden escape.

18. *Anethum graveolens* L. DILL. Annual; stem erect, 3—10 dm. high; leaves 1—3 dm. long, finely divided, smooth, dark green; flowers yellow, in clusters 8—15 cm. wide; fruit oblong, 4 mm. long. Frequently persists in gardens.

19. *Pastinaca sativa* L. WILD PARSNIP. Biennial; stem coarse, erect, 1.5—2 m. (5—7 ft.) high; lower leaves 2—3 dm. (8—12 in.) long, pinnately compound of 5—7, ovate leaflets 3—5 cm. (1½—2 in.) long, more or less toothed and lobed; flowers yellow, in clusters 1—1.5 dm. wide; fruit 5—7 mm. (¼ in.) long, nearly circular, flat, lateral ribs winged. A frequent garden escape.

20. *Heracleum lanatum* Michx. COW PARSNIP. Rough hairy perennial; stem erect, 1—2 m. high, very stout, hollow; leaves of 3, angular or rounded leaflets, 1—3 dm. wide; flowers white, in a cluster 1—2 dm. wide; fruit 1 cm. long, oblong, flat, the lateral ribs winged, each half with 4 brown marks (oil tubes) in upper half. Woods, chiefly east of Missouri River. June.

DOGWOOD FAMILY　Cornaceae

Mostly shrubs with opposite, oblong leaves and small, white flowers in dense, flat clusters; sepals 4, very small; petals 4, stamens 4, attached on top of ovary; pistil 1; fruit fleshy with a large stone. The identity of some of the species has been puzzling. Flowering Dogwood (*Cornus florida*) of southeastern U. S. has 4 large, white bracts around each cluster, as does also our Bunchberry.

Key to Species

Small, hardly woody plant; flower clusters with 4 large, white bracts.

1. *Cornus canadensis*

Shrubs; flower clusters without bracts.
　Bark often bright red; leaves oblong or ovate, abruptly pointed.

2. *Cornus stolonifera*

　Bark gray; leaves ovate or lanceolate, tapering gradually.

3. *Cornus racemosa*

1. *Cornus canadensis* L. BUNCHBERRY. DWARF CORNEL. Stems 1—1.5 dm. (4—6 in.) high from a rhizome; a cluster of flowers at top of stem, surrounded by leaves (4 on stems without flowers); flower cluster including bracts 2—3 cm. (1 in.) wide; fruit bright red, 6 mm. (¼ in.) wide, quite persistent. June. Local in woods, Pembina and Turtle Mts.

2. *Cornus stolonifera* Michx. KINNIKINNICK. RED OSIER. Shrub 1.5—3 m. (5—10 ft.) high; leaves ovate or elliptic, 5—15 cm. long, whitish below; flowers white in a flat cluster 5—10 cm. wide at ends of new branches; fruit white, 6—7 mm. wide. Along streams, drainage channels and other low places. June-Aug. The new basal shoots are ornamental in winter. The inner bark was one kind of "kinnikinnick" used by Indians for smoking. A similar species seen in plantings is European dogwood (*Cornus sanguinea* L.)

3. *Cornus racemosa* Lam. GRAY DOGWOOD. Similar to last but bark gray. In drier locations, as in woods and brush away from drainage channels. Rickett (N. Am. Fl. 28B:301, 1945) uses this name in place of *C. femina* Mill., which is of uncertain identity. Blue fruited dogwoods also occur but probably are not a different species.

HEATH FAMILY **Ericaceae**

Leaves simple, alternate; petals 5, separate or united; stamens 5 or 10, anthers often with slender appendages and opening by a small hole; pistil 1, ovary 3—10-celled, free from calyx or united with it (in blueberries). The main part of this family includes shrubs with evergreen leaves and showy pink or white flowers. They grow chiefly in acid soils, either bogs or sandy uplands. Of this group, only bearberry occurs in North Dakota. The other plants included here are sometimes put in separate families.

Key to Species

Plant not green; a white stem 1—2 dm. (4—8 in.) high with 1 flower.
                                             4. *Monotropa uniflora*
Green plants; flowers white or pink.
  Creeping, woody, evergreen; fruit a red berry. 5. *Arctostaphylos uva-ursi*
  Nearly stemless from a rhizome; not woody; fruit a capsule.
    Flowers pink; leaf blades nearly circular.   1. *Pyrola asarifolia*
    Flowers white; leaf blades usually elliptic.
      Flowers on all sides of flower stalk.   2. *Pyrola elliptica*
      Flowers all on one side of stalk.   3. *Pyrola secunda*

1. *Pyrola asarifolia* Michx. ROUND-LEAVED WINTERGREEN. (Fig. 261). Perennial with 5 or 6 thick, rounded leaves 2—3 cm. (1 in.) wide; flower stalk 2—3 dm. (8—12 in.) high with 6—10 pink or rose purple flowers 1—1.5 cm. wide. The largest and most showy species blooming in July. Local but sometimes common in moist, rich soil in aspen woods; Pembina, Benson, Rolette, Bottineau and Dunn Counties.

2. *Pyrola elliptica* Nutt. A smaller plant with greenish white flowers; same distribution.

3. *Pyrola secunda* L. Much like No. 2; same distribution, also Barnes and Richland Counties.

4. *Monotropa uniflora* L. INDIAN PIPE. GHOST PLANT. Saprophytic, living entirely on decaying vegetation. Woods. Recorded only for Ransom, Pembina and Rolette Counties.

5. *Arctostaphylos uva-ursi* (L.) Spreng. BEARBERRY. KINNIKINNICK. (Fig. 54.) Woody, creeping stems, sometimes 1 m. (3 ft.) long; leaves 2—3 cm. (1 in.) long, narrow, leathery; flowers bell shaped, pinkish, 3 mm. (⅛ in.) long; fruit red, 5 mm. wide. On stony, protected ridges; Pembina and Turtle Mts.; north sides of buttes in Bad Lands. Flowers in May; fruit persisting to next year. The leaves were used by Indians for smoking.

PRIMROSE FAMILY **Primulaceae**

Herbs with simple leaves; flowers usually showy, petals 5, partly united; stamens usually 5, opposite the corolla lobes; pistil one; fruit a 1-celled capsule with several, angular seeds. The florist's primroses belong to this family.

Key to Species

Leafy stemmed plants; flowers clustered or solitary in leaf axils.
  Low plants, 2 dm. (8 in.) or less high; flowers 3 mm. (⅛ in.) or less wide, pink.
    Plant spreading, leaves thin, flowers 1 mm. wide, petals present.
                          8. *Centunculus minimus*
    Plant upright, leaves thick, flowers 3 mm. wide, petals absent but sepals colored.   9. *Glaux maritima*
  Plants 2—8 dm. high; flowers yellow, 5—15 mm. wide or in conspicuous clusters.

Flowers 5 mm. wide in axillary spikes.　　　7.　*Lysimachia thyrsiflora*
Flowers 10—15 mm. wide, solitary on long stalks in leaf axils.
　Leaves very narrow, 2—3 cm. (1 in.) long, 3 mm. wide.
　　　　　　　　　　　　　　　　　6.　*Lysimachia longifolia*
　　Leaves 1—3 cm. wide.
　　　Leaves oblong, rounded at ends.　　　4.　*Lysimachia ciliata*
　　　Leaves lanceolate, gradually narrowed at ends.
　　　　　　　　　　　　　　　　5.　*Lysimachia verticillata*
Stem very short; flowers clustered at tip of stalk from rosette of
　leaves.
　Flowers 5—15 mm. wide; leaves 5—20 cm. long.
　　Corolla lobes spreading, united at base into a tube.　1.　*Primula incana*
　　Corolla lobes bent back, little united.　　10.　*Dodecatheon meadia*
　Flowers 2—3 mm. wide; leaves 1—3 cm. long.
　　Corolla shorter than calyx, mostly concealed. 2.　*Androsace occidentalis*
　　Corolla equalling calyx, evident when in bloom.
　　　　　　　　　　　　　　　3.　*Androsace puberulenta*

261　　　　　　　　　　　262

261.　Wintergreen (*Pyrola asarifolia*).

262.　Fairy Candelabra (*Androsace occidentalis*).

1.　*Primula incana* M. E. Jones. PRIMROSE. Small perennial with a few basal
　leaves and flower stalk 5—20 cm. (2—8 in.) high with cluster of sev-
　eral flowers, each 5—8 mm. (¼ in.) wide. Our only record of this in
　North Dakota is a specimen in grass at edge of pond from Powers Lake,
　Mountrail County, sent by Harriet McDonnell, June, 1931. In 1946, I
　visited the location with Mrs. Lester Enget, who found the plant, but
　we did not find any more. We have specimens from Carson, Sask., which
　were examined by Dr. S. F. Blake of the U. S. Dept. of Agriculture.

2.　*Androsace occidentalis* Pursh. FAIRY CANDELABRA. (Fig. 262). Winter
　annual; rosette of leaves 2—3 cm. long, bears several stalks 5—10 cm.

long with a cluster of tiny white flowers at top of each. Flowering last of April; corolla hidden in calyx. A common, little, weedy plant in fields and bare places on prairie.

3. *Androsace puberulenta* Rydb. A little larger than last, flowers distinctly larger and corolla evident. Less common but locally abundant westward. Barnes County and Park River in Walsh County are the most eastern records.

4. *Lysimachia ciliata* L. FRINGED LOOSESTRIFE. Perennial, spreading vigorously by rhizomes; stems 2—8 dm. (8—32 in.) high, leaves opposite, ovate, 3—10 cm. (1—4 in.) long, with hair fringes on petioles; flowers on long, slender stalks from leaf bases, corolla yellow, saucer shaped, deeply lobed, 15 mm. (⅗ in.) wide; pod rounded, 5 mm. wide. Common in woods or moist open ground. July. A showy plant, but weedy. The corollas fall off during the day. This species and the two following are often placed in a separate genus, *Steironema*.

5. *Lysimachia verticillata* Greene. Perennial, without long rhizomes; stems 3—6 dm. high; leaves lanceolate, 3—10 cm. long, petioles with few, short hairs. Wet places around ponds or in sloughs and ditches, apparently not common. *L. lanceolata* Walt. is said to be similar with long rhizomes, but I have been unable to recognize it in our material.

6. *Lysimachia longifolia* Pursh. Stems slender, somewhat tufted, 2—4 dm. high; leaves narrow, 2—5 cm. long. Wet, sandy soil; Richland and Ransom Counties only. This is *L. quadrifolia* Sims, but not *L. quadrifolia* L.

7. *Lysimachia thyrsiflora* L. TUFTED LOOSESTRIFE. Perennial; stems stout and leafy, 3—6 dm. high; leaves lanceolate, 5—10 cm. long; flowers 4—6 mm. wide, in dense spikes 2—5 cm. long from middle leaf axils. Swampy places, eastern half of State. Also called *Naumbergia thyrsifolia* (L.) Duby.

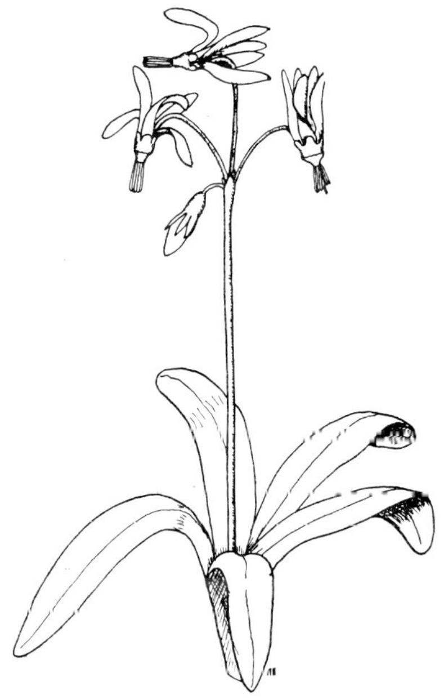

263. Shooting Star (*Dodecatheon media*).

8. *Centunculus minimus* L. Common Pimpernel. Small, spreading annual; branches 1—1.5 dm. long; leaves 4—8 mm. long, flowers very minute, solitary in leaf axils. Wet places; LaMoure, Barnes and Benson Counties.

9. *Glaux maritima* L. Sea Milkwort. Succulent perennial, 1—2 dm. high; leaves thick, narrow, 4—10 mm. long; flowers in leaf axils, pink, 3 mm. wide. On saline soil, low spots in fields, all over State but not very common.

10. *Dodecatheon meadia* L. Shooting Star. (Fig. 263). Stemless perennial with rosette of basal leaves 1—2 dm. long, 1 cm. wide; flower stalk 1—3 dm. high, cluster of 3—6 pink to purple flowers 1—1.5 cm. wide at top. On wet ground, as in semi-bare sloughs and in boggy places; Rolette, Mountrail and Bowman Counties. June.

OLIVE FAMILY Oleaceae

Mostly shrubs or trees with opposite, often compound leaves; petals usually 4, partly united; stamens 2 or 4, attached to corolla if it is present; fruit a capsule, berry or winged fruit. Lilac and privet are well known shrubs, which in addition to the olive, belong to this family. In our flora are only the ash trees.

Key to Species

Body of fruit round, winged half way to base. 1. *Fraxinus pennsylvanica*
Body of fruit somewhat flattened, winged entirely to base.
2. *Fraxinus nigra*

1. *Fraxinus pennsylvanica*, var. *lanceolata* (Borkh.) Sarg. Green Ash. Tree, 5—10 m. (17—35 ft.) high; leaves opposite, leaflets 5—7, lanceolate or ovate, the terminal one more rounded; flowers in large clusters at old leaf bases, corolla 0, a small calyx on pistillate flowers; fruit 2.5—5 cm. (1—2 in.) long (fig. 48). One of our most common trees in eastern part of State and occurring sparsely in most gullies of rough land in western part. Early May. Leaves usually have a few hairs along main veins below or occasionally all over blades (approaching typical *F. pennsylvanica* Marsh, Red Ash).

2. *Fraxinus nigra* Marsh. Black Ash. Specimens collected at Svold, Pembina County, in 1937, form the only North Dakota record of this species. Mr. H. A. Graves had noticed that the trees looked different from our common green ash. The bark is light gray instead of dark and the leaves usually have 9 leaflets. This is a swamp tree but some have been growing in several yards in Fargo for 40 years.

GENTIAN FAMILY Gentianaceae

Herbs, mostly with opposite, simple, sessile leaves; flowers often showy, petals mostly united into a narrow or wide tube with 5 lobes, sometimes with "plaits" between lobes; stamens 5, attached to corolla; pistil 1; fruit a 1-celled capsule with many small seeds attached to the side in 2 rows.

Key to Species

Leaves compound, from a rhizome; flowers white. 6. *Menyanthes trifoliata*
Leaves simple; flowers blue or yellowish.
Flowers 8—12 mm. (⅓—½ in.) long, densely clustered along the stem. 2. *Gentiana acuta*
Flowers 15—30 mm. long, fewer.
Corolla club shaped, nearly closed at top. 3. *Gentiana andrewsii*
Corolla spreading, somewhat funnel shaped.
Slender bog plant; corolla lobes rounded, longer than tube.
1. *Gentiana procera*

Prairie plants; corolla lobes pointed, shorter than tube.
  Corolla 4 cm. (1⅗ in.) long; calyx lobes spreading.
                                    5. *Gentiana puberula*
  Corolla 2—3 cm. long; calyx lobes erect.   4. *Gentiana affinis*

1. *Gentiana procera* Holm. SMALL FRINGED GENTIAN. Perennial; stems slender, 3—6 dm. (1—2 ft.) high; leaves linear, 2.5—6 cm. (1—2½ in.) long; flowers few, 3.5 cm. long, at tips of upper branches; corolla lobes with short fringes. Seepage areas by springs; eastern half of State. July, Aug. This is closely related to the well known Fringed Gentian (*G. crinita*) of eastern U. S.

2. *Gentiana acuta* Michx. NORTHERN GENTIAN. Annual; stems 3—6 dm. high; leaves lanceolate, 1—5 cm. long; flowers numerous, slender, blue, purplish or yellowish, 8—12 mm. long in a compact, terminal cluster. We have specimens from Barnes, Wells, Benson, Ramsey, Bottineau and Grant Counties, but the only place I have seen it in recent years was on old pond beaches in Turtle Mts., where it was common. Aug. Sometimes considered a variety of the Old World *G. amarella* L.

264. Closed Gentian (*Gentiana andrewsii*).

3. *Gentiana andrewsii* Griseb. CLOSED GENTIAN. (Fig. 264). Perennial, stems rather stout, 3—6 dm. high; leaves oblong—lanceolate, 3—8 cm. long; flowers blue, 2.5—3.5 cm. long, club shaped, clustered at top of stem on small plants or in middle leaf axils on large plants. Often locally abundant in low meadows. Specimens from only Richland, Ransom, Grand Forks and Benson Counties. Late Aug. Also called "bottle gentian" from the shape of the flower. One of our showy late fall flowers.

4. *Gentiana affinis* Griseb. NORTHERN GENTIAN. Perennial; stems often clustered and spreading at base, 2—4 dm. high; leaves oblong to ovate, 1—4 cm. long; flowers blue, 2.5 cm. long, clustered at upper leaf bases. Prairie, chiefly in northern part of State. Aug. I have watched for it in recent years with poor success, though near Williston it was quite common, even on rather dry hillsides. An attractive plant.

5. *Gentiana puberula* Michx. DOWNY GENTIAN. Perennial; stems usually single, 2—5 dm. high; leaves lanceolate to ovate, 2—5 cm. long; flowers 3—5 cm. long, blue, bell shaped, few at upper leaf bases. Prairie. Specimens from Ransom, LaMoure, Cass, Barnes and Ramsey Counties. Aug.

6. *Menyanthes trifoliata* L. BUCKBEAN. Perennial; leaves from stout rhizome; leaflets 3, oblong or obovate, 5—10 cm. long; flowers white or purplish, 1 cm. wide, several on a stalk from the rhizome, petals bearded. Known only from a boggy area near Sheyenne River in Ransom County south of Sheldon. Late May.

DOGBANE FAMILY   Apocynaceae

Our species are perennials with opposite, smooth, oblong leaves and milky sap; flowers small, bell shaped, 5-lobed, white or pink; stamens 5; pistils 2; fruit 2 long, slender follicles (fig. 93); seeds slender with a tuft of hairs. Oleander and vinca, grown as ornamentals, belong to this family.

Key to Species

Flowers pink, pendant; corolla lobes widely spreading.
1. *Apocynum androsaemifolium*
Flowers white, erect; corolla lobes slightly spreading.
Lower leaves widest at base, somewhat clasping. 1. *Apocynum sibiricum*
Lower leaves rounded at base with short petioles.
3. *Apocynum cannabinum*

1. *Apocynum androsaemifolium* L. SPREADING DOGBANE. (Fig. 265). Perennial, 3—6 dm. (1—2 ft.) high, usually widely branched above, with loose clusters of flowers at ends of branches, and from upper leaf axils;

265.   Spreading Dogbane (*Apocynum androsaemifolium*).

flowers 5 mm. (⅕ in.) wide. June, July. A showy plant, found near edges of woods and thickets. It is a widely distributed species but local in North Dakota.

2. *Apocynum sibiricum* Jacq. DOGBANE. INDIAN HEMP. Erect, 3—8 dm. high, often not branched, or with branches from the middle portion, a dense cluster of flowers terminating the main stem and perhaps the branches; flowers 3 mm. wide; fruit 10—15 cm. (4—6 in.) long, 4 mm. thick at base. July, Aug. A common weed in low ground, spreading freely by roots. The stem contains a good fiber of which the Indians made much use. Formerly known as *A. hypericifolium* Ait.

3. *Apocynum cannabinum* L.. Very much like the last. We have it from Billings, McLean and Mountrail Counties only.

MILKWEED FAMILY Asclepiadaceae

Plants with milky sap and opposite, simple leaves; flowers in dense umbels; fruit 2 long follicles; seeds with hair tufts as in dog-banes, but broad and flat. Flower structure not evident (fig. 90); 2 pistils surrounded by a fleshy "crown" containing 5 stamens; pollen remaining in a solid mass in each anther sac, the masses of two adjacent anthers joined by a stalk; crown with 5 tubular "hoods" erect from its base; in *Asclepias,* a slender curved horn projects from the hood; sepals and petals bent back from base after flowers open. This is a large family of mostly tropical plants. The wax plant (*Hoya*) and cactus-like *Stapelia* are often grown as house plants. All of ours are perennial.

Key to Species

Corolla hoods without horns; flowers greenish.
  Flower cluster only 1 at tip of stem; plant quite hairy.
               1. *Acerates lanuginosa*
  Flower clusters several; plant smooth or nearly so.
               2. *Acerates viridiflora*
Corolla hoods with horns; flowers purple, red or white.
  Flowers white or greenish.
    Leaves broad, oblong, 2—5 cm. (1—2 in.) wide.
              5. *Asclepias ovalifolia*
    Leaves very narrow, 2—5 mm. ($\frac{1}{12}$—⅕ in.) wide.
      Stems slender from spreading rootstocks. 8. *Asclepias verticillata*
      Stems rather stout and crowded from a thick crown.
              9. *Asclepias pumila*
  Flowers pink, red or purple; leaves 1—2 dm. (4—8 in.) long,
    2—8 cm. wide.
    Leaves with dense, fine hairs.
      Flowers 8—12 mm. wide, 25—75 in a cluster. 3. *Asclepias syriaca*
      Flowers 20—30 mm. wide, 10—20 in a cluster. 4. *Asclepias speciosa*
    Leaves not hairy; flowers 8—15 mm. wide.
      Flower stalks erect; stems 1—2 m. (3—7 ft.) high; leaves
        lanceolate.   7. *Asclepias incarnata*
      Flower stalks drooping; stems 4—8 dm. (16—32 in.) high;
        leaves narrowly ovate.   6. *Asclepias purpurascens*

1. *Acerates lanuginosa* (Nutt.) DC. Stem 2—4 dm. high; leaves hairy, oblong, 5—8 cm. long, 1 cm. wide. One specimen from Grant County (Bell 248).

2. *Acerates viridiflora* (Raf.) Eaton. GREEN MILKWEED. Stems erect or spreading, 3—5 dm. high; leaves 5—15 cm. long, quite variable in shape, broadly oblong or ovate, sometimes crinkled, lanceolate or very narrowly lanceolate (var. *ivesii* Britton). Frequent on prairie. July.

3. *Asclepias syriaca* L. COMMON MILKWEED. Stout, covered with very short hairs; stems 1—2 m. (3—7 ft.) high, usually not branched; leaves

oblong, 1—2 dm. (4—8 in.) long, 2—5 large umbels of flowers in upper leaf axils; pods 7—10 cm. (3—4 in.) long, 2 cm. thick at base, containing about 200 flat, oblong seeds. A common weed in low places or fields, eastern part of State. July. We have only one record west of Missouri River: Center, Oliver County (Stevens in 1943).

4. *Asclepias speciosa* Torr. SHOWY MILKWEED. Similar to last but hardly as tall, flowers fewer and much larger. Found all over State, especially westward, usually in slightly wet places. Rare in Red River Valley.

5. *Asclepias ovalifolia* Dec. Stems 3—6 dm. high; leaves ovate to elliptic, 4—8 cm. long; 1—3 umbels of white flowers a little smaller than in No. 3. A fairly common, low prairie plant. Eastern half of State. June.

6. *Asclepias purpurascens* L. PURPLE MILKWEED. Somewhat like a small, slender, common milkweed in appearance; leaves smooth, narrowly ovate, 10—15 cm. long. First found at Harwood, Cass County, by Gale Monson in 1926; later at several points in the county and in Richland County.

7. *Asclepias incarnata* L. SWAMP MILKWEED. A tall plant, widely branched at top, 1—2.5 m. high; leaves lanceolate, 5—15 cm. long; flowers dark red on upright stalks. Local in boggy places west to Kidder and Benson Counties.

8. *Asclepias verticillata* L. WHORLED MILKWEED. Stems slender, unbranched, 2—6 dm. high; leaves in whorls of 2—4, hardly 3 mm. wide, 2—6 cm. long; flowers white. A very common plant on fairly moist grassland. Aug.

9. *Asclepias pumila* (A. Gray) Vail. This is difficult to separate from No. 8. I watched for it with no success, until we found a dense colony along a roadside near Leal, Barnes County. A specimen of this (Stevens 468) was verified by Dr. R. W. Woodson. The late W. W. Eggleston had examined our specimens and determined two from Grant County as *pumila* (Bell 472, 1370).

MORNINGGLORY FAMILY   Convolvulaceae

Twining vines; petals united into a funnel shaped or tubular corolla; stamens 5, attached to base of corolla; pistil 1; fruit a thin walled, rounded capsule with normally 4 seeds. Contains in our area two groups which are sometimes treated as distinct families.

Key to Genera

Green, leafy plants; flowers 3—8 cm. (1—3 in.) wide.   1-2. *Convolvulus*
Yellow, leafless, parasitic plants; flowers 1—5 mm. (1/25—⅕ in.) wide.
                                                3-10. *Cuscuta*

Convolvulus   BINDWEED. "MORNINGGLORY"

Perennial vines with broad alternate leaves and large funnel shaped, white or pink flowers. The annual, purple or blue flowered morningglories (*Ipomea*) are weeds farther south but do not persist here. Bush Morningglory (*Ipomea leptophylla*), a coarse, spreading plant from a huge root, bearing large, bright pink flowers, occurs in western South Dakota.

Key to Species

Flowers 4—8 cm. (1½—3 in.) wide; 2, small, green leaves covering calyx.
Plant not hairy, leaf bases broad and widely spreading.
                                                1. *Convolvulus sepium*
Plant covered with fine hairs, leaf bases more rounded.
                                                1. *Convolvulus sepium* var. *repens*
Flowers 3—4 cm. wide; calyx not covered by a pair of leaves.
                                                2. *Convolvulus arvensis*

266   267

266.   Field Bindweed (*Convolvulus arvensis*).
267. Large Bindweed (*C. sepium*).

1. *Convolvulus sepium* L. LARGE BINDWEED. "MORNINGGLORY" (Fig. 267). Coarse vine, climbing to a height of 1—3 m. (3—10 ft.) in low ground or mostly prostrate in open fields; flowers white or pink, 4—8 cm. (1½—3 in.) wide; pods 5—8 mm. (¼ in.) wide, covered by a pair of bracts (small leaves) 1.5—2.5 cm. long; seeds black, smooth, 5 mm. long. July-Sept. Common native plant in fields, along roadsides and in brush. Spreads by horizontal rhizomes 1—2 dm. below surface of ground. These are white, of rather uniform size or thicker toward the ends. Var. *repens* L. DOWNY BINDWEED, seems a little smaller and more common in drier soils, but the relationships are not satisfactorily worked out. Its seeds are distinctly roughened.

2. *Convolvulus arvensis* L. FIELD BINDWEED. "CREEPING JENNIE." (Fig. 266). A much smaller plant, rarely 1 m. high, often forming a close mat on the ground; leaves often narrow, not much tapered, basal lobes narrow and spreading; capsules 4—5 mm. wide, often only 2-seeded; seeds gray, quite rough, 3.5 mm. long. June-Aug. This is an introduced plant, now widely distributed in the State. It spreads by irregular, horizontal root branches (not rhizomes) and develops very deep roots. It can be distinguished by the roots from the native bindweeds when no flowers are present. There seem to be two strains, one with white and one with pink flowers.

Cuscuta DODDER

Yellow, leafless vines, parasitic on other plants; flowers small (fig. 76), corolla with fringed scales attached to its inner base, one below each stamen. Plants start each year from seeds but in a few days seedlings attach themselves to a host plant from which they secure food and never develop roots. They may become quite destructive. The species are separated by details of flower structure.

Key to Species

Flowers 1—2 mm. (1/25—$\frac{1}{12}$ in.) wide, in dense, rounded heads,
    6—10 mm. wide.
  Calyx lobes longer than wide, edges not overlapping.
                              3. *Cuscuta epithymum*
  Calyx lobes wider than long, edges overlapping at base.
                              4. *Cuscuta planiflora*
Flowers 2—5 mm. wide, in looser or large clusters.
  Corolla lobes incurved at very tips.
    Corolla lobes straight but tips incurved.
      Corolla 3 mm. wide, scales few; pods flattened, often in dense
        clusters 3—8 cm. wide on large stems near ground.
                              5. *Cuscuta coryli*
      Corolla 4—5 mm. wide; scales prominent; pods not flattened.
                              6. *Cuscuta indecora*
  Corolla lobes bent back but tips incurved.
    Corolla 2 mm. wide, constricted in the middle. 7. *Cuscuta pentagona*
    Corolla 3 mm. wide, not constricted.   8. *Cuscuta campestris*
  Corolla lobes erect or spreading.
    Corolla lobes 5, widely spreading; flowers 5 mm. wide.
                           10. *Cuscuta gronovii*
  Corolla lobes 4, erect; flowers 3 mm. wide.   9. *Cuscuta cephalanthi*

3. *Cuscuta epithymum* Murr. CLOVER DODDER. Specimens of this were collected on clover plants in a lawn at Fargo about 1912, but apparently none was preserved and we have no other record. Flax Dodder (*Cuscuta epilinum*) which is similar to this and the next species, is often reported for North Dakota in weed books, but there is no record of its occurrence here.

4. *Cuscuta planiflora* Tenore. SMALL ALFALFA DODDER. The first record of this was secured in Stutsman County in 1922. It was found in Bowman County in 1926, but we have no indication that it has become generally established. The seeds are gray, greenish or yellowish, about 1 mm. long, and much resemble granules of clay or other inert materials. These species are introduced from Europe.

5. *Cuscuta coryli* Engelm. HAZEL DODDER. Vines greenish yellow, not abundant; flowers greenish white; pods 4—5 mm. wide, hardly as long, often in large clusters; seeds dark to light brown, 1.5—2 mm. wide. Common, native plant in low places, growing on various coarse weeds and shrubs. It has occurred on alfalfa in stream bottom fields but has not proved destructive. July-Sept.

6. *Cuscuta indecora* Choisy. LARGE ALFALFA DODDER. Vines yellowish, abundant; flowers white, 4—5 mm. long, often in large clusters; pods rounded, 5 mm. wide. Apparently native but rare. It grows on legumes, also on various other plants and often develops large clusters of pods. On a lake beach near Ashley, Logan County, we found it on *Ambrosia coronopifolia, Atriplex hastata, Chenopodium berlandieri, C. glaucum, Helianthus maximiliani, H. petiolaris, Lactuca scariola, Polygonum lapathifolium, Solanum triflorum* and *Xanthium.* We have specimens from Dunn and Grant Counties on alfalfa, and from Benson County on *Lotus americanus* (Bolley and Lee in 1892, previously reported as *C. arvensis.*) The seeds closely resemble those of Hazel Dodder.

7. *Cuscuta pentagona* Engelm. Vines yellowish, abundant; flowers 2 mm. wide in many small clusters. This was formerly included in *Cuscuta arvensis*. It appears to be a native plant and is usually found upon sunflowers, goldenrods and other native plants on dry soils. Specimens from Richland, Grand Forks, Logan and Barnes Counties.

8. *Cuscuta campestris* Yuncker. FIELD DODDER. Vines yellow, very abundant; flowers 3 mm. wide, in rounded clusters 1—2 cm. wide; seeds brown, 1.25—2 mm. wide. Formerly called *C. arvensis*. Dr. Yuncker regarded it as an American species, but in our area it seems clearly introduced. At Fargo, it forms extensive patches on weeds along the river edge and in similar places. It is the species which is most often troublesome in clover fields.

9. *Cuscuta cephalanthi* Engelm. This seems to be a rare native species. We have specimens from Pembina, McHenry, Ransom and Barnes Counties. One Ransom County lot was growing along river bank on *Amorpha fruticosa, Lycopus lucidus, Scutellaria galericulata, Solidago gigantea and Stachys palustris;* another in sandy soil on *Artemisia caudata*. McHenry County material was on *Artemisia frigida* and *Chrysopsis villosa*.

10. *Cuscuta gronovii* Willd. GRONOVIUS' DODDER. Vines orange yellow in large masses; flowers 5 mm. wide in rounded clusters; pods rounded or a little pointed, 5 mm. wide, seeds dark brown, 2—3 mm. wide. This dodder is one of the most conspicuous, producing large masses of coarse, orange colored stems and not blooming until early August. The flowers also are the most showy of the group. It grows on a great variety of herbs and woody plants, chiefly in wooded places. On the Missouri "bottoms" it often develops in great profusion after the original brush cover has been burned. Curiously enough, it seems not to attack alfalfa. It often occurs on various ornamentals and on potatoes, where these are grown near wooded areas. In 1916, we examined a small field of flax along the Sheyenne River in Barnes County, where both this species and *C. coryli* were abundant on the flax plants. We have another instance of its abundance on flax along the Missouri River in McLean County. Most of our material belongs to var. *curta* Engelm., which Rydberg and Yuncker have regarded as a distinct species.

PHLOX FAMILY Polemoniaceae

Annual or perennial herbs, often with conspicuous flowers; corolla with a slender tube and 5 abruptly spreading lobes; stamens 5, attached to corolla tube; pistil 1, stigmas 3; fruit a rounded, 3-celled, several seeded capsule.

Key to Species or Genera

Flowers 1—2.5 cm. (½—1 in.) wide, white, bluish or pink.
  Leaves entire, lanceolate, oblong or linear.   1-4. *Phlox*
  Leaves divided into narrow segments.   5 *Gilia congesta*
Flowers 2—3 mm. (⅛ in.) wide, white or pink.
  Stem slender, 1—3 dm. (4—12 in.) high; leaves lanceolate,
    2—5 cm. long.   6. *Collomia linearis*
  Stem 1 dm. high, wiry, widely branched; leaves narrow.
      7. *Navarettia minima*

Phlox PHLOX

Our species are perennials with opposite, simple leaves and showy flowers.

Key to Species

Stems erect, 3—6 dm. (1—2 ft.) high; flowers pink or purplish.

                1. *Phlox pilosa*

Stems creeping or 1—2 dm. high; flowers white or bluish.
 Leaves 1 cm. (⅖ in.) long, 1—2 mm. (1/25—$\frac{1}{12}$ in.) wide.
  Stems mostly creeping, green, branch tips erect. 2. *Phlox hoodii*
  Stems mostly erect, lower parts hard and whitish. 3. *Phlox andicola*
 Leaves 5—15 cm. long, 2—4 mm. wide.   4. *Phlox alyssifolia*

1. *Phlox pilosa* L. DOWNY PHLOX. Perennial; stems 3—6 dm. (1—2 ft.) high; leaves opposite, lanceolate, 3—8 cm. (1½—3 in.) long; flowers pink or purplish, 1.5—2 cm. wide in a flat cluster at top of main stem or of a few branches. Late June. We have but two records, Fargo in 1891 and Wahpeton in 1908. Possibly it was more common before the prairie was broken. It is frequent in Minnesota about 40 miles east of Fargo.

2. *Phlox hoodii* Richardson. MOSS PHLOX. HOOD'S PHLOX. (Fig. 268). Perennial, forming dense, moss-like mats 1—3 dm. wide on the ground and flowering profusely in May. A common and showy spring flower on dry prairies and hills in western part, less frequent through central part of State. Kulm, LaMoure County, is our most eastern record. Late specimens from Esmond, Benson County, July 4, had a few flowers about 5 mm. wide.

268. Moss Phlox (*Phlox hoodii*).

3. *Phlox andicola* (Britton) E. Nels. This is usually separated from *P. hoodii* by larger size of flowers but that seems an uncertain guide. It differs in habit of growth, longer leaves (1—1.5 cm.) as well as shreddy bark of lower stem. Apparently it is less common and blooms later. Flowering specimens were collected June 14, 21, 28 and one August 28, possibly an error.

4. *Phlox alyssifolia* Greene. Stems creeping, short and woody; leaves narrowly oblong, 1—2 cm. long; flowers white or bluish, 1.5—2.5 cm. wide. On hills near Buford, Williams County, in 1945 (Stevens 818). Dr. E. T. Wherry verified the identification and wrote that there was an old specimen at the New York Botanical Garden which had been collected at Buford, also one labeled "Phinney, N. D." in 1898. We do not know of such a town. The pods were well developed on June 21 and only one fresh flower was found.

5. *Gilia congesta* Hook. ROUND-HEADED GILIA. Perennial from stout root and stem base; stem erect, 1—1.5 dm. high, densely hairy; leaves alternate, divided into several, narrow segments; flowers white, 5 mm. wide, in dense, terminal clusters. Dry buttes, not common. June, July. Slope, Billings, McKenzie and Williams Counties. There are many species of *Gilia* in western U. S. and some are grown as ornamentals.

6. *Collomia linearis* Nutt. COLLOMIA. Annual; stem slender, erect, 1—4 dm. (4—16 in.) high; leaves alternate, lanceolate, 1.5 -5 cm. (½—2 in.) long; flowers pink, 4—5 mm. (⅕ in.) wide, in head-like clusters at end of main stem or branches. Bases of upper leaves just below flower clusters often are whitish or pinkish. A common, weedy plant in dry soil in bare spots on prairie, in fields and along roads. Occurs on railroad grades in Red River Valley. July. It is often included in *Gilia*.

7. *Navarettia minima* Nutt. Annual; 1 dm. high, with hard, wiry stems, short, fine leaves and dense, spiny clusters of tiny white flowers. Rather rare along small drainage channels on prairie. Grant to Mountrail County. July.

WATERLEAF FAMILY Hydrophyllaceae

Herbs with alternate leaves; corolla bell shaped, 5-lobed; stamens 5, attached to corolla, often projecting beyond it; pistil 1 with 2 stigmas or styles; fruit a capsule, often quite fleshy at first but drying and splitting when ripe; seeds usually 4, large. One striking feature of some species, also found in the next family, is the densely flowered, curved flower clusters.

Key to Species

Perennial from a thick root; flowers 1—2 cm. (⅖—⅘ in.) wide in
    terminal clusters.
  Leaves thin, smooth, divided into several broad lobes.
                        1. *Hydrophyllum virginianum*
  Leaves thick, rough hairy, oblong, not divided. 2. *Phacelia leucophylla*
Annual; flowers 5 mm. (⅕ in.) wide, solitary at leaf bases.
                          3. *Ellisia nyctelea*

1. *Hydrophyllum virginianum* L. WATERLEAF. (Fig. 269). Perennial, forming thick clumps or patches, stems 2—4 dm. (8 -16 in.) high; leaves 1—2 dm. long, deeply lobed and having pale blotches. Flowering at end of May and quite showy with masses of deep to pale lavender colored flowers which do not last long. Abundant in woods along Red River and a short distance up its tributaries. It grows readily from seed in partial shade.

2. *Phacelia leucophylla* Torr. SCORPION WEED. Stiff, rough hairy plant, 2—4 dm. high, leaves elliptic, 5 -10 cm. long; flowers pale lavender, 1 cm. wide in dense, curved, terminal clusters. On dry soil, especially steep, bare slopes. We have it from Billings and Golden Valley Counties only. June. Many species of *Phacelia* occur in western U. S. and several are grown as ornamentals.

269.   Waterleaf (*Hydrophyllum virginianum*).

3. *Ellisia nyctelea* L. WATERPOD. Annual; stems weak, spreading, 1—3 dm. high; leaves 2—4 cm. long, divided into 5—9 oblong lobes; flowers nearly white, 5 mm. wide, solitary at leaf bases; seeds spherical, black, a little larger than mustard seeds, with a honey-combed surface. Late May, drying up by July. Common weed in fields in Red River Valley, less common westward but frequent in brushy places or fields in western part of State.

BORAGE FAMILY   Boraginaceae

Herbs, with alternate, often rough hairy leaves, sometimes with smooth leaves; flowers bell shaped or funnel shaped, usually with slender tube and 5 spreading, rounded lobes; stamens 5, attached to corolla, usually short and hidden in tube; ovary divided into 4 parts, each of which ripens into a hard, 1-seeded nutlet.

This is a large group, the flowers of which much resemble those of the waterleaf family, from which they are distinguished by fruit characters. Technical classification is based largely on minute features of nutlets which we have tried to avoid in the key. The family contains a number of weeds and garden ornamentals.

Key to Species or Genera

Stems and leaves with few or no hairs.
  Flowers white, erect, star shaped; plants of saline flats.
          1. *Heliotropium curassavicum*
  Flowers blue, drooping, bell shaped; prairie plants.
          16. *Mertensia lanceolata*
Stems and leaves with many fine, soft or coarse, rough hairs.
  Flowers 10 mm. (⅖ in.) wide, yellow or red.
    Flowers yellow; nutlets smooth; prairie, spring flowering plants.
      Flowers orange yellow; leaves with fine, soft hairs.
          17. *Lithospermum canescens*
    Flowers pale yellow; stem and leaves rough.
          18. *Lithospermum incisum*

Flowers red; nutlets prickly all over; coarse weed.
        2. *Cynoglossum officinale*
Flowers 2—8 mm. ($\frac{1}{12}$—⅓ in.) wide, white, blue or yellow.
  Flowers 10 mm. long, slender, lobes not spreading; nutlets smooth.
        19. *Onosmodium occidentale*
  Flowers 3—5 mm. long, lobes spreading; nutlets rough or spiny.
    Flowers yellow; plants prickly with coarse hairs which
      break off easily.
      Calyx in fruit 10 mm. long.   14. *Amsinckia idahoensis*
      Calyx in fruit 5 mm. long.   15. *Amsinckia menziesii*
    Flowers blue or white; plants rough but hardly prickly.
      Flowers white, 8 mm. wide; stem stiff, 1—3 dm. high.
        11. *Cryptantha bradburiana*
      Flowers white or blue, 2—4 mm. wide; stems various.
        Nutlets without prickles, usually with ridges or tubercles.
          Flowers blue; stems covered with hooked hairs.
        10. *Asperugo procumbens*
          Flowers white; hairs of stems not hooked.
            Plants erect, gray; leaves 3—5 cm. long.
        12. *Cryptantha calycosa*
            Plants spreading, green; leaves 1—3 cm. long.
        13. *Plagiobothrys scopulorum*
      Nutlets with hooked hairs or prickles, forming burs.
        Nutlets with hooked hairs all over their backs.
        3. *Cynoglossum boreale*
        Nutlets with 1 or 2 rows of stiff, hooked spines around
          edges and sometimes over whole back.
        4-9. *Hackelia* and *Lappula*

1. *Heliotropium curassavicum* L. Smooth, rather fleshy perennial; stems spreading, 1—3 dm. (4—12 in.) high; leaves oblong, 2—4 cm. long; flowers white, 5-lobed, 5 mm. (⅕ in) wide, in 3—5 slender, curved clusters at ends of branches. On saline flats, central part of State. June-Sept. Dr. I. M. Johnston called our plants var. *obovatum* DC.

2. *Cynoglossum officinale* L. HOUND'S TONGUE. Coarse biennial, 6—10 dm. high; leaves alternate, oblong or lanceolate, 1—3 dm. long, velvety; entire top of plant becoming a large, branching flower cluster; flowers dark red, 7—10 mm. wide; nutlets 5—7 mm. wide, rounded but flattened. An introduced weed, a nuisance on account of its burs. Has a pronounced "doggy" odor. We have it from Ransom, Barnes and Steele Counties only.

3. *Cynoglossum boreale* Fern. Slender, native, woodland plant with a few, rather rough, oblong leaves, 1—3 dm. long and a few blue flowers, 6—8 mm. wide on a stalk 3—6 dm. high. Woods, Turtle Mts., Bottineau and Rolette Counties. June.

Hackelia and Lappula   STICKSEED

Mostly annuals with rough leaves, small, blue or white flowers and nutlets forming burs with hooked spines. Species of *Hackelia* are often included in *Lappula* from which they differ in the bracted flowers and less elongated attachment of nutlets. The burs are often called sticktights or beggar's lice, names applied also to various other burs.

Key to Species

Lower leaves 1—3 cm. (⅖—1⅕ in.) wide; flowers blue.
  Leaves oblong or elliptic; nutlets with slender prickles around
    edges and on backs.   4. *Hackelia americana*
  Leaves lanceolate or oblanceolate; nutlets with 1 row of prickles
    which are broad and flat at base.   5. *Hackelia floribunda*
Lower leaves 5—8 mm. (¼ in.) wide; flowers blue or white.

Nutlets with 2 rows of prickles on edges.
Nutlets with a row of barbless prickles on middle of back.
                                    8. *Lappula cenchrusoides*
Nutlets without prickles on back.   6. *Lappula echinata*
Nutlets with only 1 row of prickles.
Prickles well separated, not united.   7. *Lappula redowski*
Prickles of some or all nutlets more or less united at base.
Base of prickes forming a prominent, rounded ridge.
                          9. *Lappula texana*, var. *homosperma*
Prickles flat at base, forming a flaring cup.
                          9. *Lappula texana*, var. *heterosperma*

4. *Hackelia americana* (A. Gray) FERN. WOOD STICKSEED. Native biennial, 3—9 dm. (1—3 ft.) high, flowering top often widely branched; basal leaves ovate, stem leaves elliptic or lanceolate, 2—10 cm. (1—4 in.) long, soft hairy. Common in woods and thickets. July, Aug. This and the following have often been regarded as varieties of the European *H. deflexa*.

5. *Hackelia floribunda* (Lehm.) Johnst. Biennial, 5—10 dm. high, with few, long branches; leaves 8—10 cm. long, 1—2 cm. wide, soft hairy; flowers 4 mm. wide. Rare, native plant in woods. July, Aug. We have it from Benson and Bottineau Counties only (fig. 21).

6. *Lappula echinata* Gilib. BLUE STICKSEED. (Fig. 270). Annual or winter annual, 3—6 dm. (1—2 ft.) high, widely branched above, entire plant rough hairy; leaves narrowly oblong, 2—5 cm. (1—2 in.) long; flowers blue, 2—3 mm. (⅛ in.) wide; nutlets with 2 rows of crowded spines. A common, introduced weed, often abundant in pastures and yards. June-Aug. A great nuisance on account of the burs.

7. *Lappula redowski* (Hornem.) Greene, var. *occidentalis* S. Wats. Low STICKSEED. Annual, 1—3 dm. high, widely branched from base; flowers white or blue; nutlets with only a few spines in one row. A very a-bundant, native, prairie plant, western part of State, also along dry roadsides in extreme east. May-July.

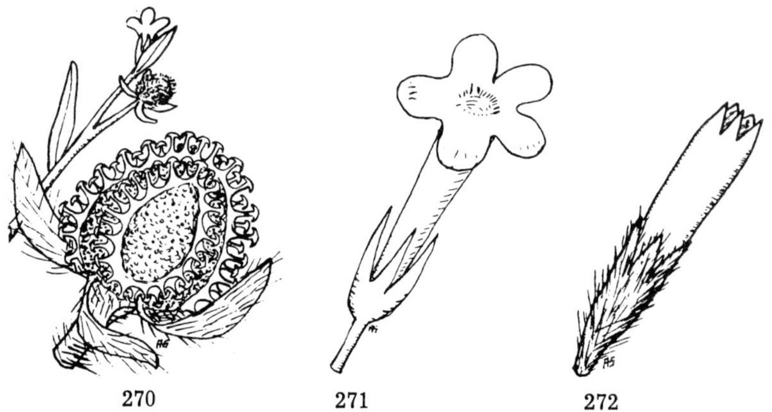

270   271   272

8. *Lappula cenchrusoides* A. Nels. Similar to Nos. 6, differing chiefly in the nutlets. Annual, native or introduced from farther west. We have it from Bowman, Billings and Williams Counties.

9. *Lappula texana* (Scheele) Britton. Similar to No. 7. Two varieties col-lected in sandy river bottom at Marmarth, Slope County, July 4, 1918. I have watched for it since without success. Possibly it was a chance introduction there from the west.
Var. *heterosperma* (Greene) Nels. & Macbr. Usually 3 of the 4 nut-lets have a wide spreading border; spines of fourth nutlet little modi-fied, and in some flowers on same plant, no spines modified. The flow-ers were noted as blue in fresh material.

Var. *homosperma* (Nels.) Nels. & Macbr. Nutlet with a thick, rounded rim. A few slender, fully mature plants were found among those of No. 7 at Marmarth. Why a few plants of such forms as these two varieties should appear among quantities of the common one is still a puzzle.

10. *Asperugo procumbens* L. CATCHWEED. Annual; stems weak, spreading, 3—6 dm. long, covered with short, stiff, recurved hairs; leaves oblong, 2—8 cm. long; flowers blue, 2 mm. wide, 1—3 at upper leaf bases; calyx enlarged, saucer shaped in fruit, 1 cm. wide. First received from the late A. R. Palmer in 1937 with a statement that he saw it first in 1924. On visiting his farmstead at foot of Killdeer Mts., it was found abundant in partial shade around yards and buildings. He reported that livestock ate it readily. A specimen was received from Dickinson in 1946.

11. *Cryptantha bradburiana* Payson. BUTTE CANDLE. Biennial; stem stiff, 1—3 dm. (4—12 in.) high, usually only short branches at top, forming a narrowly oblong or branching flower cluster 5—20 cm. (2—8 in.) long; whole plant very rough and gray with coarse hairs; leaves oblong or oblanceolate, flowers white, 7—10 mm. (⅓ in.) wide. One of the conspicuous spring flowers on buttes and dry hills. Our only records east of Missouri River are near its valley. We have suggested the name "butte candle" because of the spike-like, white, flower clusters. Formerly called *Oreocarya glomerata* (Pursh) Greene.

12. *Cryptantha calycosa* (Torr.) Rydb. Annual, somewhat like No. 7, but more slender. Found under the side of some "scoria" blocks on Twin Buttes at Bowman, in June, 1918. It was still there in 1946 (Stevens 890).

13. *Plagiobothrys scopulorum* (Greene) Johnst. Small, native annual; stems spreading, hardly 1 dm. high; leaves green, narrow, rough, 1—3 cm. long; flowers white, 2 mm. wide. Occasionally common on low, saline spots, where there is little vegetation and water often stands for a time. Bottineau, Grant, Morton and Hettinger Counties are represented in our collections, but it probably occurs in most of the cen-

273. Wild Forget-me-not (*Mertensia lanceolata*).

tral and western counties. A widely distributed species, the only one of numerous western ones which extends this far east. Previously recorded as *Allocarya scopulorum* Greene.

14. *Amsinckia idahoensis* M. E. Jones. Biennial; stems much branched, 3—6 dm. high; leaves lanceolate or narrower, 3—8 cm. long; flowers yellow, 3 mm. wide, in long, slender clusters at ends of branches. Whole plant covered with coarse, stiff hairs which stick into skin and clothing and easily break off. A weed from western U. S.; along railroad tracks at Pembina in 1912 and Rugby in 1918, but not known to be well established.

15. *Amsinckia menziesii* (Lehm.) Nels. & Macbr. Plants collected at Fargo in 1942 (Stevens, 629) are referred to this form on account of the short calyx. They were making a vigorous growth on a street corner between highway and railroad but may have been eliminated by a Victory garden the following summer.

16. *Mertensia lanceolata* (Pursh) DC. LUNGWORT. WILD FORGET-ME-NOT. "BLUEBELLS". CHIMING BELLS. (Fig. 273). Smooth plant from a stout, perennial root; stems 1—2 dm. (4—8 in.) high with a few branches at top bearing drooping, blue flowers, 1 cm. (⅖ in.) long; leaves narrowly oblong to ovate, 2—8 cm. long. A very attractive, early spring flower on hills in western North Dakota and one which responds well to cultivation. Occurs locally as far east as hills of Sheyenne River. Flower has a slender tube which widens abruptly about the middle. It becomes purplish as it fades. Leaves have a few, very short "puffy" hairs on upper side. The previous record of *M. paniculata* (Ait.) Don at Williston (Bell 123) was based on an unusually large plant of *M. lanceolata*. Dr. L. O. Williams has examined the specimen.

17. *Lithospermum canescens* (Michx.) Lehm. HAIRY PUCCOON. (Fig. 271). Perennial from a thick root; stems 2—4 dm. high, a few, curved flowering branches at top; leaves oblong, 2—5 cm. long, soft hairy; flowers 1 cm. wide, bright orange colored. A common prairie plant. June.

18. *Lithospermum incisum* Lehm. NARROW-LEAVED PUCCOON. Perennial; stems erect at first, widely branched and spreading in summer; leaves quite narrow, 2—5 cm. long, and whole plant rough hairy; flowers lemon yellow, edges of corolla lobes "erose" (as if gnawed by an insect). Common, blooms a little earlier than the preceding and is inclined to drier and stony soils. Sometimes dwarfed to a cushion-like form. Formerly known as *L. linearifolium* Goldie.

19. *Onosmodium occidentale* Mack. FALSE GROMWELL. (Fig. 272). Perennial; very rough with a flat mat of coarse hairs; several stout stems 4—10 dm. high; leaves elliptic, 5—10 cm. long; flowers 1 cm. long, yellowish white, in clusters at ends of stems, not showy because the small, triangular lobes do not spread out; nutlets white, smooth, rounded, 3—4 mm. long, persisting on stems more or less all winter. A fairly common prairie plant. July.

VERVAIN or VERBENA FAMILY Verbenaceae

Annual or perennial herbs with opposite leaves and small flowers in spikes or heads; corolla partly united to form a slender tube with 5 spreading, slightly unequal lobes; stamens 5, attached to corolla tube and enclosed in it; ovary 4-seeded, splitting at maturity into 4 nutlets which are oblong, nearly square in cross-section, roughened on inner faces and faintly ridged or netted on outer side (fig. 121). None of our species have the rounded clusters of large flowers as in the garden verbenas.

Key to Species

Stems erect; flowers in leafless, terminal spikes.
  Flowers white, in very slender, loose spikes.   1. *Verbena urticifolia*
  Flowers purple, in dense spikes.
    Flowers 3 mm. (⅛ in.) wide; spikes 2—10 cm. (1—4 in.) long
      on several upper branches.   2. *Verbena hastata*
    Flowers 5 mm. wide; often only 1 spike 2—3 dm. (8—12 in.) long
      at end of main stem.   3. *Verbena stricta*
Stems creeping or ascending; flowers scattered among small leaves
  along stem.   4. *Verbena bracteata*

1. *Verbena urticifolia* L. NETTLE-LEAVED VERVAIN. Annual; stems 4—12 dm. (1½—4 ft.) high, with long, slender branches; leaves ovate, 5—15 cm. (2—6 in.) long, 2 cm. wide, sharply toothed; flowers white, 2—3 mm. ($\frac{1}{10}$ in.) wide, distinctly separated, in very slender spikes 1—2 dm. long from upper leaf bases. Uncommon in woods. Recorded for Richland, Ransom, Dickey, Cass, Stutsman, Nelson and Pembina Counties. July-Sept.

2. *Verbena hastata* L. SWAMP VERVAIN. Perennial; stem erect, 3—10 dm. high, not much branched except on top; leaves lanceolate, 5—15 cm. long, 2—3 cm. wide, sharply toothed; flowers purple, 3 mm. wide, crowded in spikes 2—10 cm. long. Common in wet meadows and along streams. July-Sept.

3. *Verbena stricta* Vent. HOARY VERVAIN. Perennial, gray and rough with coarse hairs; stems stiff, erect, 3—10 dm. high, 1 or several stout spikes at top of plant; leaves ovate or rounded, 2—6 cm. long, sharply toothed; flowers purple, 5 mm. wide. Frequent in southeastern part of State to eastern Dickey and southern Barnes County. July-Sept. A Fargo record is from plants established in railroad yards. Common pasture weed farther south.

4. *Verbena bracteata* Lag. & Rodr. BRACTED VERVAIN. (Fig. 274). Annual; usually forming mats 3—8 dm. wide; leaves 2—4 cm. long, broadest toward tip, deeply toothed or lobed; flowers purplish blue, 3 mm. wide, in crowded clusters along branch ends, a narrow bract 1—1.5 cm. long below each flower. Common in dry soil. July-Sept. Formerly called *V. bracteosa* Michx.

    A number of plants of *Verbena rigida* Spreng. were found on the Agricultural College grounds at Fargo in 1943. It is a South American plant which has become well established in southern U. S. Similar to No. 2 but has still shorter spikes, leaves tending to be narrower and tapered toward base.

MINT FAMILY Lamiaceae or Labiatae

    Herbs, often aromatic, with opposite, simple leaves; corolla usually a slender tube divided above into two "lips," the upper of which is 2-lobed, the lower 3-lobed; stamens 2 or 4 on corolla; pistil 1, ovary 4-lobed, developing 4 nutlets enclosed in persistent calyx.

    Shape of corolla, opposite leaves and often square stems are characteristic of this group, but these are also found in some plants of the figwort family. The mint family is a large one and technical characters depend much on details of stamens and nutlets. The leaves are much alike in many of the species.

Key to Species

Corolla without distinct upper and lower lip.
  Corolla 10 mm. (⅖ in.) long, no upper lip but stamens protruding
    in its place.   1. *Teucrium occidentale*
  Corolla 2—5 mm. long, lobes short, nearly equal.
    Stamens 4; plant aromatic.   26. *Mentha arvensis*
    Stamens 2; plants not aromatic.

Calyx teeth rounded, shorter than mature nutlets.
23. *Lycopus uniflorus*
Calyx teeth pointed, longer than nutlets.
Leaves with long, slender teeth. 24. *Lycopus americanus*
Leaves with short, broad teeth. 25. *Lycopus asper*
Corolla with a distinct upper and lower lip.
Calyx with a crest on upper lip, (fig. 275), but no sharp teeth.
Flowers in slender, axillary and terminal clusters.
2. *Scutellaria lateriflora*
Flowers solitary in leaf axils.
Flowers 2 cm. (⅘ in.) long; plants 2—4 dm. (8—16 in.) high.
3. *Scutellaria galericulata*
Flowers 5—8 mm. (¼ in.) long; plants 1—1.5 dm. high.
4. *Scutellaria leonardi*
Calyx with sharp teeth, not crested.
Leaves 3 mm. or less wide.
Flowers nearly white; plants slender below, branching above.
22. *Pycnanthemum virginianum*
Flowers bluish purple; plant branching near base.
Annual; flowers 5 mm. long. 20. *Hedeoma hispida*
Perennial; flowers 10—12 mm. long. 21. *Hedeoma drummondii*
Leaves 5 mm. or more wide.
Stamens only 2.
Flowers blue or purple, in elongated clusters.
Flowers purple, 1 cm. long; leaves white hairy below.
17. *Salvia sylvestris*
Flowers blue, 5—8 mm. long; leaves green, not hairy.
18. *Salvia reflexa*
Flowers lilac colored, in heads 3—8 cm. (1—3 in.) wide.
19. *Monarda fistulosa*
Stamens 4.
Stems creeping or spreading, not over 3 dm. (1 ft.) high.
Stems creeping, rooting at joints. 7. *Nepeta hederacea*
Stems spreading, not rooting.
Flowers in terminal clusters; leaves oblong or ovate.
10. *Prunella vulgaris*
Flowers at leaf bases; leaves rounded.
13. *Lamium amplexicaule*
Stems erect, 3—15 dm. high.
Plants smooth or not distinctly hairy.
Leaves and flowers decidedly prickly when old.
8. *Dracocephalum parviflorum*
Leaves and flower clusters not prickly.
Flowers blue, 6—10 mm. (⅓ in.) long; leaves ovate.
5. *Agastache anethiodora*
Flowers purple, 10—15 mm. long; leaves lanceolate.
11. *Physostegia parviflora*
Plants well covered with hairs.
Plant soft hairy, fragrant; flowers white. 6. *Nepeta cataria*
Plant coarse or rough hairy; flowers pink or purple.
Sepals in fruit hooked and spiny (fig. 277); flowers
hairy. 12. *Leonurus cardiaca*
Sepals not hooked; flowers not hairy.
Corolla tube twice as long as calyx. 16. *Galeopsis tetrahit*
Corolla tube not longer than calyx.
Stems hairy all over. 14. *Stachys palustris*
Stems with hairs on corners only. 15. *Stachys aspera*

1. *Teucrium occidentale* A. Gray. GERMANDER. Perennial by rhizomes; plant soft hairy, stem usually not much branched, 3—6 dm. (1—2 ft.) high; leaves oblong to lanceolate, 5—10 cm. (2—4 in.) long, sharply toothed; flowers pale purple, 8—12 mm. (⅓—½ in.) long in a terminal spike 1—3 dm. long; nutlets rounded, light brown, with a coarse net-

work of ridges and a scar which covers most of inner side. Low places in neglected fields or pastures, forming patches. July, Aug. Resembles hedge nettle (Nos. 14, 15) but flowers are paler in color and more densely grouped in a terminal spike with none at upper leaf bases.

2. *Scutellaria lateriflora* L. BLUE SKULLCAP. Perennial; stems rather weak, spreading, 3—6 dm. high; leaves ovate, thin, 3—6 cm. long; flowers blue, 5—7 mm. long, in slender clusters from upper leaf bases. River banks and other wet places. July-Sept. Specimens from Richland, Cass, Ransom and Barnes Counties only. The flowers are rather pretty though small.

3. *Scutellaria galericulata* L. MARSH SKULLCAP. (Fig. 275). Perennial; stems slender, 3—6 dm. high, leaves lanceolate, 2—5 cm. long, with short, rounded teeth; flowers blue, 2 cm. long, solitary at upper leaf bases. Stream banks and other wet places, eastern part of State. June-Sept. Flowers are fairly showy.

4. *Scutellaria leonardi* Epling. SMALL SKULLCAP. Perennial by tubers; stems often unbranched, 1—1.5 dm. high; leaves ovate, 1—2 cm. long, entire; flowers 5—8 mm. long, solitary at upper leaf bases. Moist, sandy fields and prairies. June. We have this only from Richland and Ransom to Walsh Counties. It seems not common, but may be more widely distributed for it is small and matures early, so that it could easily be overlooked. Formerly included in *S. parvula* Michx.

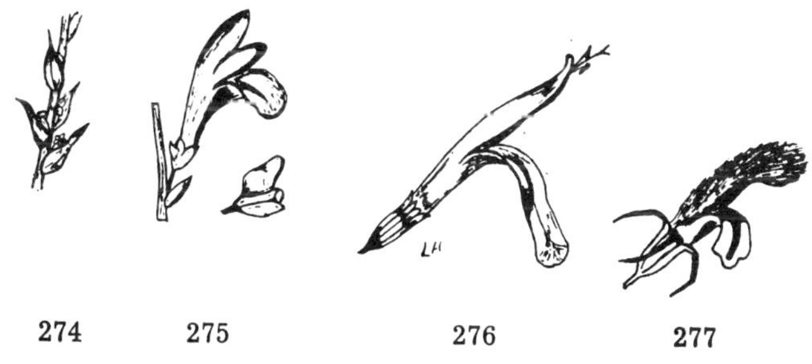

274   275   276   277

5. *Agastache anethiodora* (Nutt.) Britton. FRAGRANT GIANT HYSSOP. FALSE ANISE. Perennial; stems 1—2 m. (3—7 ft.) high, square, sharply angled; leaves ovate, 5—10 cm. (2—4 in.) long and nearly as wide, teeth rounded; flowers blue, 6—8 mm. (¼ in.) long in terminal spikes or somewhat clustered farther down the stem. Woods and thickets all over State. July-Sept. First discovered in the Fort Mandan area. It has an anise-like fragrance when bruised for which it was used by the Indians. Not related to true anise.

6. *Nepeta cataria* L. CATNIP. Silvery gray, soft hairy perennial, large plants becoming widely branched, 1—1.5 m. high; leaves rounded ovate, 2—8 cm. long, teeth rounded; flowers 8—10 mm. long, white with a few dark spots, in somewhat rounded clusters toward ends of branches; nutlets black, 1.5 mm. long with two white spots on scar. A well known, introduced plant, established in a number of places.

7. *Nepeta hederacea* L. GROUND IVY. GILL OVER-THE-GROUND. Perennial; stems slender, creeping, rooting at joints; leaves rounded, 1—3 cm. long, teeth short, rounded; flowers blue, 1—2 cm. long, in groups of 2 or more at leaf bases. Specimens from Richland and Barnes Counties. Often becomes a weed in shaded places. Also known as *Glechoma hederacea*. Sometimes called "creeping charley," a name applied to various plants of creeping habit.

8. *Dracocephalum parviflorum* Nutt. DRAGONHEAD. (Fig. 278). Biennial; stems becoming widely branched, 1—1.5 m. (3—5 ft.) high; leaves oblong to lanceolate, 5—10 cm (2—4 in.) long, with sharp, somewhat spiny teeth; flowers pale blue, 1 cm. long, nearly concealed by spiny calyces and bracts of dense flower clusters at ends of branches; nutlets black, 2—3 mm. long, with a prominent scar extending across both

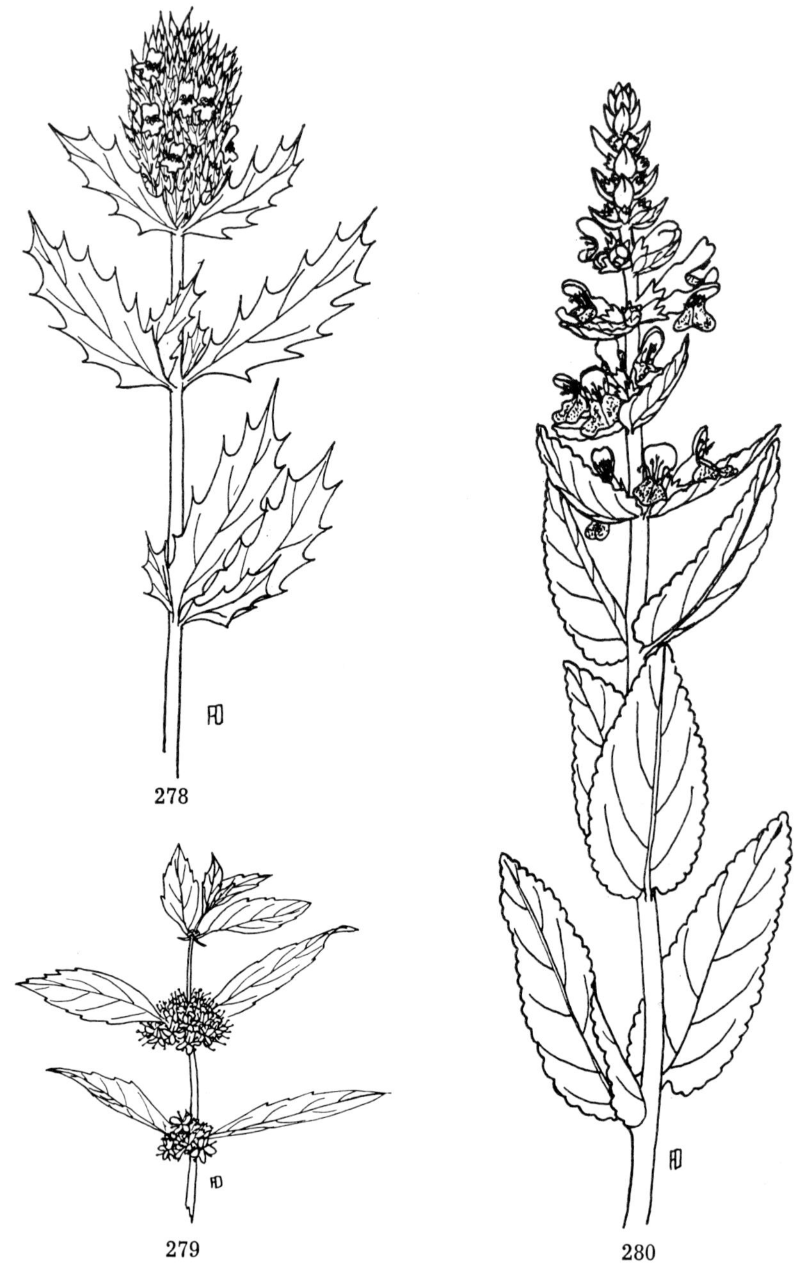

278. Dragonhead (*Dracocephalum parviflorum*).
279. Wild Mint (*Mentha arvensis*).
280. Hedge Nettle (*Stachys palustris*).

sides of base. July, Aug. This coarse, native weed has attracted attention in recent years because the nutlets are a conspicuous impurity in alfalfa and sweet clover seed. It grows especially in clearings and along edges of woods or brushy areas and seems not to spread into open fields to any extent. The dry seed clusters are quite spiny. It is sometimes abundant locally, and we have it from widely separated localities in Richland, Cass, Walsh, Pembina, Ramsey, Hettinger, Stark and Dunn Counties.

9. *Dracocephalum thymiflorum* L. Perennial; stems more slender than the preceding; leaves 2—5 cm. long, not spiny; flowers clustered at nearly all leaf bases. This European plant was collected in a field of *Bromus inermis* near Belfield, Stark County, in 1914 and again near Mott in Hettinger County, in 1938 (Mrs. Meissner). We do not know whether it has become estabished as a weed.

10. *Prunella vulgaris* L. SELFHEAL. Perennial; stem erect, 1—3 dm. high; leaves ovate, oblong or lanceolate, 2—5 cm. long; flowers purple, 8—10 mm. long, in oblong clusters at ends of branches. A widely distributed plant, common in many parts of U. S. and Europe. I have seen it in considerable abundance on low prairie at different points in Minnesota, 15—25 miles from Fargo, where it seemed native. At Harlow, Benson County, in 1930, I found it in a prairie coulee. We have only one other record, at Fargo in 1910, where it has not been seen since. The nutlets are a frequent impurity in seeds of white clover and this probably was the source of the Fargo specimen.

11. *Physostegia parviflora* Nutt. OBEDIENT PLANT. (Fig. 281). Perennial; stems smooth, upright, 3—15 dm. (1—5 ft.) high; leaves lanceolate or elliptic, 5—15 cm. (2—6 in.) long, sharply toothed; flowers rose purple, 1—1.5 cm. long, in spikes 1—2 dm. long at ends of branches. Along river and creek banks, sometimes abundant. We have it only from the southeast and from Pembina to McLean County, but it may occur in any county where conditions are suitable. It is also called false dragonhead and lion's heart, but the name obedient plant is especially suitable because when flowers are pushed to one side, they remain in that position. A closely related species of eastern U. S. is commonly grown as an ornamental.

12. *Leonurus cardiaca* L. LION'S TAIL. (Fig. 277). Perennial; stem 3—8 dm. high, widely branched above; lower leaves 3—5 palmately cleft and toothed, 2—5 cm. long and nearly as wide, on long petioles; upper leaves lanceolate, few toothed; flowers pink, 1 cm. long, quite hairy, several at each leaf base along upper branches; calyx teeth becoming hooked and spiny in fruit. Introduced in wooded or brushy areas. We have it from Richland, Cass, LaMoure, Walsh and Bottineau Counties. July, Aug. A very undesirable plant on account of the spiny branches. It seems to thrive well and reproduce freely by seed. The name "motherwort" is more commonly used for it, but is surely less appropriate.

13. *Lamium amplexicaule* L. DEAD NETTLE. Annual with weak, spreading stems, 1—2 dm. high; leaves 1—4 cm. long, rounded and with rounded teeth; flowers pink, slender, 1—1.5 cm. long at leaf bases. I found this quite abundant in a dooryard at Langdon in 1918. It is an introduced weed found in most parts of U. S. The lower leaves have long petioles, the upper clasp the stem.

14. *Stachys palustris* L. HEDGE NETTLE. (Fig. 280). Perennial by rhizomes; stems 3—8 dm. (1—2½ ft.) high, usually not branched; leaves oblong or lanceolate, 5—15 cm. (2—6 in.) long, sparsely or densely hairy; flowers purplish, 1.5 cm. long in a terminal, interrupted spike, also at upper leaf bases; nutlets black, 2 mm. long and nearly as wide, with a small, rounded scar. Common in ditches, low places in fields, pond and river edges. The flowers are quite showy and the plant is frequently one of the most showy ones around low, wet spots in July. It is quite rough with hairs but does not sting like a nettle.

15. *Stachys aspera* Michx. A more slender, smaller plant than the preceding, with only a few hairs on angles of stem. It is less frequent and we have it only from Ransom, Cass, Barnes, Pembina and Bottineau Counties.

16. *Galeopsis tetrahit* L. HEMP NETTLE. Annual; stems widely branched, 3—6 dm. high, covered with coarse, prickly hairs; leaves ovate to lanceolate, 3—8 cm. long; flowers purple, white spotted, 1—1.5 cm. long, clustered at upper leaf bases and branch tips; calyx enlarged in fruit and spiny; nutlets flattened, much narrowed toward base, 3 mm. long and nearly as wide at upper end, mottled gray. An introduced weed found in 1913 at Perth, Towner County. July, Aug.

17. *Salvia sylvestris* L. Perennial; stems 3—8 dm. high; leaves oblong—lanceolate, 5—10 cm. long, lower surface gray with short hairs, edges with small, rounded teeth; flowers purple, 1 cm. long, in terminal, spike-like racemes 1—3 dm. long. Collected by Mrs. Anna Meissner near the HT Ranch near Amidon, Slope County, in 1941 and identified by O. M. Freeman, U. S. Dept. Agr. I visited the place in 1945 and found a patch about 50 yds. wide, presumably from an old planting.

18. *Salvia reflexa* Hornem. LANCE-LEAVED SAGE. Annual; stems 2—6 dm. high; leaves oblong or lanceolate, with few if any hairs or teeth; flowers blue, 8—12 mm. long, in slender clusters at ends of upper branches; nutlets smooth, tan colored. Probably introduced here from farther south. It is common from Nebraska to Texas. We have it from Sargent, Cass, Foster, Morton and Stark Counties. Formerly called *S. lanceolata* Willd.

19. *Monarda fistulosa* L. WILD BERGAMOT. (Fig. 276). Perennial, developing large clumps; stems 5—10 dm. (1½—3 ft.) high with few, erect branches; leaves lanceolate, 5—10 cm. (2—4 in.) long, sharply toothed; flowers in heads 3—8 cm. wide at ends of branches; corolla lilac colored, 3 cm. long, the tube only 2—3 mm. wide, dividing above the middle into two, narrow, widely spreading "lips," stamens projecting beyond corolla. A handsome plant, occurring all over State but common only locally, usually in valleys or coulees, on lower hillsides, in open wooded or bushy places. July. Dr. M. R. Gilmore said that the Indians recognized four varieties which had different odors.

20. *Hedeoma hispida* Pursh. ROUGH PENNYROYAL. Annual; stems simple or widely branched, 1—1.5 dm. high; leaves 1—1.5 cm. long, 1—2 mm. wide; flowers bluish purple, 5 mm. long, several at most leaf bases. Common in dry soil on prairie or other uncultivated places. Probably all over State, though we have only Pembina County recorded north of Wells County. June-Aug.

21. *Hedeoma drummondii* Benth. Perennial from a stout, woody root; stems widely branched, 1—1.5 dm. high; leaves as in preceding, but flowers longer, 10—12 mm. long. Stony hills and buttes west of Missouri River. An attractive, fragrant plant, worthy of consideration for a rock garden. Previously recorded as *H. nana* Nutt.

22. *Pycnanthemum virginianum* (L.) Durand & Jackson. MOUNTAIN MINT. Perennial; stems 3—6 dm .high; leaves 2—5 cm. long, 2—3 mm. wide; flowers nearly white, 5 mm. wide. Boggy, sandy soil, Richland County only. Aug. Branches, leaves and flowers are crowded into a rounded or flat, dense cluster at top of plant. Fragrant.

23. *Lycopus uniflorus* Michx. BUGLE-WEED. Perennial from a tuberous base; stem 2—5 dm. high; leaves lanceolate, 2—6 cm. long, teeth rather short and rounded; flowers white, 2 mm. wide, closely crowded at leaf bases; calyx lobes blunt. Local in shaded, boggy places. We have it from Ransom, Barnes, Benson, Pembina and Bottineau Counties (fig. 15).

24. *Lycopus americanus* Muhl. WATER HOARHOUND. Similar to preceding, but stouter, often widely branched, leaves with long, sharp teeth; calyx teeth sharp; corolla white, 3 mm. wide. Common in moist soil in low fields, ditches, stream banks, either in partial shade or in the open. July-Sept. It has considerable resemblance to wild mint but is not fragrant. *Lycopus* species have 3-sided, angular nutlets with thick, corky margins.

281. Obedient Plant (*Physostegia parviflora*).

25. *Lycopus asper* Greene. WESTERN WATER HOARHOUND. Perennial from rhizomes; stems stiff, usually not branched; leaves thick, oblong or lanceolate, sharply toothed, 2.5—10 cm. long. Common in saline, wet, open places. This plant has also been called *L. lucidus* Turz., a species of Asia.

26. *Mentha arvensis* L. WILD MINT. (Fig. 279). Perennial, spreading freely by rhizomes; stems 2—6 dm. high; leaves elliptic, ovate or lanceolate; flowers lavender or nearly white, 3 mm. wide, clustered at upper leaf bases; corolla lobes spreading, acute, nearly equal in size; nutlets rounded, brown, 1 mm. long. Common in moist, either open or shaded places. The plants may be short and stiff in open, dry places, or weak and spreading among a rank growth of grasses and other plants. They also vary from nearly smooth to quite hairy. The lower sides of the leaves are covered with glandular dots, where the oils are secreted, which give the plant its characteristic odor. Our specimens seem mainly var. *villosa*, f. *glabrata* (Benth.) Stewart, often called *M. canadensis* L., which is not considered distinct from the European *M. arvensis*. Peppermint and spearmint do not occur in North Dakota. They have flowers in terminal, not axillary, clusters.

Phlomis tuberosa L. JERUSALEM SAGE. Stout perennial, 3—10 dm. high, with heart shaped, ovate or lanceolate leaves 5—10 cm. long, and head-like clusters of flowers at upper leaf bases, flowers 2 cm. long with dense hairs projecting from underside of the large, rounded upper lip of corolla. A specimen from Barlow, Foster County, received through G. B. Schmid, July, 1929, was said to have been found in a field.

Elsholtzia cristata Willd. Annual, 3—5 dm. high; leaves ovate to lanceolate, 2—10 cm. long; flowers light blue, 2 mm. long, in terminal, dense, 1-sided clusters. Collected in 1930 at Fannie M. Heath's place near Grand Forks, where it had run wild in the grove and gave promise of becoming a weed. It looks something like wild mint but flowers are in dense clusters at ends of branches and calyx enlarges in fruit.

NIGHTSHADE FAMILY   Solanaceae

Herbs or shrubs with alternate leaves; sepals 5, separate or united; petals united into a saucer shaped or funnel shaped corolla often with pointed or rounded lobes; stamens 5, attached to base of corolla; pistil 1; fruit a many seeded berry or pod. A large and important family, containing many drug plants, food plants such as potato, tomato, egg plant, pepper and ornamentals such as petunia.

Key to Species or Genera

Fruit a berry; seeds in a soft pulp.
  Herbs, not thorny; flowers yellow or white.
    Berry loosely enclosed by enlarged calyx.   1-3. *Physalis*
    Berry not enclosed by calyx.
      Flowers white; leaves entire or toothed.   4-5. *Solanum*
      Flowers yellow; leaves much divided.   7. *Lycopersicon esculentum*
  Woody, thorny plant; flowers purple.   8. *Lycium halimifolium*
Fruit a pod; dry, splitting open when ripe.
  Pod not prickly, opening by a lid.   9. *Hyoscyamus niger*
  Pod prickly, splitting down the sides.
    Leaves prickly; flowers yellow, 2 cm. (¾ in.) wide.
                6. *Solanum rostratum*
    Leaves not prickly; flowers white, 7—10 cm. wide.
              10. *Datura stramonium*

Physalis GROUNDCHERRY

Annuals or perennials with simple leaves and yellow flowers solitary at leaf bases; calyx and corolla cup shaped, calyx 5-toothed, much enlarged in fruit, loosely enclosing a berry. An interesting form, grown as an ornamental, is Chinese-lantern Plant (*P. alkekengi*), the calyx of which becomes bright red.

Key to Species

Plant closely covered with short, sticky hairs.   1. *Physalis heterophylla*
Plant not sticky; hairs fewer.
   Fruiting calyx sharply 5–angled.   2.   *Physalis virginiana*
   Fruiting calyx rounded.   3.   *Physalis lanceolata*

1. *Physalis heterophylla* Nees. Perennial, stems and leaves closely covered with sticky hairs; stem stout, widely spreading, 2—3 dm. (8—12 in.) high; leaves ovate or rounded, 2—8 cm. (1—3 in.) long; flowers 1—2 cm. wide, yellow with purple center; berry yellow, 1 cm. wide; fruiting calyx rounded, 2 cm. long. Sandy soil, southeastern part of State, also Grant and Stark Counties. June-Sept. This species is sometimes grown for the fruit.

2. *Physalis virginiana* Mill. (Fig. 282). Similar to preceding but not sticky; leaves ovate to lanceolate, entire or toothed; fruiting calyx 5-angled, sunken at base, 3 cm. long. This seems not restricted to sandy soil. We have it in the east to LaMoure and Ramsey Counties, also in Grant County. It is probably more widely distributed.

3. *Physalis lanceolata* Michx. Perennial; stems slender, 2—6 dm. high, leaves lanceolate with wavy or toothed edges, 2—5 cm. long; flowers 2 cm. wide, yellow with brownish center; berry yellow, 1 cm. wide; calyx rounded, 2 cm. long. Sandy prairies. Specimens from Richland to Barnes and Dickey and from Grand Forks, Pembina, Grant and Golden Valley Counties. *P. ixocarpa*, a groundcherry planted for the fruit, resembles this species, but has a flower with purple center and a purplish berry. It is an annual and will volunteer from seed.

Solanum NIGHTSHADE

Annuals or perennials with simple or compound leaves; flowers usually in small clusters opposite leaves; corolla shallow, with 5 pointed lobes; fruit a berry or pod. There are many species. Ours are all annuals. Potato, egg-plant and Jerusalem-cherry, which is grown as a house plant, are other species.

Key to Species

Flowers white; leaves simple; plant not prickly.
   Leaves nearly entire; mature fruit black.   4.   *Solanum nigrum*
   Leaves sharply toothed; mature fruit green.   5.   *Solanum triflorum*
Flowers yellow, leaves divided; plant very prickly.   6.   *Solanum rostratum*

4. *Solanum nigrum* L. BLACK OR GARDEN NIGHTSHADE. Annual; stem erect or spreading, becoming widely branched, 3—8 dm. (1—2½ ft.) high; leaves ovate, 2—8 cm. (1—3 in.) long, edges wavy; flowers white, star shaped, 6—8 mm. (¼ in.) wide, in clusters of 3—10; berry black when mature, 8—10 mm. wide. Fields, waste ground, woods or thickets; widely distributed but not very common. July-Sept. The plant is usually smooth. We have a few specimens from the western part of the State which are densely covered with short hairs (*S. villosum* (L.) Mill. or *S. interius* Rydb.?). The berries are often used for sauce, jam or pie. The plant called wonderberry, stubbleberry and garden huckleberry, is a stouter form of this species.

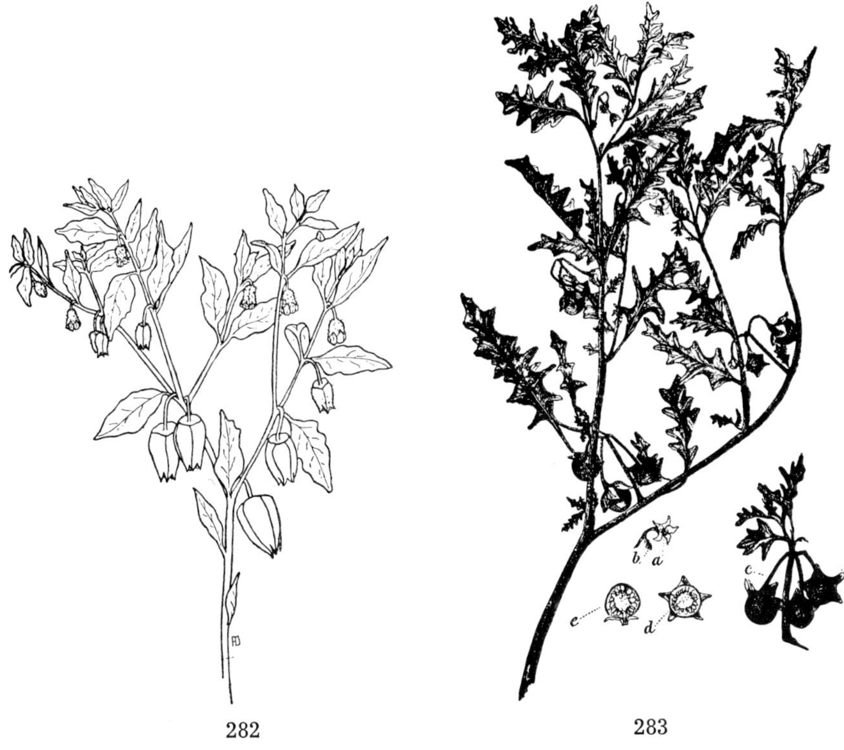

282　　　　　　　　　　　283

282.　Groundcherry (*Physalis virginiana*).

283.　Cut-leaved Nightshade (*Solanum triflorum*).

5.　*Solanum triflorum* Nutt. CUT-LEAVED NIGHTSHADE. (Fig. 283). Annual; stems spreading along ground, 3—10 dm. long; leaves oblong, 2—6 cm. long, divided about half way to middle into several teeth or lobes; flowers white, 8—10 mm. wide, in clusters of 1—3 at leaf bases; berry 1—1.5 cm. wide, green in color when ripe. Dry, bare soils, common west, infrequent east. June-Aug. This plant was first described from specimens collected in the gardens of the Mandan Indians. We had many inquiries about the edibility of the berries and were unable to find any record of it. Trials showed they had a very persistent "bite" which made them unpalatable.

6.　*Solanum rostratum Dunal.* BUFFALO BUR. (Fig. 284). Annual, prickly plant; stem erect or widely branched, 2—8 dm. high; leaves ovate, 1—3 dm. long, divided into rounded lobes (somewhat as in watermelon); flowers 2—3 cm. wide, bright yellow; clusters becoming stout in fruit with rounded, spiny burs 1—2 cm. wide, which are fleshy at first, then dry and split open; seeds black, flat and rough, 3 mm. wide. Roadsides, pastures and feed lots; abundant in some localities, chiefly in Missouri River Valley. Specimens are received each year from various parts of the State where seeds were doubtless introduced in crop seed from farther south. It is a common native weed in Kansas and Colorado but does not seem to persist in North Dakota except in the southern part. The potato beetle feeds freely upon it. The plant is a nuisance on account of its prickly nature but is easily destroyed by cutting.

7. *Lycopersicon esculentum* Mill. TOMATO. This is scarcely established as a wild plant though it comes up freely from scattered seeds and plants from seeds left in the garden may mature fruit.

8. *Lycium halimifolium* Mill. MATRIMONY VINE. Spreading or trailing shrub, 1—2 m. (3—7 ft.) high, with slender, gray stems; leaves oblong or lanceolate, 2—8 cm. (1—3 in.) long, a stout spine 1 cm. long at base of some of them; flowers purple, 1 cm. wide, in clusters at leaf bases; fruit a red, oblong berry 1 cm. long. Often planted as an ornamental, chiefly for the berries. We have received several specimens in recent years which suggest that it has become wild. In one nursery at Bismarck, it was offering considerable resistance to eradication.

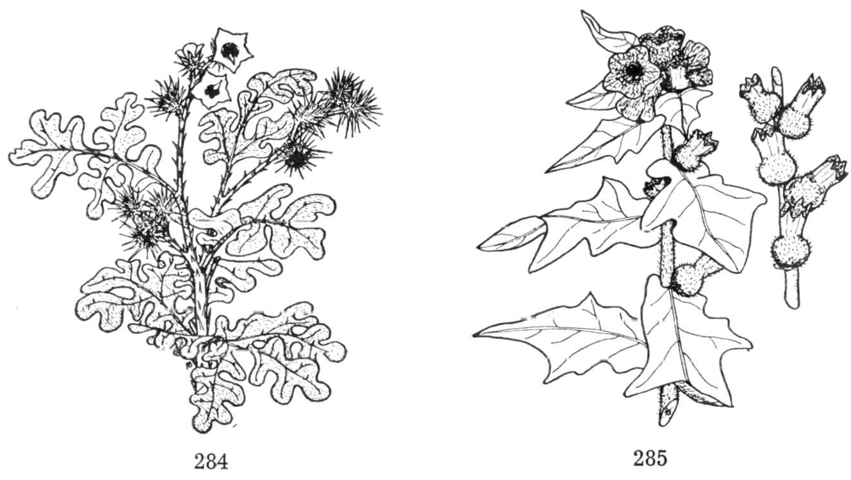

284 285

284.   Buffalo Bur (*Solanum rostratum*).

285.   Henbane (*Hyosscyamus niger*).

9. *Hyoscyamus niger* L. HENBANE. (Fig. 285). Introduced biennial; stem stout, 1—2 m. high, forking above into widely spreading branches with a flower at each small leaf; leaves angular-ovate, 1—2 dm. long, with dense, short, sticky hairs; flowers 2 cm. wide, brownish yellow with purple center and purple veins; pod oblong, 2 cm. long, erect, enclosed by calyx and opening by a lid; seeds many, yellowish, rough, 1—1.5 mm. wide. Dump grounds, other waste ground and around buildings. Specimens from Barnes, LaMoure, Logan, Grant, Morton, Oliver, McLean and Towner Counties. Mr. E. C. Moran found a quantity of it along the Little Missouri River below Medora a few years ago. It is a very poisonous plant, containing the alkaloids hyoscyamine and scopalamine, but the foliage is not likely to be eaten. The plant produces the first year a thick root and a few large leaves. Several years ago, two children in Hettinger County were poisoned, one fatally, from eating roots. A more recent report is of children poisoned by eating the seeds.

10. *Datura stramonium* L. JIMSON WEED. Very stout, smooth annual, 1—2 m. high; leaves angular-ovate, 1—3 dm. long, rank scented; flowers white, trumpet shaped, 1 dm. long, solitary in forks of branches; fruit a spiny, rounded pod, 2—3 cm. long, splitting in 4 pieces; seeds similar to those of Buffalo Bur. We have this from Christian Hanson, Mayville, Traill County, in 1025, when frost came late. It is a common barnyard weed in eastern and central U. S., but not this far north. A

related form with velvety, rounded leaves, is often grown as an ornamental (Angel's Trumpet). These plants contain the alkaloids hyoscyamine and atropine. Some years ago two children near Fargo were seriously ill from eating some of the seeds.

One kind of tobacco (*Nicotiana quadrivalvis* Pursh) was grown by the Indians and is mentioned by Rydberg (Fl. Pl. Pr.) as "escaped in N. D.", but we have no record of it.

FIGWORT FAMILY Scrophulariaceae

Annuals or perennials with alternate or opposite leaves; corolla united in a tube with spreading lobes, usually more or less 2-lipped; stamens attached to corolla tube, 4 or 2 (5 in*Verbascum*); pistil 1, ovary 2-celled; fruit a capsule with many small seeds.

This is a large group of plants, including many ornamentals, some weeds and many of no importance. Some species seem to have the general characters of mints, but are to be recognized by the different type of fruit. Most of our specimens were examined by Dr. F. W. Pennell during preparation of his monograph.[1]

Key to Species or Genera

Stamens 5; corolla nearly regular, a shallow cup.
　Leaves with a thick covering of long, branching hairs.
　　　　　　　　　　　　　　　　　　1. *Verbascum thapsus*
　Leaves with only some short, sticky hairs. 　2. *Verbascum blattaria*
Stamens 4 or 2; corolla usually irregular, tubular, 2-lipped.
　Corolla with a slender spur 1 cm. (⅖ in.) long from lower end.
　　　　　　　　　　　　　　　　　　3. *Linaria vulgaris*
　Corolla without such spur.
　　Stems creeping or spreading; mud or water plants.
　　　Flowers yellow, 8—12 mm. (⅓—½ in.) long. 　14. *Mimulus geyeri*
　　　Flowers white, blue or purple, 3—5 mm. long.
　　　　Flowers blue; stems floating or spreading.
　　　　　Leaves oblong, with short petioles. 　20. *Veronica americana*
　　　　　Leaves lanceolate, without petioles, often clasping at
　　　　　　base. 　　　　　　　　21. *Veronica connata*
　　　　Flowers white; stems creeping, rooting at nodes.
　　　　　Leaves 3—8 mm. wide, long petioled, appearing tufted.
　　　　　　　　　　　　　　16. *Limosella aquatica*
　　　　　Leaves 10—15 mm. wide, sessile, opposite.
　　　　　　　　　　　15. *Hydranthelium rotundifolium*
　　Stems upright, but often low and widely branched.
　　　Flowers yellow, white or yellowish white.
　　　　Flowers bright yellow.
　　　　　Leaves 3—5 mm. (⅙ in.) wide, usually entire.
　　　　　　　　　　　　　30. *Orthocarpus luteus*
　　　　　Leaves 10—20 mm. wide, toothed and curled.
　　　　　　Stems tufted, 2—3 dm. (8—12 in.) high.
　　　　　　　　　　　　31. *Pedicularis canadensis*
　　　　　　Stems not tufted, 3—6 dm. high. 32. *Pedicularis lanceolata*
　　　　Flowers white or yellowish white.
　　　　　Stems stout, little branched, 2—8 dm. high.
　　　　　　Plant gray, 2—3 dm. high; flowers 2.5—5 cm. (1—2 in.)
　　　　　　　long. 　　　　　　28. *Castilleja sessiliflora*
　　　　　　Plant green, 3—8 dm. high; flowers 5 mm. long.
　　　　　　　　　　　25. *Veronicastrum virginicum*
　　　　　Stems slender, often low branching, 1—2 dm. high.
　　　　　　Flowers 2 mm. long, not tubular; capsule flattened.
　　　　　　　　　　　24. *Veronica peregrina*

---

[1]Pennell, F. W. The Scrophulariaceae of eastern temperate North America. Acad. Nat. Sci. Phila. Monograph No. 1. 650 pp. 1935.

Flowers 10 mm. long, tubular; capsule rounded.
                                 17. *Gratiola neglecta*
Flowers some shade of purple, red or blue.
   Flowers red or dark reddish purple.
      Flowers red, in a spike-like cluster.   29. *Castilleja miniata*
      Flowers (fig. 288) greenish purple in a large, branching
        cluster.   5. *Scrophularia lanceolata*
   Flowers pink to purple, lavender or blue.
      Flowers 12—25 mm. (½—1 in.) long.
        Flowers solitary in leaf axils.
          Corolla strongly 2-lipped, throat nearly closed.
                        13. *Mimulus ringens*
          Corolla lobes nearly equal, throat open. (fig. 290).
            Corolla 2—2.5 cm. long, rose purple. 26. *Gerardia aspera*
            Corolla 10 mm. long, light purple.
                      27. *Gerardia tenuifolia*
      Flowers in large, terminal clusters.   6-12. *Penstemon*
    Flowers 5—10 mm. long.
      Flowers in slender, terminal clusters. 23. *Veronica longifolia*
      Flowers at leaf bases.
        Corolla tube much shorter than the narrow lobes.
          Leaves 3—6 cm. long; stamens 2. 22. *Veronica scutellata*
          Leaves 1—3 cm. long; stamens 4. 4. *Collinsia parviflora*
        Corolla tube much longer than the rounded lobes.
          Lower leaves narrowed at base.   18. *Lindernia dubia*
          Lower leaves not narrowed at base.
                    19. *Lindernia anagallidea*

1. *Verbascum thapsus* L. MULLEIN. Biennial; stem stiff, thick, not branched, 1—2 m. (3—7 ft.) high; leaves alternate, oblanceolate, 1—4 dm. (4—16 in.) long, thickly covered with yellowish, felted, branched hairs; flowers yellow, 1—2.5 cm. (½—1 in.) wide, in a thick, terminal spike. Uncommon, introduced weed. July, Aug. Specimens from Cass, Sargent and Rolette Counties. In the Turtle Mts. in Rolette County, I saw a considerable amount of it at one place along the roadside in 1942. It is a familiar weed of eastern and central U. S. and even in Minnesota within 50 miles of Fargo. It grows especially on stony hills, in pastures and in dry, sandy, neglected fields.

2. *Verbascum blattaria* L. MOTH MULLEIN. Biennial; stem 4—12 dm. high; slender branched, with few hairs except short sticky ones on upper branches; leaves oblong or ovate, 2—10 cm. long, lower ones toothed or lobed, upper clasping; flowers white or yellow, 2—3 cm. wide, scattered along upper branches. This was found at Fargo in 1901 but has not been seen since.

3. *Linaria vulgaris* L. TOADFLAX. BUTTER AND EGGS. Strong scented perennial; stem leafy, unbranched, 3—10 dm. high; leaves 2—8 cm. long, 3—5 mm. wide; flowers yellow, 3 cm. long, in long, terminal clusters; capsules rounded, 5 mm. wide; seeds black, flattened, 1 mm. wide with a thin wing all around the edge. A persistent, introduced perennial, spreading by roots and forming dense patches. It has been most abundant in southern Barnes County, but found at various other places and is often planted as an ornamental. The flowers have the same shape as those of snapdragon, except that they have a slender, pointed spur 1—2 cm. long, projecting downward from the base.

4. *Collinsia parviflora* Dougl. BLUE-EYED MARY. Annual; stem erect or spreading, 1—3 dm. high; leaves opposite, oblong to linear, 1—3 cm. long; flowers blue, 4—6 mm. long at leaf bases. I found this in considerable quantity in a sparsely wooded coulee on north side of Killdeer Mts. in 1938 (Stevens 345). It is a species chiefly of western U. S. and whether or not recently introduced at this place is unknown. Dr. Pennell verified the identification.

5. *Scrophularia lanceolata* Pursh. FIGWORT. (Fig. 288). Large, native perennial, 1.5—2 m. high; leaves opposite, ovate to lanceolate, 5—20 cm. long, sharply toothed; flowers purplish, 8—10 mm. long, in a large branching top; pods ovate, pointed, 8—10 mm. long. Frequent in open woods, edges of woods and thickets, coulees or open roadsides. June, July. Previously called *S. leporella* Bickn.

Penstemon BEARDTONGUE

Perennials with opposite, usually entire, thick leaves and showy flowers in terminal clusters; stamens 4, the fifth without an anther but with a brush of hairs; capsules rounded or pointed, sometimes quite woody; seeds irregularly angled, brown or black.

This large, distinctive genus is most abundant in western U. S. It is characterized by the fifth stamen having its filament well developed, without an anther, but bearing a tuft of coarse hairs. Some western species have bright red flowers and a few have woody stems. Many are cultivated as ornamentals. The main flowering period is June.

Key to Species

Flowers blue or decidedly bluish.
  Flowers 3 cm. (1⅕ in.) long; leaves little waxy. 6. *Penstemon glaber*
  Flowers 1—2 cm. long; deep blue; leaves thick, waxy.
    Upper leaves ovate. 8. *Penstemon nitidus*
    Upper leaves linear or lanceolate. 9. *Penstemon angustifolius*
Flowers lavender, purple, pink or white.
  Flowers white with some dark spots. 11. *Penstemon albidus*
  Flowers lavender, purple or pinkish.
    Flowers with yellow hairs protruding from inside. 10. *Penstemon eriantherus*
    Flowers without such hairs.
      Flowers 4 cm. long, pinkish lavender; leaves rounded, not toothed. 7. *Penstemon grandiflorus*
      Flowers 2 cm. long, lilac; leaves narrow, toothed. 12. *Penstemon gracilis*

6. *Penstemon glaber* Pursh. SMOOTH BEARDTONGUE. Stems 3—6 dm. (1—2 ft.) high; leaves oblong or lanceolate, 5—10 cm. (2—4 in.) long; flowers nearly blue, 3 cm. long. Our only record is a specimen from W. F. Blume of Dickinson in 1922, with the statement that it had been present in the native sod in his yard. It seems strange that such a showy plant could have been overlooked by early collectors, especially in this area which was quite thoroughly explored. Dr. Pennell writes that it should occur in the State. In 1941, I saw two selections growing in the Oscar H. Will Co. nursery. One had leaves about 1 dm. long and 2 cm. wide, the width quite uniform, flowers blue. The other had ovate leaves, 4—5 cm. wide at base, flowers tending to become purplish with age. Specimens of each were sent to Dr. Pennell.

7. *Penstemon grandiflorus* Nutt. LARGE BEARDTONGUE. (Fig. 286). Stout, usually unbranched, waxy plant, 3—10 dm. high; leaves thick, broad ovate to nearly circular, 3—8 cm. long; flowers pinkish lavender, 4 cm. long, in a loose, terminal cluster 2—4 dm. long. Our most showy species and commonly cultivated. Abundant in very sandy soil along Missouri and lower Cannon Ball Rivers; in dune area in Richland and Ransom Counties. We have no specimens north of Stutsman and Morton Counties.

8. *Penstemon nitidus* Dougl. Stocky plant, 1—4 dm. high; leaves ovate to lanceolate, 2—8 cm. long, dark green but somewhat waxy; flowers blue, 1.5—2 cm. long, in late May or early June. This has often been regarded as a variety of *P. angustifolius*. It was first brought to our

286. Large Beardtongue (*Penstemon grandiflorus*).

attention by a specimen from Bowman in 1921 sent by Marie Jorgenson and identified by Dr. Pennell. It proves to be fairly common on hills and buttes in the west and we now have specimens from Slope, Golden Valley, Dunn, Mountrail and McKenzie Counties. It is quite a handsome plant and the Oscar H. Will Co. of Bismarck had been growing it as different from *P. angustifolius* before we recognized it as such.

9. *Penstemon angustifolius* Pursh. NARROW-LEAVED BEARDTONGUE. Similar to *P. nitidus*, perhaps a little taller, rather slender with narrow leaves and flowers in a narrow, dense cluster 1—2 dm. long. Hills and buttes from Missouri River west; locally near Medina and Dawson in Stutsman and Kidder Counties and along Sheyenne River near Valley City, Barnes County.

10. *Penstemon eriantherus* Pursh. CRESTED BEARDTONGUE. Stem stout, 2—4 dm. high; leaves oblong to linear; flowers lavender, 2—3 cm. long, the tube 1 cm. wide, nearly closed by a crest on lower lip, coarse yellow hairs projecting. A showy species, quite common west of Missouri River on bare clay slopes (fig. 18). Also known as *P. cristatus* Nutt., an earlier name which was not published in approved form.

287.   Slender Beardtongue (*Penstemon gracilis*).

11. *Penstemon albidus* Nutt. WHITE BEARDTONGUE. Stocky, somewhat sticky plant, 2—4 dm. high; leaves lanceolate, 5—8 cm. long; flowers 2 cm. long, white, spotted with purple. Prairies, especially stony hills. One of the commonest species, beginning to flower in late May.

12. *Penstemon gracilis* Nutt. SLENDER BEARDTONGUE. (Fig. 287). Stems slender, 2—6 dm. high, upper part branching and flowers scattered; leaves 5—8 cm. long, lanceolate to linear with fine teeth; flowers lilac, 1.5—2 cm. long, tube 5 mm. wide. This is perhaps more common than *P. albidus*, but grows in good or sandy prairie soils, especially in coulees where moisture supply is better than average. Flowers in late June, the latest of the Penstemons.

13. *Mimulus ringens* L. MONKEYFLOWER. (Fig. 289). Perennial, dark green, erect plant; stem angled, 2—8 dm. (8—32 in.) high; leaves nearly clasping, lanceolate, 3—8 cm. (1—3 in.) long with a few sharp teeth; flowers solitary at upper leaf bases, violet, 2—3 cm. long; calyx tubular, 1—2 cm. long, with 5 lengthwise plaits ending in slender teeth; corolla with a slender tube and widely flaring lips, the upper 2-lobed, rounded, erect, the lower horizontal; throat nearly closed by a raised portion on lower lip; capsule oblong, 1 cm. long, with many, very minute seeds. In wet places along river and pond banks, west to Stutsman County. Aug. A handsome flower.

14. *Mimulus geyeri* Torr. YELLOW MONKEYFLOWER. Perennial, stems mostly horizontal in shallow water or on wet ground, 1—4 dm. long; leaves rounded, as wide as long, 8—25 mm. long; flowers yellow, 8—12 mm. long. July, Aug. We have it only from Richland and Barnes Counties but it probably occurs elsewhere in suitable locations. First named from plants collected by the Nicollet Expedition at Devils Lake. Later referred to *M. glabratus. var. fremontii* (Benth.) Grant.

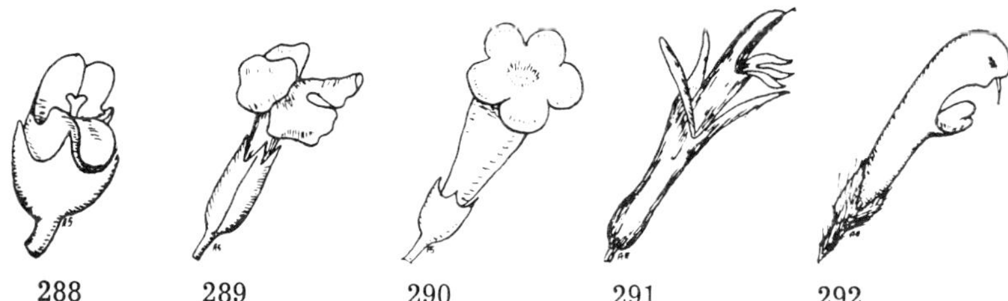

288   289   290   291   292

15. *Hydranthelium rotundifolium* (Michx.) Pennell. WATER HYSSOP. Perennial; stems floating in water or creeping on mud, rooting at joints, 2—6 dm. long; leaves opposite, obovate, 1—2 cm. long; flowers white or yellowish, 5—6 mm. long, solitary at leaf bases. We have it from Cass, Ransom, Grant and Morton Counties. This plant has passed under various names, *Monniera, Bacopa, Macuillamia.*

16. *Limosella aquatica* L. MUDWORT. Annual but spreading by runners, no erect stems; leaves tufted, 1—3 cm. long, 2—4 mm. wide on slender petioles; flowers white or purplish, 2 mm. long, each on a slender stalk from a leaf base. On mud along ponds or ditches. We have it from Cass, LaMoure, Walsh, Pembina, Morton and Grant Counties. An insignificant plant; probably occurs in every part of State.

17. *Gratiola neglecta* Torr. HEDGE HYSSOP. Annual; stem finely glandular, 1—3 dm (4—12 in.) high; leaves opposite, oblong to lanceolate, 1—3 cm. (⅖—1⅕ in.) long; flowers yellowish white, 8—10 mm. (⅓ in.) long, tubular with short, rounded lobes; capsule 3—6 mm. long, equaling the sepals. Along ditch and pond edges and in low spots in fields where water has stood for a time. We have it from Grant, Morton and Cass to Bottineau Counties. Probably it could be found in any county in suitable places. Previously referred to *G. virginiana* L. Pennell records a specimen of *Gratiola lutea* Raf. from Grand Forks. It is perennial by rhizomes; capsule much shorter than sepals, upper leaves clasping at bases.

18. *Lindernia dubia* (L.) Pennell. FALSE PIMPERNEL. Annual; stem becoming widely branched, 1—2 dm. high; leaves ovate, entire or nearly so, 1—2 cm. long; flowers lavender, 8—10 mm. long, solitary at leaf bases. On wet soil along river banks or in low places in fields. Aug. Specimens from Fargo, our only record, were determined as var. *typica* Pennell.

19. *Lindernia anagallidea* (Michx.) Pennell. Specimens from Cass and from Richland (Bell 357, mixed with *Gratiola*) Counties have been determined by Pennell as this species. He cites "Stevens 992," but that number was not originally assigned to it. They are insignificant but rather pretty little mud plants, previously recorded as *Ilysanthes dubia* (L.) Barnhart.

20. *Veronica americana* Schwein. SPEEDWELL. BROOKLIME. Stems erect, spreading, or floating in water, 1—6 dm. long; leaves opposite, ovate or oblong-lanceolate, 1—3 cm. long; flowers blue, 4—5 mm. wide, in slender axillary clusters. In flowing water or in boggy places. One record from Killdeer Mts., Dunn County, others from Richland, Ransom, Barnes, Cass and Pembina Counties. Probably it will be found in other localities where conditions are suitable.

21. *Veronica connata* Raf. Very similar to preceding in appearance and in distribution; leaves more elongated lanceolate, 2—7 cm. long. Previously referred to *V. anagallis-aquatica* L., which Pennell regards as a distinct European species which has been introduced into America in various places. Our specimens he refers to *V. connata,* var. *typica* Pennell, but some collected by Lunell in Benson and Rolette Counties, he recorded as var. *glaberrima* Pennell. These lack the very small glandular hairs which are present on terminal branches and flower stalks of var. *typica.*

22. *Veronica scutellata* L. MARSH SPEEDWELL. Perennial; stems slender, 1—5 dm. high; leaves linear or lanceolate; flowers lavender, 6—10 mm. wide, on slender stalks from upper leaf bases. Collected in 1946 near Lake Metigoshe, Turtle Mts., in a roadside ditch. Pennell recorded specimens collected by Lunell at Leeds, and by Geyer at Devils Lake.

23. *Veronica longifolia* L. CLUMP SPEEDWELL. Perennial, stems erect, 3—8 dm. high; leaves lanceolate, sharply toothed, 5—10 cm. long; flowers lilac, 3 mm. wide, in slender, terminal spikes. Specimens were received from F. W. Schmidt Jr., of Kongsberg, Wells County, in 1918 and 1920. It was no doubt an escape from cultivation. I have had no opportunity to check further on its persistence. The name *V. maritima* L. has also been used for this plant.

24. *Veronica peregrina* L. PURSLANE SPEEDWELL. Sticky, aromatic annual, 1—2 dm. high; leaves alternate, narrowly oblong, 1—2 cm. long; flowers white, 2 mm. wide at leaf bases; capsule wider than long, notched at top, 4 mm. long. A very common little weed in fields, gardens, etc. especially in heavy soils where water may stand for a time. May, June. It occurs all over the U. S., and in parts of Canada and Mexico. Our form is var. *xalapensis* (HBK.) Pennell, originally described from Xalapa, Mexico. Var. *typica* Pennell, which does not have glandular hairs, occurs all through eastern U. S. and might be found in eastern North Dakota.

25. *Veronicastrum virginicum* (L.) Farwell. CULVER'S ROOT. Perennial; stem erect, 3—10 dm. (1—3 ft.) high; leaves lanceolate, sharply toothed, 5—15 cm, (2—6 in.) long, usually in whorls of 5; flowers white, 5 mm. (⅕ in.) long, in a dense spike 1—3 dm. long at end of main stem and usually on several, upper branches. Our only record is a specimen received in 1904 from Miss Ella Olafson, Hallson, Pembina County. It is to be expected in that area, but some search for it in recent years has failed to locate any plants. It is quite common in Minnesota within 40 miles of our border. It grows in low grassland, usually near edges of woods. It is a striking plant, the long, slender flower clusters appearing in late July. Leaves in whorls of 5 make an unusual arrangement, which is found also in Joe-Pye Weed. The species is often included in *Veronica.*

26. *Gerardia aspera* Dougl. Rough Gerardia. (Fig. 290). Annual; stems slender, simple or widely branched, 2—6 dm. high; leaves opposite, 1—3 cm. long, 1—2 mm. wide; flowers rose purple, 2—2.5 cm. long, 1—1.5 cm. wide, solitary at leaf bases; capsule rounded, 5 mm. long. Low prairie. Aug. We have it from scattered localities east of Missouri River. It is a small but striking plant, because of the large flowers and inconspicuous leaves. The plants turn black when pressed and dried, so they are a disappointment to a collector. The plant is named for Thomas Gerard, author of the "Great Herbal" (1597), one of the famous early botanical books.

27. *Gerardia tenuifolia* Vahl. Slender Gerardia. Very similar to preceding but more slender, flowers smaller, about 1 cm. long. In boggy or decidedly wet places. We have it from the four southeastern counties and from Rolette and McHenry Counties. Pennell referred our material to var. *parviflora* Nutt.

28. *Castilleja sessiliflora* Pursh. Downy Painted Cup. Stocky, gray perennial, 1—3 dm. high; leaves crowded, 3—5 cm. long, lanceolate or divided into several narrow lobes; flowers pale yellow, slender, 3—5 cm. long, crowded in a terminal cluster. Common on dry prairies and hills. June.

29. *Castilleja miniata* Dougl. (Fig. 291). Perennial; stem erect, 3—6 dm. high; leaves alternate, lanceolate, 5—10 cm. long; flowers bright red, clustered at tip of stem. On July 15, 1938, Mrs. Dana Wright of St. John, N. D., sent two stalks, which were identified by Dr. Pennell. She wrote that she found these north of Walhalla, Pembina County, and saw many more across the boundary in Manitoba. This is one of the numerous Rocky Mountain species, which would not have been expected in this area. The very similar Scarlet Painted Cup, or Indian Paintbrush (*C. coccinea*) occurs in Minnesota within 15 miles of Fargo, but has not been found in North Dakota. It and the above species owe their striking color to red bracts in the flower cluster rather than to the corolla itself.

30. *Orthocarpus luteus* Nutt. Owl Clover. Annual; stem erect, usually unbranched, 1—3 dm. (4—12 in.) high; leaves lanceolate or linear, 3—6 cm. (1—2½ in.) long; flowers yellow, 10—15 mm. (½ in.) long, in a thick, terminal spike. Prairie, all over State. July. This was first discovered by Nuttall in the Fort Mandan area. He was especially pleased with it because it represented a new genus. Many species, some with red flowers, were later described from the Pacific Coast region. The origin of the name "owl clover" is not known.

31. *Pedicularis canadensis* L. Lousewort. (Fig. 292). Perennial; stems tufted, 1—3 dm. high; leaves mostly near the ground, oblong, 8—15 cm. long, "crinkly", edges cut into close, rounded lobes; flowers yellow, 2—2.5 cm. long, in a terminal, head-like cluster. We have it only from Richland County where it is locally abundant on low, sandy prairie, blooming in mid-May. One specimen (Bergman 1420), previously referred to the following, belongs to this species. It is a showy plant which I propose to call "bumblebeewort" because the flowers are visited so freely by these insects.

32. *Pedicularis lanceolata* Michx. Taller plant than the preceding, stems rather bare below, 3—6 dm. high. Aug. We have it from low, sandy prairie in the southeast, also Walsh, Pembina and McHenry Counties

BROOMRAPE FAMILY Orobanchaceae

Low, fleshy, pinkish or purplish plants, parasitic on roots of other plants; stems with pinkish scales in place of leaves; flowers purplish, tubular, 2-lipped; stamens 4; pistil 1; fruit a pod with dust-like seeds. Members of this group are more abundant in Europe where some of them infest crop plants.

Key to Species

Each flower with a scale at its base.　　1. *Orobanche ludoviciana*
Flower clusters without scales.
　Stems 1—3-flowered; woodland plants.　　2. *Orobanche uniflora*
　Stems many flowered; prairie plants.　　3. *Orobanche fasciculata*

1. *Orobanche ludoviciana* Nutt. (Fig. 77). Stem stout, 1—3 dm. (4—12 in.) high; flowers 12—25 mm. (½—1 in.) long in a thick cluster at top of stem. Prairie; Pembina, McHenry and Billings Counties are our only records. It probably can be expected in any part of the State. Aug.

2. *Orobanche uniflora* (L.) Endl. CANCERROOT. Stem short, often not coming above ground; each flower on a stalk 5—15 cm. long from top of stem; corolla yellowish or purple, 10—20 mm. long. Our only record is from aspen woods at Fargo.

3. *Orobanche fasciculata* Nutt. Similar to preceding, but with several or many flower stalks; flowers 20—30 mm. long. Fairly common on prairie all over State, especially in sandy soils. June-Aug. Parasitic on wormwoods and various other plants.

BLADDERWORT FAMILY　Utriculariaceae

Aquatic plants, stems lying in water, leaves finely divided; flowers 1—several on erect branches; corolla of united petals, 2-lipped, a short spur or rounded projection from the base; stamens 2, attached to upper side of corolla; fruit a capsule with many small seeds.

On the leaves or stems are many tiny "bladders," sacks 1—3 mm. wide. These capture small aquatic animals which are used by the plant as food. There are nearly 100 known species. Some grow in wet places out of water and have broad leaves. For a recent account of these and other carnivorous plants, see the book by F. F. Lloyd, "Carnivorous Plants", 1943.

1. *Utricularia vulgaris* L. BLADDERWORT. (Fig. 64). Floating plant with long stems and finely divided leaves; flower stalks 1—2 dm. (4—8 in.) high with 1 or 2 flowers 15—20 mm. (⅜—⅘ in.) long. The bright yellow flowers in thick masses in July and August are conspicuous. Hardly common, but we have records from all parts of State. Some of the best displays I have seen in recent years have been along the main highways just west of Tappen in Stutsman County and near Doyon in Ramsey County. Our plants have not been carefully studied and more than one species may be represented.

LOPSEED FAMILY　Phrymaceae

This family contains a single species. The flower is similar to many in the mint and figwort families, but the fruit is 1-seeded.

1. *Phryma leptostachya* L. LOPSEED. (Fig. 118). Perennial; stem erect, 3—10 dm. (1—3 ft.) high; leaves opposite, ovate, 3—15 cm. (1—6 in.) long, very thin, with rounded teeth; flowers white, tinged with purple, 8 mm. (⅓ in.) long, scattered along slender, upper, leafless branches; fruiting calyx bent sharply downward, the teeth spiny. Woods. July, Aug. We have it from the southeast, where it is quite common, also from Pembina and Morton Counties.

PLANTAIN FAMILY **Plantaginaceae**

Tufted plants, mostly perennials, with very short stems and a few, spreading leaves; flowers very small in a dense spike on one or more bare stems from center of leaf cluster (fig. 85); calyx and corolla 4-lobed; stamens 4; pistil 1; fruit a 2—several-seeded capsule which splits around the middle.

Key to Species

Flower cluster thick and short, about 1 in. long; rare weed in
   lawns.   1. *Plantago lanceolata*
Flower cluster slender, flowering on most of its length.
   Very small plants 3—10 cm. (1—4 in.) high; leaves green, hardly
      3 mm. (⅛ in.) wide.   2. *Plantago elongata*
   Larger plants 1—5 dm. (4—20 in.) high; or if small, gray hairy
      and leaves 6—8 mm. wide.
      Leaves gray, hairy, 6—8 mm. wide; plants of dry prairie.
         Spikes with a slender bract, 6—12 mm. long below each
            flower.   4. *Plantago spinulosa*
         Spikes without such bracts.   3. *Plantago purshii*
      Leaves green, 3—7 cm. wide; plants of rather moist soil.
         Leaves about 3 cm. wide, tapering at ends.   5. *Plantago eriopoda*
         Leaves 7 cm. wide, rounded.
            Leaves green, smooth or quite rough with short hairs.
                      6. *Plantago major*
           Leaf stalks purple; blades not hairy.   7. *Plantago rugelii*

1. *Plantago lanceolata* L. LANCE-LEAVED PLANTAIN. BUCKHORN. Perennial; leaves elliptic, 5—30 cm. (2—12 in.) long; flowering stalks 3—5 dm. (1—1½ ft.) high, bare except for short flower spike at top. Seeds are a common impurity in clovers. It has been found a few times in lawns but seems not to persist. It is a troublesome weed in meadows in eastern U. S.

2. *Plantago elongata* Pursh. Small annual, 3—10 cm. (1—4 in.) high. Frequent in western North Dakota on saline flats. July.

3. *Plantago purshii* R. & S. PURSH'S PLANTAIN. PRAIRIE PLANTAIN. Gray, hairy, native annual, 1—3 dm. high; leaves linear, 3—10 cm. long; flower spikes 5—30 cm. (2—12 in.) high. Common, especially west of Missouri River, often forming gray patches.

4. *Plantago spinulosa* DC. BRACTED PLANTAIN. Resembles the last but is less common. Previously listed by error as *P. aristata* Michx.

5. *Plantago eriopoda* Torr. ALKALI PLANTAIN. Native perennial from a short, thick crown with long, brown hairs; leaves green, elliptic, 5—20 cm. long; flower stalks 5—30 cm. high. Frequent on saline flats. June, July.

6. *Plantago major* L. COMMON PLANTAIN. "RIBGRASS." Perennial from a short, thick crown; leaves rounded or ovate, 1—2 dm. long; flower spike 1—3 dm. high, the flowers crowded. Common in sloughs, along roads and in waste ground. July-Sept. Probably both native and introduced. The leaves of some plants are quite hairy.

7. *Plantago rugelii* Decne. RUGEL'S PLANTAIN. Similar to last; leaves with purple petioles; spikes more slender, flowers not crowded. Rather rare, usually in wooded areas, eastern part of State. More common farther south.

MADDER FAMILY **Rubiaceae**

Plants of various habits; petals 4, partly united and borne on top of 2-celled ovary; stamens 4; fruit various. A large family of mostly tropical, woody plants. Madder (*Rubia tinctoria*) is a European plant, formerly much used for dye. Coffee and quinine belong to this family. Commercial coffee is the seed from a red berry.

Key to Species

Leaves opposite; flowers funnel shaped, purplish white.

1. *Houstonia longifolia*

Leaves in whorls of 4—8; flowers flat, white; fruit a bur.

Leaves in whorls of 6—8; stems weak, spreading or climbing.

Leaves in 8's; angles of stem very rough.   2. *Galium aparine*

Leaves in 6's; stem only slightly rough.   4. *Galium triflorum*

Leaves in whorls of 4.

Prairie plants, mostly upright.   3. *Galium boreale*

Small, swamp plants, mostly creeping.   5. *Galium trifidum*

1. *Houstonia longifolia* Gaertn. BLUETS. Erect, tufted perennial, about 1 dm. (4 in.) high; leaves 1—2 cm. (⅖--⅘ in.) long, 1—2 mm. wide; flowers purplish white, 6—8 mm. (¼ in.) long, in few flowered clusters at tips of stems; fruit a rounded pod. Low places on prairie. June, July. We have it from Benson, Pierce, Cavalier and Pembina Counties. Several species of bluets are familiar and attractive flowers in eastern U. S.

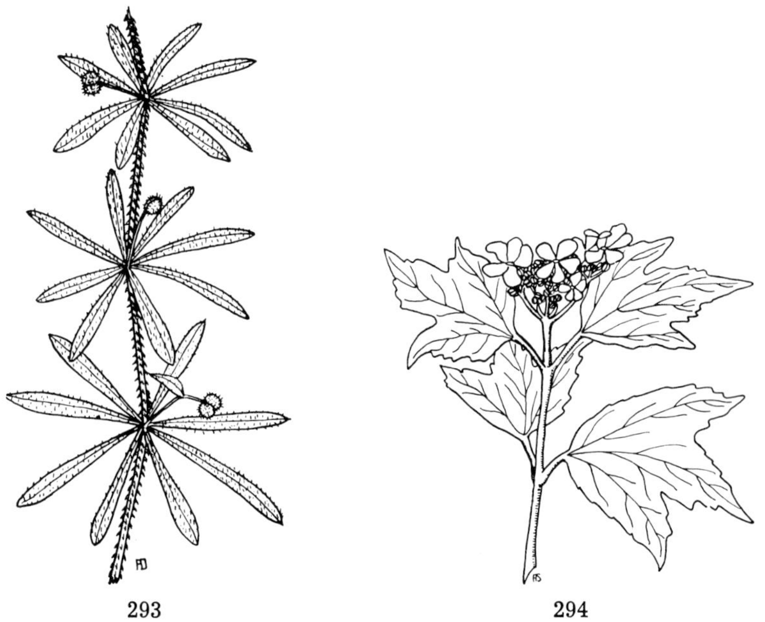

293

294

293. Cleavers (*Galium aparine*).

294. Pembina (*Viburnum trilobum*).

2. *Galium aparine* L. Cleavers. GOOSEGRASS. (Fig. 293). Annual with weak stems and few, shallow roots; stem climbing over other plants to 1 m. (3 ft.) high, the sharp angles covered with short, stiff hairs projecting backward; leaves oblanceolate, rough, 3—6 cm. long, 5—10 mm. wide; flowers white, 3 mm. wide in loose clusters on slender branches; fruit 2 rounded burs, 3—5 mm. wide. Open woods and brushy places. May. We have it from southeast counties and from Pembina, Stark and Billings, but it probably occurs elsewhere in wooded places. It is an annoying weed because the whole plant clings to clothing and the weak roots easily break away from the ground.

3. *Galium boreale* L. NORTHERN BEDSTRAW. Perennial; stems erect, often
   tufted, 2—4 dm. high; leaves 2—5 cm. long, 2—4 mm. wide; flowers
   white, 2 mm. wide, in large, branching, terminal clusters; fruit rounded, 2 mm. wide, hairy but hardly a bur. Common in open woods through
   the east, lower prairies, coulees and slopes west. An attractive wild
   flower, sometimes called "wild baby's breath".

4. *Galium triflorum* Michx. Perennial; stems usually creeping, 3—10 dm.
   long; leaves elliptic, 3—5 cm. long; flowers in loose clusters on slender
   stalks from leaf bases; fruit 3—5 mm. wide, somewhat bristly. Woods,
   usually in shady places. We have it from eastern third of State. It is
   called "sweet-scented bedstraw", but we fail to recognize much fragrance.

5. *Galium trifidum* L. SMALL BEDSTRAW. Perennial; stems very slender and
   weak, creeping, 1—3 dm. long; leaves 1—2 cm. long, 1—2 mm. wide;
   flowers 1 mm. wide, 1—3 on slender stalks from upper leaf bases;
   fruit 1 mm. wide, smooth. Wet meadows or swampy woods, eastern
   third of State.

HONEYSUCKLE FAMILY   Caprifoliaceae

Mostly woody plants with opposite leaves; corolla on top of
ovary, petals mostly or partly united; sepals very small; stamens 5,
attached to corolla; pistil 1; fruit usually fleshy.

Key to Species

Small, creeping plant, slightly woody.   1. *Linnaea borealis*
Shrubs, 1—3 m. (3—10 ft.) high.
   Corolla 4—8 mm. (¼ in.) long, tube very short.
      Corolla flat; flowers in large, flat, terminal clusters.
         Leaves 3-lobed; fruit red.   2. *Viburnum trilobum*
         Leaves not lobed; fruit black.
            Leaves finely toothed, smooth.   3. *Viburnum lentago*
            Leaves coarsely toothed, finely hairy.   4. *Viburnum affine*
      Corolla bell shaped; flowers in small, axillary clusters.
        Flowers 6—8 in each cluster; leaves 2—5 cm. (1—2 in.) long.
              5. *Symphoricarpos occidentalis*
        Flowers 1—3 in each cluster; leaves 1—2 cm. long.
              6. *Symphoricarpos albus*
   Corolla 15—25 mm. long, tubular with spreading, oblong lobes.
      Trailing or vine-like shrub; flowers in dense group at tip
        of branch.   7. *Lonicera dioica*
      Upright shrub; flowers in pairs from upper leaf bases.
              8. *Lonicera tatarica*

1. *Linnaea borealis* Gron. TWINFLOWER. Stems slender, creeping, 3—6 dm.
   (1—2 ft.) long; leaves rounded, 5—10 mm. (⅕—⅖ in.) long; flowers
   pink, funnel shaped, nodding, 8—10 mm. long, in pairs on slender
   stalks 2.5—10 cm. (1—4 in.) long; fruit a 1-seeded pod. Our only record
   is a specimen from Turtle Mts. near Bottineau, collected by Neil Hotchkiss of the Fish and Wildlife Service, U S Dept. of Interior, in 1921.
   It might occur in Pembina Mts. June.

2. *Viburnum trilobum* Marsh. PEMBINA. HIGHBUSH CRANBERRY. (Fig. 294).
   Spreading shrub 1—3 m. (3—10 ft.) high; leaves broadly ovate, or widest above middle, 7—15 cm. (3—6 in.) long, lobes of lower leaves
   short, triangular, but upper usually long and tapering; ripe fruit
   drooping, bright red, 1 cm. long, with 1 large, flat seed. Swampy wooded places, Richland to Pembina and Bottineau Counties. June. The fruit
   remains on the bushes all winter and is eaten by some winter birds,
   especially Bohemian waxwings. The fruit is used for jelly. Pembina
   was the Indian name for the plant and we have the town and county
   named from this source. This plant was formerly called *V opulus* L.
   or var. *americanum* Mill., but is now considered a distinct species.

The European *V. opulus* is often planted an ornamental, usually non-fruiting with rounded flower heads (Snowball). The fruits of it have an unpleasant taste. The two plants may be distinguished as follows:

V. opulus. Lobes of leaves narrowed to a short, wide tip, petiole with about two pairs of red protuberances 2 mm. wide; fruit bright scarlet.

V. trilobum. Lobes of leaves tapering to slender tips, petiole with smaller protuberances; fruit dark red.

3. *Viburnum lentago* L. BLACK HAW. NANNYBERRY. (Fig. 295). Shrub 2—5 m. high; leaves oblong or ovate with a tapering tip, glossy and finely toothed, 5—15 cm. long; flowers white, 5 mm. wide in flat clusters 5—15 cm. wide; fruit black, 10—15 mm. long. Frequent in open woods and along streams through eastern third of State; records also from Bottineau, McLean and Dunn Counties. The large clusters of white flowers in June are showy and the leaves often become bright red in fall. The fruit is sweet when fully mature but there is little flesh on the flat seed.

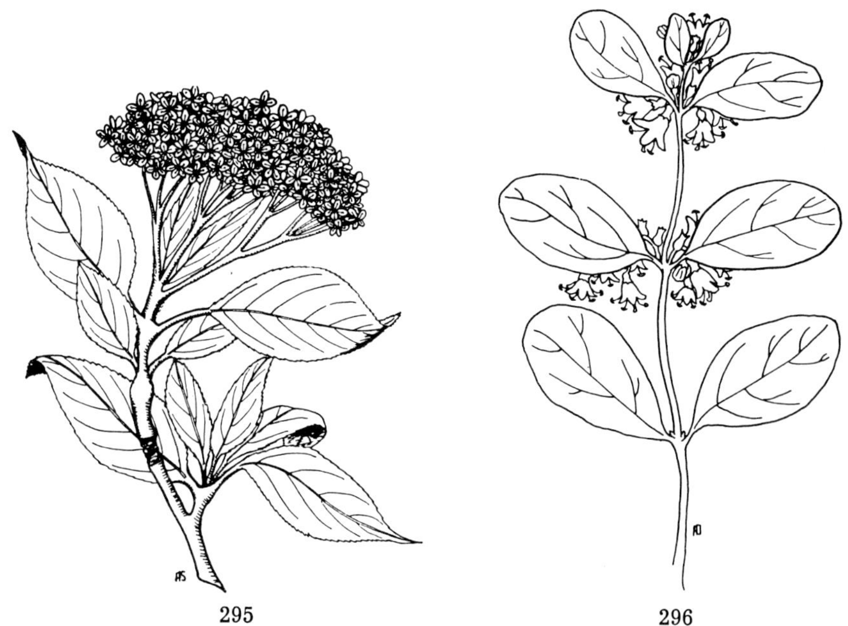

295

296

295. Black Haw (*Viburnum lentago*).

296. Wolfberry (*Symphoricarpos occidentalis*).

4. *Viburnum affine* Bush. DOWNY-LEAVED ARROWWOOD. Shrub 1—2 m. high; leaves ovate, 4—8 cm. long, coarsely toothed, covered with fine hairs; fruit black, 1 cm. long, 8 mm. thick, not pleasant to taste. Recorded only at Fargo, in Pembina, eastern Cavalier and Bottineau Counties. Our plant is var. *hypomalacum* Blake, and was formerly referred to *V. pubescens* Pursh.

5. *Symphoricarpos occidentalis* Hook. WOLFBERRY. (Fig. 296). Upright or spreading bush, 3—10 dm. (1—3 ft.) high; leaves oblong or rounded, 2—5 cm. (1—2 in.) long, larger ones sometimes with a few teeth or short lobes; flowers pinkish, 6 mm. (¼ in.) wide, at leaf bases along upper parts of branches; fruit white, later turing dark, oblong, 5 mm. long, 2-seeded. A very common bush, becoming over 1 m. high in open

woods; forming dense patches in lower places on the prairie or in coulees. July. The fruits remain on the plants over winter and are an important food for game birds. It is commonly called "buckbrush", a name applied to various small shrubs.

6. *Symphoricarpos albus* L. Small shrub, hardly distinguishable from small specimens of the preceding, but occurs only on thickly wooded slopes in certain localities. Records from Ramsey, Pembina, Rolette, Bottineau, Hettinger, Dunn and McKenzie Counties. The snowberry, often planted for its large, white fruits, is a form of the same species. Formerly called *S. racemosus* Michx.

7. *Lonicera dioica* L. WILD HONEYSUCKLE. Trailing or clambering shrub, sometimes more or less erect with drooping branches and 1 m. or more high; leaves commonly pale, oblong, 3—10 cm. long, the uppermost grown together at base around stem; flowers pinkish or yellow, 15—25 mm. long in a cluster above these joined leaves; fruit orange red, 5—8 mm. wide. Occasional and locally common in wooded areas. Late May. A rather attractive plant which could be grown as an ornamental by using a low trellis. The abundant, early flowers are attractive to hummingbirds. Our plants apparently should be called var. *glaucesens* (Rydb.) Butters, but I cannot separate them satisfactorily from the species.

8. *Lonicera tatarica* L. TARTARIAN HONEYSUCKLE. Shrub 1—3 m. high with fine, spreading branches; leaves oblong or ovate, 3—7 cm. long; flowers pinkish or yellowish, 12—15 mm. long, in pairs from leaf bases; berry scarlet, 5—6 mm. wide. June. Birds scatter the seeds of this commonly planted ornamental. It has become well established at Fargo and perhaps elsewhere. First noted in 1921.

TEASEL FAMILY   Dipsacaceae

Herbs with opposite leaves; flowers small, in involucrate heads (fig. 116); corolla funnel shaped with 4 or 5 lobes; stamens 4 or 2, attached to corolla which is on top of ovary; calyx cup shaped or funnel shaped, with bristle pointed lobes; fruit dry, 1-seeded.

This is a small group to which belong Teasel (*Dipsacus sylvestris*), a weed in eastern U. S., and Mourning Bride (*Scabiosa atropurpurea*), a common garden flower. The spiny heads of teasel were formerly used for carding wool. Specimens of this group would run out to aster family in our "short key". They are closely related to asters but differ in having a true calyx.

1. *Knautia arvensis* (L.) F. Coult. FIELD SCABIOUS. Rough perennial; stems 6—10 dm. (2—3 ft.) high, not much branched; lower leaves 1—2 dm. (4—8 in.) long, oblong, entire or pinnatifid; middle leaves usually with 1—3 pairs of lanceolate lobes; flower heads lilac colored, 2—5 cm. (1—2 in.) wide, at ends of few, long branches. Found at Barton, Pierce County, in 1903; in a bromegrass field at Northwood, Grand Forks County, in 1917, and in a similar location at Minot in 1911. It is of interest and probably significant that Harry Graves saw it again at Northwood in 1942. The seeds were apparently introduced in bromegrass seed.

GOURD FAMILY   Cucurbitaceae

Herbaceous vines with alternate, usually broad, lobed leaves; flowers at leaf bases, solitary or clustered; corolla on top of ovary, short or long bell shaped with 5—6 lobes, stamens commonly 3; fruit fleshy, dry, or fleshy becoming dry. A large group of mostly tropical plants, including squashes, melons, cucumbers and several ornamentals. Stamens and pistils are usually produced in separate flowers, therefore, it is impossible for staminate flowers to "set" fruits.

Key to Species

Fruits solitary, 2.5—5 cm. (1—2 in.) long, watery, many seeded.
1. *Echinocystis lobata*
Fruits clustered, 1—1.5 cm. long, dry, 1-seeded. 2. *Sicyos angulatus*

1. *Echinocystis lobata* (Michx.) T. & G. WILD CUCUMBER. (Fig. 299). Annual vine climbing by tendrils, 2—10 m. (7—34 ft.); leaves 1—2 dm. (4—8 in.) long, usually cut about half way into 5 triangular, spreading lobes; staminate flowers yellowish, 4—5 mm. (⅕ in.) wide, forming clusters 1—1.5 dm. long; pistillate flowers few; fruits broad oblong, 2.5—5 cm. (1—2 in.) long, covered with soft prickles, eventually becoming dry, seeds falling from lower end and only network partitions remaining inside wall of fruit. Common along streams in east, less frequent as far west as Missouri River, usually climbing over bushes in open places or at edges of woods. July-Sept. It is much planted as a porch vine and so more widely distributed.

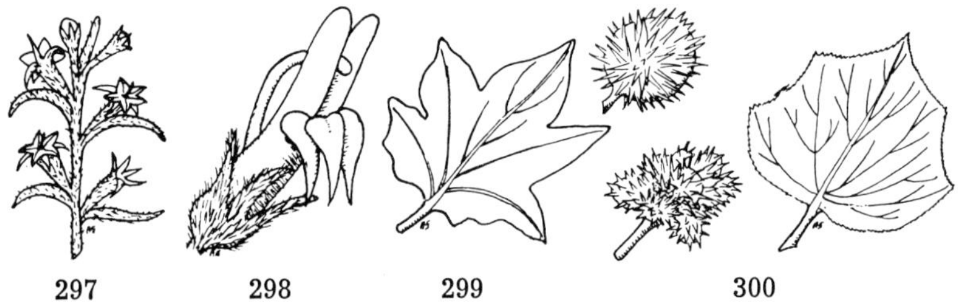

297   298   299   300

2. *Sicyos angulatus* L. BUR CUCUMBER. (Fig. 300). Similar in habit to preceding but leaves more rounded; staminate flowers 10 mm. wide, saucer shaped, in smaller, less conspicuous clusters; fruit oblong, 1—1.5 cm. long, with 1 large seed. Found by Reynold Shunk near Anselm, Ransom County, about 1916; not seen by me until 1946 (Stevens 894). It is a common plant in central U. S. where the vines extend into corn fields and become a nuisance because the coarse hairs on the fruits are quite prickly and break off easily.

BELLFLOWER FAMILY Campanulaceae

Annuals or perennials with simple, alternate leaves; petals partly or mostly united into a bell shaped or saucer shaped corolla on top of ovary; stamens 5, attached to corolla; pistil 1, stigmas usually 3; fruit a 3-celled capsule, splitting or opening by small holes. A large group of plants, represented by the common bluebell and by species of *Campanula* and *Platycodon* in garden flowers.

Key to Species

Corolla bell shaped with short, wide lobes.
    Corolla 15—25 mm. long, purple; stems smooth.
        Stems slender, 2—4 dm. (8—16 in.) high. 1. *Campanula rotundifolia*
        Stems stout, 10—15 dm. high. 2. *Campanula rapunculoides*
    Corolla 5—10 mm. long, white; stems rough. 3. *Campanula aparinoides*
Corolla saucer shaped, lobes narrow, longer than wide.
    Leaves rounded, clasping. 4. *Specularia perfoliata*
    Leaves narrow, not clasping. 5. *Specularia leptocarpa*

1. *Campanula rotundifolia* L. BLUEBELL. HAREBELL. (Fig. 301). Perennial; stems slender, 2—4 dm. (8—16 in.) high; lowest leaves ovate or rounded 5—15 mm. (⅕—⅗ in.) long, upper ones very narrow, 1.5—8 cm. (½— 3 in.) long; flowers bluish purple, nodding, in loose clusters at top of stems; capsule 5 mm. long, opening by 3 small holes at base (upper end as capsule hangs down). Common on prairie, sometimes abundant, especially in sandy soils. June, July. It grows under quite a range of conditions. Some authors call the American plant *C. petiolata* A. DC., regarding it as distinct from the plant of Europe.

2. *Campanula rapunculoides* L. CREEPING BELLFLOWER. Perennial, spreading by roots and forming thick, radish-like roots; stems erect, 6—10 dm. high; lowest leaves cordate, upper lanceolate; flowers purple, 2 cm. long, at upper leaf bases. July-Sept. Often planted and must be established at some places though we have no definite records.

3. *Campanula aparinoides* Pursh. MARSH BELLFLOWER. Perennial; stems weak, 3—6 dm. high, the angles with hooked hairs which catch on hands and clothing; flowers white or bluish, 5—10 mm. wide. Our only record is from the northwestern corner of Richland County (Bell 720). It is to be expected in that area but I have not seen it. It is quite common in boggy places, growing among grasses, in Minnesota within 50 miles of our eastern border.

   Campanula glomerata L., a rough, stout perennial with ovate, finely toothed leaves and blue flowers in terminal, head-like clusters and at bases of upper leaves, was found by Harry Graves in a flower bed at Hillsboro, Traill County, in 1942. It had apparently come up there by accident.

4. *Specularia perfoliata* (L.) A. DC. VENUS' LOOKINGGLASS. Annual; stem slender, not branched, 3—6 dm. high; leaves rounded, toothed, clasping, 5—15 mm. wide; flowers solitary at leaf bases, bluish purple, nearly flat, 7—10 mm. wide; capsule oblong, 6—8 mm. long. Our only record is from Grant County in 1907 (Bell 459). I remember the plant as a common weed in our orchard in northeastern Kansas.

301. Bluebell (*Campanula rotundifolia*).

5. *Specularia leptocarpa* (Nutt.) A. Gray. (Fig. 297). Annual; stem rough, usually unbranched, 3—6 dm. high; leaves lanceolate or narrower, 1—3 cm. long; flowers solitary at nearly all leaf bases, broad funnel shaped, 6—8 mm. long; capsule 1—2 cm. long, splitting at top, the persistent sepals 3—8 mm. long. First detected in a specimen which I collected at Selfridge, Sioux County, in 1935. Only a few days later, Mr. V. T. Heidenreich of the Indian Service, U. S. Dept. of Int., who has contributed many specimens, found it in Dunn County near Elbowoods. In 1943, I found it near Glen Ullin, Morton County, and at North Roosevelt Park, McKenzie County (fig. 20). In a broad coulee near the Little Missouri River, it was so abundant that it colored the ground. Apparently it is easily missed unless seen when in bloom, July 10—30.

LOBELIA FAMILY Lobeliaceae

Annuals or perennials with alternate leaves and showy flowers in terminal spikes; corolla on top of ovary, gradually widening from base, then spreading more abruptly into 2 short, triangular upper lobes and 3 lower ones; corolla tube split down upper side nearly to base; stamens 5, filaments grown together into a tube and little attached to corolla; ovary 2-celled; fruit a capsule with many small seeds.

This group is often placed in the bellflower family, though the shape of the corolla is different and the united stamens are a peculiar feature. Several species are grown as ornamentals, especially *Lobelia erinus* as a low border plant. Cardinal Flower (*L. cardinalis*) which resembles our largest species (No. 1) but has flaming red flowers, is a well known wild flower of central and southern U. S.

Key to Species

Corolla 15—25 mm. (⅗—1 in.) long; plants 3—6 dm. (1—2 ft.) high.
                                        1. *Lobelia siphilitica*
Corolla 5—10 mm. long; plants 1—3 dm. high.
  Stem leaves 2—4 mm. wide; 3—8 flowers in a loose cluster.
                                        3. *Lobelia kalmii*
  Stem leaves 5—10 mm. wide; flowers many in a spike.
                                        2. *Lobelia spicata*

1. *Lobelia siphilitica* L. GREAT BLUE LOBELIA. (Fig. 298). Stem stout, not branched, 3—6 dm. (1—2 ft.) high; leaves numerous, lanceolate or oblong, 5—15 cm. (2—6 in.) long; flowers deep blue, 15—25 mm. (⅗—1 in. long), in a terminal spike 2—4 dm. long. Along streams or in wet meadows. We have it only from the sandhills area in Richland and Ransom Counties where it is frequent. A striking plant. Aug.

2. *Lobelia spicata* Lam. PALESPIKE LOBELIA. Perennial; stems slender, not or but little branched, 2—6 dm. high; leaves oblong, 3—5 cm. long; flowers bluish white, 5—10 mm. long, in a spike 1—3 dm. long. A common, often abundant plant in fairly moist prairie. The flowers, which appear in late July, are rather pretty though small. Our specimens seem mostly var. *hirtella* A. Gray, in which the calyx has many short hairs, but there is much variation even in different specimens of the same lot.

3. *Lobelia kalmii* R. & S. KALM'S LOBELIA. Biennial; stem very slender, usually about 2 dm. high; lower leaves spatulate or oblanceolate, 1—2.5 cm. long; flowers blue, 8—10 mm. long, solitary at upper leaf bases on slender stalks or 1 or 2 on branches 1—5 cm. long. Local in boggy places. Specimens from Nelson, Benson, McHenry, Bottineau and Kidder Counties. Quite a pretty little plant, flowering in July. Named for the Swedish explorer, Peter Kalm, who did some of the first scientific bontanizing in America in 1732-34.

ASTER FAMILY **Asteraceae**

Flowers very small, few to many in a dense head, surrounded by small leaves or scales (bracts, forming an involucre); corolla attached to top of ovary, tubular with 5 short lobes (disk flower) or flat and extended (ray flower); stamens 5, included in corolla; anthers united by their edges; pistil 1, stigmas 2; fruit an achene; calyx much modified (pappus), persisting on top of achene as hairs, bristles or spines, or none at all. (Fig. 302).

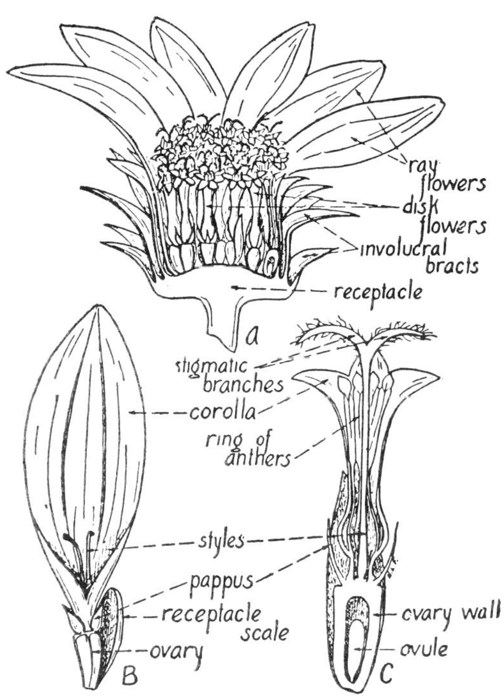

302. Sunflower head and detail of flowers.

This is the largest of all plant families, usually easily recognized by the heads as in asters, sunflowers and thistles. In some cases, the heads or flowers are highly modified, as in cocklebur, where 2 pistillate flowers are enclosed by a woody involucre, developing into a bur. Heads may have only disk flowers as in thistle, ironweed and blazing star, or only ray flowers as in dandelion and lettuce; often both are present but sometimes rays are small. In some species only disk flowers or only ray flowers produce achenes. Most of our species are fall blooming. Many are weeds and many are ornamentals.

Key to Species or Genera

A. Heads with only tubular flowers, no ray flowers. (If rays
   present go to **AA.**)
   Outer flowers enlarged but same shape as others.   129-131. *Centaurea*

Bracts hooked; lower leaves broad, not spiny.　121.　*Arctium minus*
Outer flowers not enlarged.
　　Bracts of involucre spiny tipped or hooked.
　　　Bracts hooked; lower leaves broad, not spiny.　121.　*Arctium minus*
　　　Bracts sharp, spiny; leaves toothed and spiny.
　　　　　　　　　　　　　122-128.　*Cirsium* and *Carduus*
　　Bracts green or chaffy, not spiny.
　　　Plants 5—20 cm. (2—8 in.) high, very white wooly.
　　　　Leafy stemmed; leaves narrow, 3—6 mm. (⅛—¼ in.)
　　　　　wide.　　　　　　59.　*Gnaphalium palustre*
　　　　Leaves basal, rounded, 5—15 mm. wide.　　55-58.　*Antennaria*
　　　Plants usually over 2 dm. (8 in.) tall or not white wooly.
　　　　Flowers bright reddish purple.
　　　　　Heads in flat topped clusters.
　　　　　　Leaves alternate.　　　　1.　*Vernonia fasciculata*
　　　　　　Leaves in whorls of 5.　　2.　*Eupatorium maculatum*
　　　　　Heads in long, slender clusters.　　6-8.　*Liatris*
　　　　Flowers yellow, white or whitish.
　　　　　Flowers yellow.
　　　　　　Shrubs with very narrow leaves.　12-13.　*Chrysothamnus*
　　　　　　Herbs.
　　　　　　　Leaves spiny; stiff plant 1—3 dm. high.
　　　　　　　　　　　　17.　*Aplopappus nuttallii*
　　　　　　　Leaves not spiny; plants 3—15 dm. high.
　　　　　　　　Leaves opposite; pappus of 2 barbed spines.
　　　　　　　　　　　　　82-85.　*Bidens*
　　　　　　　　Leaves alternate; pappus none.　102.　*Tanacetum vulgare*
　　　　　Flowers white, greenish, gray (or yellowish from pollen).
　　　　　　Leaves opposite.
　　　　　　　Flowers white, in flat-topped clusters.　3-4.　*Eupatorium*
　　　　　　　Flowers greenish, monoecious, not so clustered.
　　　　　　　　Pistillate and staminate flowers in same head.
　　　　　　　　　　　　　60-62.　*Iva*
　　　　　　　　Pistillate flowers few, separate; involucre woody.
　　　　　　　　　　　　　63-65.　*Ambrosia*
　　　　　　Leaves alternate.
　　　　　　　Pistillate heads forming a bur or hardened fruit.
　　　　　　　　Leaves not lobed; bur 1—3 cm. (⅖—1⅕ in.)
　　　　　　　　　long.　　　　　　67.　*Xanthium*
　　　　　　　　Leaves much lobed; bur 5—8 mm. (¼ in.) long.
　　　　　　　　　　　　66.　*Franseria acanthicarpa*
　　　　　　　Heads all similar, flowers not monoecious.
　　　　　　　　Pappus a tuft of hairs.
　　　　　　　　　Leaves toothed; heads in flat topped clusters.
　　　　　　　　　　　　5.　*Kuhnia eupatorioides*
　　　　　　　　　Leaves not toothed; heads scattered.
　　　　　　　　　　　　44.　*Aster brachyactis*

　　　　　　　　Pappus none.
　　　　　　　　　Heads 5—10 mm. (⅕—⅖ in.) wide.
　　　　　　　　　　Plant 3—6 dm. (1—2 ft.) high; western
　　　　　　　　　　　native.　　88.　*Hymenopappus tenuifolius*
　　　　　　　　　　Plant 1—2 dm. high; introduced in towns.
　　　　　　　　　　　　99.　*Matricaria matricarioides*
　　　　　　　　　Heads 2—3 mm. wide.　　103-111.　*Artemisia*
AA. Heads with ray flowers but these sometimes very short.
　B. Heads with both ray and disk flowers. (If only ray, go to BB).
　　C. Rays yellow. (If white, blue or purple, go to CC).
　　　Pappus of many slender bristles.
　　　　Bracts of involucre in 1 row, scarcely overlapping.
　　　　　Leaves opposite; heads 2.5—5 cm. (1—2 in.) wide.
　　　　　　　　　　　113.　*Arnica fulgens*

　　　　　Leaves alternate; heads 1—2 cm. wide.　114-120.　*Senecio*
　　　　Bracts overlapping, in 2—5 rows.
　　　　　Heads 2—5 mm. (1/12—⅕ in.) wide.　　18-26.　*Solidago*

Heads 10—20 mm. wide.
Leaves divided into narrow segments with sharp tips.
14. *Aplopappus spinulosus*
Leaves narrow, not divided.
Stems leafy, rough hairy.   11. *Chrysopsis villosa*
Leaves mostly basal, smooth.   14-17. *Aplopappus*
Pappus of a few scales, awns or none.
Receptacle chaffy, a scale below each flower.
Leaves opposite, 1—2 dm. (4—8 in.) long, grown
together at base.   68. *Silphium perfoliatum*
Leaves not so united.
Pappus of scales or awns.
Pappus of 2 scales which fall off easily.
74-80. *Helianthus*

Pappus of persistent scales or awns.
Pappus of 2, stiff, barbed awns.   82-85. *Bidens*
Pappus of several, delicate, lanceolate scales.
Heads 3—8 cm. (1—3 in.) wide; scales 5, not
fringed.   93. *Gaillardia aristata*
Heads 5—8 mm. (¼ in.) wide; scales 6—10,
fringed.   94. *Dyssodia papposa*
Pappus absent or indistinct.
Disk of head cylindrical, 2—5 cm. long
72. *Ratibida columnifera*
Disk of head rounded, not greatly elongated.
Leaves opposite.
Leaves simple, ovate; achenes 4-angled.
69. *Heliopsis helianthoides*
Leaves pinnate; achenes flattened.
81. *Coreopsis tinctoria*
Leaves alternate.
Heads slender, 8—10 mm. (⅓ in.) wide.
86. *Madia glomerata*
Heads broad, 5—10 cm. (2—4 in.) wide.
Leaves 5—15 mm. wide, entire, rough.
70. *Rudbeckia serotina*
Leaves 1—2 dm. wide, deeply lobed, smooth.
71. *Rudbeckia laciniata*
Receptacle not chaffy or with some slender scales.
Heads 2—3 mm. wide in clusters of 2—5.
9. *Gutierrezia sarothrae*
Heads 5 mm. or more wide, usually solitary on stalk.
Leaves divided into narrow segments.
Leaves opposite; widely branched plants.
89. *Bahia oppositifolia*
Leaves alternate; erect plants.
91. *Hymenoxys richardsonii*
Leaves not divided.
Heads gummy; leaves narrowly oblong, thick.
10. *Grindelia squarrosa*
Heads not gummy; leaves thin, lanceolate.
Leaves all basal.   90. *Actinella acaulis*
Leafy stemmed plant   92. *Helenium autumnale*
CC. Rays white, blue, purple or pinkish.
Disk prickly with stiff brown chaff: rays pink or purple.
73. *Brauneria angustifolia*
Disk not prickly.
Leaves all basal.
Leaves ovate, 1—3 dm. (4—12 in.) wide; swamp plant.
112. *Petasites sagittata*
Leaves linear, 2—10 mm. ($\frac{1}{12}$—⅖ in.) wide; prairie plant.
27. *Townsendia exscapa*
Leafy stemmed plants, but stem sometimes quite short.
Leaves finely divided or lobed; pappus none.

Heads 5—6 mm. wide in a dense, flat topped cluster.
  Leaves divided into numerous, fine segments.
                        95.   *Achillea lanulosa*
  Leaves deeply cut but not finely divided.
                        96.   *Achillea multiflora*
Heads 2—4 cm. wide, not so clustered.
  Plant strong scented; achenes tuberculate.
                        98.   *Anthemis cotula*
  Plant not strong scented; achenes 3-ribbed.
                        100.   *Matricaria inodora*
Leaves rarely divided; pappus usually present.
  Pappus none; rare, introduced weed.
                        97.   *Chrysanthemum leucanthemum*
  Pappus of hairs, bristles or scales.
    Pappus a few, stiff bristles or scales.
      Pappus of 2—5 bristles; tall, slender, native.
                        28.   *Boltonia latisquama*
      Pappus of several fringed scales; low, spreading
        weed.   87.   *Galinsoga ciliata*
    Pappus a dense tuft of hairs.
      Ray flowers usually 50 or more.   45-54.   *Erigeron*
      Ray flowers usually 15—30.   29-44.   *Aster*
BB. Heads with only ray flowers; plants usually with milky juice.
  Pappus of short, blunt scales or none; rare introduced weeds.
    Flowers blue or whitish; pappus of blunt scales.
                        134.   *Cichorium intybus*
    Flowers yellow; pappus none.   135.   *Lapsana communis*
  Pappus a tuft of fine hairs.
    Pappus hairs branched.
      Leaves long, narrow, smooth; achenes 2—3 cm. (1 in.) long.
                        133.   *Tragopogon dubius*
      Leaves broad, spiny; achenes 5—6 mm. (¼ in.) long.
                        132.   *Picris echioides*
    Pappus hairs not branched.
      Bracts in 2 rows, the lower widely spreading. 136-137.   *Taraxacum*
      Bracts mostly in 1 row, not widely spreading.
        Achenes flattened.
          Achenes beaked, usually black or gray.   151-155.   *Lactuca*
          Achenes not beaked, red.   148-150.   *Sonchus*
        Achenes cylindrical or angled, not flattened.
          Leaves small, often scale-like; heads 5-flowered.
                        143-144.   *Lygodesmia*
          Leaves ordinary; heads 25—50-flowered.
            Leafy stemmed plants.
              Heads yellow; leaves lanceolate.
                        145.   *Hieracium canadense*
              Heads white or purplish; leaves broad.
                        146-147.   *Prenanthes*
            Leaves mostly basal; heads yellow.
              Heads solitary on basal stalks.   138-139.   *Agoseris*
              Heads on branching stems.   140-142.   *Crepis*
1. *Vernonia fasciculata* Michx. IRONWEED. Perennial; stem clustered, erect,
6—12 dm. (2—4 ft.) high, branched only at top; leaves many, lanceolate, finely toothed, 7—15 cm. (3—6 in.) long; heads deep reddish
purple, 10—15 mm. (½ in.) wide, in a flat topped cluster 6—15 dm.
wide. Along ditches or in wet grassland. Records chiefly from southern half of State, especially eastern. Aug. One specimen from Kidder
County (Perrine 1323) with ovate leaves 4—6 cm. long, seems to fit
Rydberg's description of *V. corymbosa* Schwein.

Eupatorium THOROUGHWORT

Perennials, usually with opposite leaves; flower heads small,
white or purplish, of disk flowers only, in dense, terminal, flat
topped clusters; achenes 5-angled, pappus of fine bristles.

Key to Species

Leaves in whorls of 3—6; flowers purplish.   2. *Eupatorium maculatum*
Leaves opposite; flowers white.
  Plant densely fine hairy; leaves not stalked. 3. *Eupatorium perfoliatum*
  Plant not hairy; leaves long stalked.   4. *Eupatorium rugosum*

2. *Eupatorium maculatum* L. JOE-PYE WEED. Stem 1—2 m. (3—7 ft.)
   high; leaves ovate to lanceolate, sharply toothed, 1—2 dm. (4—8 in.)
   long; heads pink or purplish, the cluster 1—2 dm. wide, flat or rounded. Occasional in swampy places, eastern half of State (fig. 14). Aug.
   A showy plant. Often referred to as *E. purpureum* L.

3. *Eupatorium perfoliatum* L. BONESET. Stem 6—15 dm. high, very hairy;
   leaves often united at base, triangular-ovate, 1—2 dm. long, wrinkled,
   teeth rounded; heads white, in a flat cluster 6—15 cm. wide. Occasional
   in boggy places, Richland and Ransom Counties. Aug. This is the
   source of "boneset tea."

4. *Eupatorium rugosum* Houtt. WHITE SNAKEROOT. Stem erect, often widely branched, 3—10 dm. high; leaves broadly ovate, 5—10 cm. long, very
   thin, petioles nearly as long as blades; flowers white, usually in many
   small clusters 5—10 cm. wide. Local in woods. Richland and Cass
   Counties. Formerly called *E. ageratoides* L. f. and *E. urticaefolium*
   Reich. Poisonous to cattle, causing "milk sickness" which is transmit-
   ted to humans.

5. *Kuhnia eupatorioides* L. FALSE BONESET. Perennial; stems tufted, 3—8
   dm. high; leaves numerous, lanceolate, 2—7 cm. long, sharply toothed;
   heads white, 5—8 mm. wide, in flat or rounded clusters 3—8 cm.
   wide at ends of branches; achenes cylindrical, ribbed, 5 mm. long.
   Prairie, especially hills along valleys, southern part of State. July,
   Aug.

Liatris   BLAZING STAR

Perennials from short, thick, upright, underground stems; leafy
stem erect, stiff, not branched; leaves alternate, narrow, thick,
rough; flower heads rose purple in a long, terminal cluster; achenes
elongated, ribbed, hairy.

Key to Species

Heads 10—40-flowered; bracts broadly rounded.   6. *Liatris scariosa*
Heads 3—6-flowered; bracts pointed.
  Plant 2—6 dm. (8—24 in.) high; common on dry prairie.
                                           7. *Liatris punctata*
  Plant 5—12 dm. high; local in wet meadows. 8. *Liatris pycnostachya*

6. *Liatris scariosa* (L.) Willd. ROUND-HEADED BLAZING STAR. Stem 6—10
   dm. (2—3 ft.) high; upper leaves narrowly oblong to oblanceolate, 5—
   10 cm. (2—4 in.) long, the lowest 1—3 dm. long; heads 1—2 cm. wide in
   a loose cluster 1—3 dm. long. Low, arable prairie. Aug, Sept. Recent
   authors limit *L. scariosa* to southeastern U. S. Our plants have been
   separated into *L. aspera* Michx. in the east and *L. ligulistylis* A. Nels.
   in the west.

7. *Liatris punctata* Hook. NARROW-LEAVED BLAZING STAR. Stem 3—6 dm.
   high; leaves linear, 5—15 cm. long, 2—4 mm. (⅙ in.) wide, often twisted; heads 5—10 mm. wide in a dense cluster 1—2 dm. long. Dry, and especially stony, prairie. Aug., Sept.

8. *Liatris pycnostachya* Michx. TALL BLAZING STAR. GAYFEATHER. Stem 5—
   12 dm. high; leaves lanceolate, narrowed at both ends, 1—3 dm. long;
   heads 5—10 mm. wide in a dense cluster 2—4 dm. long. Low grassland,
   southeast 4 counties (fig. 19). Late July and Aug. Often planted as an
   ornamental.

9. *Gutierrezia sarothrae* (Pursh) Britt. & Rusby. BROOMWEED. Perennial;
   stems tufted, 2—4 dm. high; leaves alternate, linear, 2—4 cm. long;

heads yellow, 2—3 mm. wide, forming flat topped clusters 2—5 cm. wide. Dry prairie and hills. Aug., Sept. Resembles a small goldenrod. The common name is from its use as brooms by Indians.

10. *Grindelia squarrosa* (Pursh) Dunal. GUMWEED. Biennial; stem erect, often widely branched, 3—6 dm. (1—2 ft.) high; leaves alternate, narrowly oblong, 2—4 cm. (1—1½ in.) long, finely toothed, thick and stiff; heads yellow, very gummy, 2—3 cm. wide, forming somewhat flat topped clusters; bracts numerous, linear, curved; achenes 2—4 mm. (⅛ in.) long, straw colored, smooth, usually somewhat curved and angled. Common on prairie especially along roadsides and in overgrazed pastures. Aug., Sept.

11. *Chrysopsis villosa* (Pursh) Nutt. GOLDEN ASTER. Perennial; stems rough hairy, erect or spreading, widely branched, 2—6 dm. high; leaves alternate, oblong or oblanceolate, 2—4 cm. long; heads yellow, 2—3 cm. wide; achenes hairy, 2—3 mm. long. Common on dry prairie and hills. Aug., Sept.

Chrysothamnus   RABBITBRUSH

Low shrubs with narrow, alternate leaves and flower heads resembling those of goldenrod; heads rather long and narrow, bracts stiff, yellowish, apparently in vertical rows.

Key to Species

Leaves dark green, not hairy.   12. *Chrysothamnus graveolens*
Leaves white with fine hairs.   13. *Chrysothamnus nauseosus*

12. *Chrysothamnus graveolens* (Nutt.) Greene. Low rounded shrub, 1— 1.5 m. (3—5 ft.) high; leaves numerous, linear, 4—7 cm. (1½—3 in.) long; heads yellow, 3—4 mm. (⅛ in.) wide, in somewhat flat topped clusters 5—10 cm. wide. One of the most characteristic plants of bare, clay buttes, more abundant and larger on sides of deep, narrow gullies. West of Missouri River, occasionally on east side where clay buttes occur (fig. 18). Aug., Sept. Flowers resemble those of goldenrods.

13. *Chrysothamnus nauseosus* (Pursh) Britton. Usually quite low, widely spreading, 1—2 dm. high; leaves white wooly, 2—3 cm. long. Specimens from Billings, McKenzie, Mountrail and Williams Counties. It seems less common than the preceding but is less conspicuous, and may have been overlooked. The name *nauseosus* is older, and *graveolens* is sometimes considered a variety of it.

Aplopappus

Yellow flowered shrubs or tough herbs with alternate leaves. Our species are rather diverse in appearance but are low plants, not obviously shrubby. Some authors separate the group into several genera.

Key to Species

Heads 2—3 cm. (1 in.) wide, with rays; leaves not spiny.
  Leaves gray, divided into narrow lobes.   14. *Aplopappus spinulosus*
  Leaves not gray, not divided.
    Leaves mostly basal, not toothed.   15. *Aplopappus armerioides*
    Leafy stemmed; leaves with a few, sharp teeth.
      16. *Aplopappus lanceolatus*
Heads 1 cm. wide without rays; leaves thick, with spiny teeth.
      17. *Aplopappus nuttallii*

14. *Aplopappus spinulosus* (Pursh) DC. Perennial; stems widely branched, 2—4 dm. (8—16 in.) high; leaves gray, hairy, oblong, 1—3 cm. (⅖—1⅕ in.) long, pinnately divided into linear lobes which have spiny tips; heads yellow, 1.5—2 cm. wide. Frequent on dry prairie. Aug. The flower heads resemble those of *Chrysopsis*. Also called *Sideranthus spinulosus* (Pursh) Sweet.

15. *Aplopappus armerioides* (Nutt.) Gray. Perennial; stem very short, woody; leaves mostly basal, stiff, narrow, 3—7 cm. long; heads yellow, 2 cm. wide, 1—3 on a stalk 5—15 cm. high. Hills and buttes from Adams and Dunn Counties west. June. Also called *Stenotus armerioides* Nutt.

16. *Aplopappus lanceolatus* (Hook.) T. & G. Perennial; stem erect, 2—4 dm. high with few short branches above, each bearing a flower head; leaves linear to lanceolate, 5—20 cm. long with a few sharp teeth; heads yellow, 2 cm. wide, bracts green on upper half. Apparently only on saline flats; Ward, Mountrail and Williams Counties. Aug. Also called *Pyrrocoma lanceolata* (Hook.) Greene.

17. *Aplopappus nuttallii* T. & G. Perennial; stem erect, stiff, tufted, 5—30 cm. high; leaves oblong-lanceolate to spatulate, 1—2.5 cm. long, thick, spiny toothed; heads yellow, 1 cm. wide, rays short or none. Dry clay or stony soil on hills, Bowman to McKenzie Counties. July, Aug. Also called *Sideranthus grindelioides* (Nutt.) Britt.

Solidago GOLDENROD

Erect perennials with numerous, alternate leaves; heads small, many clusters of a few heads forming a large, terminal cluster; rays yellow, short; bracts usually yellowish, pappus a tuft of fine hairs.

Key to Species

Leaves rounded ovate; heads in small clusters at upper leaf bases.
                18. *Solidago flexicaulis*
Leaves linear, lanceolate or oblong; heads in a large, terminal cluster.
  Leaves oblong, 2—5 cm. (1—2 in.) wide, rough; flower cluster
    flat topped.   25. *Solidago rigida*
  Leaves rarely over 2 cm. wide, usually not rough.
    Leaves oblong or ovate, 2—5 cm. long, soft hairy.
             23. *Solidago mollis*
    Leaves elongated, linear or lanceolate, 5—15 cm. long.
      Flower cluster flat topped; leaves very narrow or elongated.
        Leaves few, the lower often 2 dm. (8 in.) long.
            24. *Solidago riddellii*
        Leaves many, not especially long.   26. *Solidago graminifolia*
      Flower cluster usually not flat topped nor leaves long and
        narrow.
        Stem and leaves with fine hairs.
          Flower clusters short and dense, 10—15 cm. wide.
            19. *Solidago altissima*
          Flower cluster elongated, narrow, 5—10 cm. wide.
            22. *Solidago nemoralis*
        Stems and leaves not hairy.
          Plants 3—6 dm. (1—2 ft.) high on prairies.
            21. *Solidago missouriensis*
          Plants 6—15 dm. high on low ground.   22. *Solidago gigantea*

18. *Solidago flexicaulis* L. BROAD-LEAVED GOLDENROD. Stems 3—10 dm. (1—3 ft.) high; leaves mostly rounded ovate, 5—10 cm. (2—4 in.) long, thin, sharply serrate, abruptly narrowed at base into a winged petiole; heads in small groups forming a terminal cluster 5—20 cm. long. We have only two localities, Fargo, Cass County, and near Leonard in Ransom County. The plant is common in woods a little farther east.

19. *Solidago altissima* L. TALL GOLDENROD. Stems usually in clumps, 3—10 dm. high; finely hairy; leaves numerous, lanceolate, 5—10 cm. long, finely toothed, short hairy; heads on spreading branches 5—10 cm. long, forming a large, terminal cluster 10—20 cm. long, usually pyramidal, sometimes flattened on top. Common. This had been called *S. canadensis* L. which as now understood does not occur here. Our specimens seem all or mostly var. *gilvocanescens* Rydb. (*S. pruinosa* Greene). Five specimens, from widely separated localities, have linear-lanceolate leaves, but these are probably only extreme examples.

20. *Solidago gigantea* Ait. TALL SMOOTH GOLDENROD. Stems smooth, 1—1.5 m. high from long rhizomes; leaves numerous, 5—15 cm. long, smooth, dark green, lanceolate, sharply toothed. Common in low ground, forming large patches. Resembles the preceding but is taller, dark green, smooth stemmed and begins to bloom a little earlier. *S. serotina* Ait. is another name for the same plant.

21. *Solidago missouriensis* Nutt. EARLY GOLDENROD. Stems smooth, 3—6 dm. high; leaves narrowly oblanceolate, smooth, thick, the lower ones elongated to 1 or 1.5 dm.; flower cluster narrowly or broadly pyramidal, 5—15 cm. long, heads rather large, 5 mm. wide. Common on prairie. The earliest of our goldenrods, beginning to bloom in late July.

22. *Solidago nemoralis* Ait. GRAY GOLDENROD. Stems gray with fine hairs, 3—6 dm. high from a thick crown; leaves oblanceolate, 5—10 cm. long; flower cluster slender, curving, 1-sided, 5—20 cm. long. Common, especially on dry hillsides.

23. *Solidago mollis* Bartl. SOFT GOLDENROD. Stems gray hairy, 3—6 dm. high; leaves ovate to obovate, 3—6 cm. long, velvety with dense hairs; flower cluster usually dense, narrow or short and flat, 5—15 cm. long. Common on dry prairie, the most common species in western N. D.

24. *Solidago riddellii* Frank. RIDDELL'S GOLDENROD. Stems 3—6 dm. high, smooth; leaves unusually long, linear or lanceolate, the lower often 2—3 dm. long; heads in a flat topped cluster 1—2 dm. wide. Low, sandy prairie, Walcott, Richland County, in 1947 (Stevens 1040).

303. False Aster (*Boltonia latisquama*).

25. *Solidago rigida* L. STIFF GOLDENROD. Stems stout, rough, 3—8 dm. high; leaves rough, broadly oblong to elliptic, the larger ones 2—3 dm. long, 2—5 cm. wide, flower cluster usually flat topped, 1—2 dm. wide; heads 6—8 mm. wide. Common on prairie, often increasing in heavily grazed pastures.

26. *Solidago graminifolia* (L.) Salisb. NARROW-LEAVED GOLDENROD. Stems slender, 4—8 dm. high; leaves numerous, 4—10 cm. long, 3—5 cm. wide, smooth; flower cluster usually flat topped, 3—10 cm. wide. Local in sandy, usually moist soil.

27. *Townsendia exscapa* (Rich.) Porter. EASTER DAISY. Tufted perennial; leaves oblanceolate, 2—4 cm. long; heads 2—3 cm. wide, purplish white, appearing among the leaves. Rather rare on prairie; records from Stark and Billings Counties only. Late May. This is the nearest to a true daisy which we have.

28. *Boltonia latisquama* A. Gray. FALSE ASTER. (Fig. 303). Perennial, the slender rhizomes inconspicuous; stem erect, 4—10 dm. (16—40 in.) high, branching above to a rounded or flattened top; leaves alternate, linear to oblanceolate, 5—10 cm. (2—5 in.) long, quite smooth; heads 2 cm. wide, white; achenes flattened, wedge shaped, with a few, irregular bristles (fig. 308). Roadside ditches, edges of ponds, etc. Probably widely distributed, but our specimens are only from southeast and north central parts of State. It is an attractive plant, often cultivated under the name *Boltonia*. It resembles *Aster paniculatus* and grows in the same places but begins to bloom 10 days earlier, about August 20. The leaves are narrow at base, lacking the broad, fringed base of the aster. The more rounded top of the plant will often distinguish it at a distance. Previously recorded as *B. asterioides* (L.) L'Her.

Aster   ASTERS

Mostly perennials with alternate, simple leaves; flower heads numerous; rays many, white, blue or purple, never yellow; disk flowers yellowish from pollen; achenes cylindrical, somewhat angled or ribbed; pappus of fine, branched hairs. A large, variable and difficult group of showy plants. Many of our species are grown as ornamentals, a few are weedy. Flowering Aug.-Oct.

Key to Species

Rays absent or little developed; annual in wet soil. 44. *Aster brachyactis*
Rays well developed; perennials or biennials.
  Rays blue, violet or purple.
    Leaves broad, with long, winged petioles.   29. *Aster sagittifolius*
    Stem leaves without petioles.
      Leaves obviously hairy.
        Leaves silvery with long, fine hairs lying against surface.
            36. *Aster sericeus*
        Leaves green, not silvery.
          Leaves strongly clasping, 2—5 cm. (1—2 in.) long.
             30. *Aster novae-angliae*
          Leaves not or little clasping.
            Tall, swamp plant; leaves 5—15 cm. long. 31. *Aster puniceus*
            Low, prairie plant; leaves 1—3 cm. long.
              32. *Aster oblongifolius*
      Leaves fringed, hairy along veins, or with short hairs.
        Lower leaves ovate or cordate, smooth.   33. *Aster laevis*
        All leaves lanceolate to linear.
          Leaves lanceolate, hairs only on edges and veins.
             37. *Aster coerulescens*
         Leaves linear, covered with fine hairs.   34. *Aster canescens*
  Rays white, rarely bluish or purplish.

Heads few, about 5—15.
 Upland plant; bracts not hairy.   41. *Aster ptarmicoides*
 Saline marsh plant; bracts glandular.   42. *Aster pauciflorus*
Heads many.
 Tall plant, 1—2 m. (3—7 ft.) high; leaves mostly 2—3 cm.
  (1 in.) wide.   35. *Aster umbellatus*
 Medium to low plants, 3—10 dm. (1—3 ft.) high; leaves rarely
  over 1—2 cm. wide.
  Leaves 5—15 cm. long, smooth or nearly so.
   Leaves 10—20 mm. (⅖—⅘ in.) wide. 37. *Aster coerulescens*
  Leaves 3—10 mm. wide.
   Stems green, 1—1.5 m. high.   38. *Aster longifolius*
   Stems usually red, slender, 4—6 dm. high. 39. *Aster junceus*
  Leaves 2—4 cm. long, covered with short hairs.
   Heads 7—10 mm. wide.   40. *Aster ericoides*
   Heads 12—15 mm. wide.   41. *Aster commutatus*

29. *Aster sagittifolius* Wedem. ARROW-LEAVED ASTER. Stem 6—10 dm. (2—3 ft.) high; upper leaves lanceolate to ovate, 3—10 cm. (1—4 in.) long, with prominent, winged petioles; lower leaves ovate, cordate or sagittate, 10—15 cm. long, not so distinctly wing petioled; heads blue or purple, 2 cm. wide. Local in woods; Cass, Pembina, Benson and Rolette Counties.

304   305

304. Smooth Blue Aster (*Aster laevis*).

305. White Prairie Aster (*Aster ericoides*).

30. *Aster novae-angliae* L. NEW ENGLAND ASTER. Stem very leafy, branched mostly at top, 1—2 m. (3—7 ft.) high; leaves oblong to lanceolate, 3—8 cm. long, rough with short hairs; heads purple or rose colored, 2—2.5 cm. wide. Local in wet grassland, eastern third of State; often grown as an ornamental.

31. *Aster puniceus* L. SWAMP ASTER. Stem with coarse hairs, 6—15 dm. high, widely branched above; leaves lanceolate or oblong-lanceolate, 5—15 cm. long; heads violet or pale purple, 2—3 cm. wide. Local in swamps; Richland, Ransom, Benson and Bottineau Counties. A striking plant, our largest flowered aster.

32. *Aster oblongifolius* Nutt. AROMATIC ASTER. Stem widely branched, 1—6 dm. high; leaves oblong, 1—3 cm. long, thick, rough; heads 1.5—2.5 cm. wide, violet to pink. Common, especially on dry prairies and hillsides westward. Sometimes grown in gardens. It is an attractive plant for dry, stony soils.

33. *Aster laevis* L. SMOOTH BLUE ASTER. (Fig. 304). Stem 3—10 dm. high, often widely branched; upper leaves ovate to lanceolate, more or less clasping, 3—10 cm. long, lowest leaves with long, somewhat winged petioles; heads blue purple or rose colored, 1.5—2.5 cm. wide. Common all over State, lower prairie, woods and brush. Commonly cultivated and very showy, especially because it blooms from September to late October.

34. *Aster canescens* Pursh. HOARY ASTER. Biennial; stem widely branched, 3—6 dm. high; leaves 2—6 cm. long, mostly linear, entire, the lowest spatulate and toothed; heads 2—3 cm. wide, dark blue. A striking plant on bare clay slopes, western edge of State (fig. 18). Blooms late, though we have one specimen of June 12 and one July 2. It was listed as *Machaeranthera pulverulenta* by Bergman but proves to be *canescens*. A more western species with divided leaves, *tanacetifolius*, Tansy Aster, is often grown as an ornamental.

35. *Aster umbellatus* Mill. FLAT-TOP ASTER. Stems 1—2 m. high, branched only at top; leaves elliptic or lance-elliptic, 7—15 cm. long, 2—3 cm. wide; heads 1—2 cm. wide, white, in a flat or rounded cluster 1—3 dm. wide. Local in swampy areas, Richland, Barnes, Benson, Pembina, Rolette and Bottineau Counties (fig. 14).

36. *Aster sericeus* Vent. SILKY ASTER. Stems 2—4 dm. high, lower part very smooth and leafless at flowering time; leaves oblong, 2—4 cm. long, very silvery silky; heads purple, 2—2.5 cm. wide. Prairie, recorded only from Richland, Cass and Ransom Counties (fig. 19). A striking plant because of the silvery leaves.

37. *Aster coerulescens* DC. TALL WHITE ASTER. Spreading freely by rhizomes; stems 3—10 dm. (1—3 ft.) high, leafy, often widely branched; leaves lanceolate, 5—15 cm. (2—6 in.) long, mostly smooth, a few bristly hairs along edges of lower ones; heads white or bluish, 1—2 cm. wide in a large, branching, leafy top. Very common in low ground, showy but decidedly weedy. This is what has often been called *A. salicifolius* Lam. or has passed as *A. paniculatus* Lam. In 1936, Dr. K. M. Wiegand examined our specimens and referred them to *coerulescens* except for specimens from Rolette, Grant and Slope Counties, which he retained as *A paniculatus*, var. *simplex* (Willd.) Burgess (St. John, L. R. Waldron 1758; Pretty Rock, Bell 1278). Wiegand (Rhodora 35.30, 1933) listed *A. luetevirens* Greene as occurring in the Bad Lands of N. D. It is another of this group, said to differ from *coerulescens* in smaller heads (10—16 mm. wide compared to 12—25 mm. for *coerulescens* and *paniculatus*).

38. *Aster longifolius* Lam. Resembles the last except for narrower leaves. Three specimens from Richland, Barnes and Pembina Counties were doubtfully referred to this species by Dr. Wiegand.

39. *Aster junceus* Ait. RUSH ASTER. Stem slender, smooth, often red, 3—8 dm. high; leaves linear, 4—8 cm. long; heads white, 15—20 mm. wide. Boggy ground; Ransom, Barnes, Kidder, Benson and Pembina Counties. This is our smallest and most slender plant in this group. Several

of our specimens were examined by Dr. S. F. Blake in 1935. Some specimens previously referred to this species were labeled *coerulescens* by Dr. Wiegand.

40. *Aster ericoides* L. WHITE PRAIRIE ASTER. (Fig. 305). Stems erect or spreading, 3—8 dm. (1—2½ ft.) high; leaves linear, 2—5 cm. (1—2 in.) long, rough hairy; heads numerous, white, 8—15 mm. (⅓—⅖ in.) wide. Our commonest prairie aster, formerly referred to *A. multiflorus* Ait. It is highly variable and several species recognized by Rydberg can be found in our plants.

41. *Aster commutatus* (T. & G.) Gray. I am not able to separate this satisfactorily from the last. Key characters given by Rydberg, "bracts almost equal in length", seem to apply to several specimens (Rolla, Miss Lovell in 1905 and L. R. Waldron No. 1729; Walhalla, L. R. W. 1665; Minot, L. R. W. 1851a; Denbigh, Bergman in 1909; Williston, Bell 467) from northern edge of State, but the usual interpretation of *commutatus* seems a less branched, coarser plant, common westward.

A Fargo specimen (Stevens 299) was identified by J. A. Steyermark in 1938 as *commutatus*, var. *crassulus* (Rydb.) Blake. L. H. Shinners in 1942 regarded this as *adsurgens* Greene. Among this lot were a few stems which had heads with purplish rays. These probably were from a different plant, but may have changed color in pressing since they were not noted when the specimens were collected as average Fargo material.

42. *Aster ptarmicoides* (Nees) T. & G. WHITE UPLAND ASTER. Stems slender, wiry, smooth, 2—4 dm. high; leaves linear or oblanceolate, 5—15 mm. long, smooth, not toothed; heads white, 1—1.5 cm. wide, usually 6—10, often but one on each branch from top of stem. Common on rocky hills and dry prairie. Quite different from the last in its smooth, wiry stems and few, larger heads.

43. *Aster pauciflorus* Nutt. FEW-FLOWERED ASTER. Stem 3—6 dm. high; leaves thick, linear to oblanceolate, 3—10 cm. long; heads white, 2—3 cm. wide, few, each on an upper, glandular hairy, spreading branch; lower stems smooth. Wet, saline soil; Bottineau, Grant, Billings, Williams and McKenzie Counties.

44. *Aster brachyactis* Blake. RAYLESS ASTER. Annual; stem 1—6 dm. high, leaves linear, 3—10 cm. long; heads whitish, 6—15 mm. wide, rays usually lacking, bracts 1 cm. long. Edges of ponds or streams. Specimens from southeast, southwest and north central parts of State. It is probably generally distributed but easily overlooked. It blooms late but we have two specimens well developed in July. Listed by some authors as *Brachyactis angusta* (Lindl.) Britt., but Dr. S. F. Blake found *angustus* could not be used.

Erigeron FLEABANE. DAISY FLEABANE

Annual, biennial or perennial with alternate leaves; heads with many disk and ray flowers, the latter white or pink; achenes very small, pappus of fine hairs. Flowering mainly in June.

Key to Species

Rays evident, 5—10 mm. (⅕—⅖ in.) long.
  Perennial from a thickened base.
    Leaves finely divided; plant 5—10 cm. (2—4 in.) high.
45. Erigeron compositus

    Leaves entire or only toothed.
      Stems many, spreading.   46. *Erigeron pumilus*
      Stems few, erect.
        Middle and upper stem leaves much smaller than lowest.
47. Erigeron glabellus

        Stem leaves gradually smaller, middle ones as large or
          larger than lowest.

Main root 5—10 mm. thick, 5—10 cm. long.
                          48. *Erigeron caespitosus*
Roots slender from a thick, short base.
                          49. *Erigeron subtrinervis*
Annual or biennial; crown less well developed.
   Stem leaves 5—15 mm. wide, clasping at base.
                          50. *Erigeron philadelphicus*
   Stem leaves 2—5 mm. wide, not clasping.   51. *Erigeron strigosus*
Rays usually less than 5 mm. long.
   Stem stout, 5—15 dm. (1½—5 ft.) high; leaves very numerous.
                          53. *Erigeron canadensis*
   Stem slender, 2—8 dm. high; leaves not crowded.
      Heads 1 cm. or more wide; stem 2—6 dm. high.
                          52. *Erigeron lonchophyllus*
      Heads 4—5 mm. wide; stem 1—3 dm. high.   54. *Erigeron divaricatus*

45. *Erigeron compositus* Pursh. Perennial from a branched crown; leaves
    finely divided, 2—5 cm. (1—2 in.) long, nearly smooth, clustered on a
    branched, thickened root; heads white or pink, 1—2 cm. wide, solitary
    on single stalks 1—2 dm. (4—8 in.) high from the leaf tufts. We have a
    single record from Minot. In 1946, I saw a specimen which Mrs. Alvina
    Thorgrimson collected on White Earth Creek in eastern Williams
    County, but I did not find plants myself. It is common in the Rocky
    Mountain region where it grows on gravelly or rocky slopes, banks
    or hills. Bergman listed it as *multifidus* Rydb., but Rydberg himself
    later decided that form was not distinct from *compositus*.

46. *Erigeron pumilus* Nutt. Perennial from a branched crown; stems often
    many, widely spreading, 1—3 dm. long, shaggy hairy; stem leaves
    linear, 1—3 cm. long, the lowest oblanceolate, 2—10 cm.; heads 2—3
    cm. wide. Common on dry prairie and hills from Missouri River west.

47. *Erigeron glabellus* Nutt. Perennial from a short, irregular base; stems
    2—4 dm. high, rough hairy; stem leaves linear to lanceolate, 3—5 cm.
    long, lowest oblanceolate, 3—10 cm.; heads white, 2—3 cm. wide,
    solitary or 2—4 on slender, upper branches. Common on lower prairie
    and north slopes, all over State. Some authors recognize *E. asper*
    Nutt. as a separate species. Dr. A. R. Cronquist united them (Bull.
    Torrey Cl. 70:273, 1943) and recognized ssp. *typicus* Cronquist, with
    appressed hairs on the stems and ssp. *pubescens* (Hook.) Cronquist,
    with spreading hairs. Both seem about equally well represented in
    our specimens.

48. *Erigeron caespitosus* Nutt. Perennial from a thick root; stem 1—2 dm.
    high, finely hairy; leaves oblanceolate, 5—10 cm. long; heads 1.5—2 cm.
    wide. Bergman referred a single specimen from Gambetta, Williams
    County, to this western species. Another from Killdeer Mts. was
    identified as this by Dr. S. F. Blake in 1936.

49. *Erigeron subtrinervis* Rydb. Perennial from a short crown; stems 2—4
    dm. high; lower leaves oblanceolate, 5—15 cm. long, stem leaves lanceo-
    late, gradually shorter; heads 1.5—2.5 cm. wide, rays white or pink.
    One specimen from Leeds was referred to this form by Lunell and
    Bergman had placed one from Dickinson here. Three others, Leeds, Wil-
    liston (Bell 497) and Schafer may belong to it, but all may be *E. glabel-
    lus*.

50. *Erigeron philadelphicus* L. Rather coarse but soft biennial; stem 3—8
    dm. (1—2½ ft.) high, branching above; lowest leaves 5—15 cm. (2 -
    6 in.) long, 1—4 cm. wide, usually toothed, the upper lanceolate, clasp-
    ing; heads many, 1—2 cm. wide; rays white or pale violet. Our speci-
    mens, except one from Pembina County, are from southeastern 8
    counties, where it is common in low ground. June-Sept.

51. *Erigeron strigosus* Muhl. Annual; stem 3—8 dm. high, widely branched
    to a flat top; leaves linear to oblanceolate, 2—8 cm. long; heads many,
    1 cm. wide, rays white. Frequent on prairie at least in southern part
    of State. This had been called *E. ramosus* (Walt.) BSP

52. *Erigeron lonchophyllus* Hook. Slender biennial; stem 3—6 dm. high, the few branches erect; leaves linear, 5—15 cm. long; heads 1.5 cm. wide, rays white, 2—3 mm. long, erect. We have this northern swamp plant from Turtle Mts. and Devils Lake (fig. 21). Lunell collected it at Leeds, Benson County.

53. *Erigeron canadensis* L. HORSEWEED. Annual; stem stout, 4—15 dm. high, rough hairy, usually branched only in the flowering top; leaves very numerous, oblanceolate to linear, 5—15 cm. long, the lower often with a few teeth toward tip; flowering top 1—6 dm. long, a dense, oblong mass of branches and tiny heads 3—4 mm. wide; rays white, very short. Common weed, especially in abandoned fields. This and the next species are placed by some authors in a separate genus, *Leptilon.*

54. *Erigeron divaricatus* Michx. LOW HORSEWEED. Annual; stem very slender, widely branched, 1—3 dm. high; leaves narrowly linear, 1—2.5 cm. long; heads 3—4 mm. wide, rays very short, purple. An insignificant weed, frequent farther south. One record from southwestern Grant County (fig. 20).

Antennaria CATSFOOT

Small, white hairy perennials, spreading by stolons, blooming in early spring; leaves rounded, mostly basal; heads dioecious, white or pinkish, few in a cluster at top of a short flower stalk which elongates considerably at maturity; achenes very small, pappus a tuft of fine hairs. This is a very difficult group and the status of species is uncertain. The following treatment follows determinations of our specimens in 1946 by Cronquist.

Key to Species

Basal leaves 2—5 cm. (⅘—2 in.) long, 1—3 cm. wide, 3-nerved.
                55. *Antennaria plantaginifolia*
Basal leaves 1—4 cm. long, 3—15 mm. (⅛—⅗ in.) wide, 1-nerved.
  Upper side of leaf green, less white than lower. 56. *Antennaria neglecta*
  Upper side of leaf as white as lower.
  Leaves 10—20 mm. long; plants in loose colonies.
                57. *Antennaria parvifolia*
  Leaves 5—15 mm. long; plants in dense colonies.
                58. *Antennaria microphylla*

55. *Antennaria plantaginifolia* (L.) Hook. Leaves ovate, elliptic or spatulate, 2—5 cm. (⅘—2 in.) long, usually greenish above, obviously 3-nerved or veins branched; flower stalks about 1 dm. (4 in.) high, becoming 3—4 dm. in fruit. One specimen from Richland County and one from Pembina County. Others from Sioux, Billings and McKenzie Counties, referred to var. *ambigens* (Greene) Cronquist, have leaves more white above, heads larger, 8—10 mm. in fruit. This is called *A. fallax* Greene by some authors.

56. *Antennaria neglecta* Greene. Basal leaves obovate or spatulate, 1—3 cm. long, usually quite green above; flower stalks 2—10 cm. high, elongating to 2—3 dm. in fruit. Specimens from eastern third of State; also Grant and Stark Counties were referred by Cronquist to var. *attenuata* (Fern.) Cronquist. This form seems larger, the leaves whiter above with a small, sharp tip. It has also been called *A. neodioica* Greene. Some of our specimens had previously been named *A. campestris* Rydb. This plant is one of the early spring flowers, the clusters first showing as little furry balls unfolding, whence the name "catsfoot" or "pussytoes".

57. *Antennaria parvifolia* Nutt. Basal leaves spatulate, 2—5 cm. long, white above, usually somewhat folded; flowers sometimes pinkish. Specimens from southeastern and western counties but it may occur all over the State. It resembles the preceding but the leaves are whiter,

narrowed gradually from the tip (more rounded in *neglecta*), usually somewhat folded lengthwise. It had previously been called *A. aprica* Greene and *parviflora* had been considered the same as *microphylla* Greene.

58. *Antennaria microphylla* Greene. Plants growing in dense mats, tight to the ground, 1—6 dm. wide; basal leaves spatulate, 5—10 mm. long; flower stalks 1—2 dm. high, becoming 3—4 dm. in fruit. Common in coulees and low prairie, blooming in early June. The dense mats of fine leaves are striking and the plant is worth considering as an ornamental.

59. *Gnaphalium palustre* Nutt. EVERLASTING. Small, white wooly annual; stem widely branched, 5—20 cm. high; leaves spatulate, oblong or lanceolate, 1—2 cm. long; heads 3—4 mm. long, in small, leafy clusters. Our only specimen from Miss Hazel A. Sandberg of Larson, Burke County, in 1932, was verified by Dr. S. F. Blake.

Iva   MARSH ELDER.   POVERTY WEED

Native annuals or perennials with opposite. rough leaves and small, greenish flower heads; staminate and pistillate flowers mixed in each head; achenes wedge-shaped, black, without pappus.

Key to Species

Heads in a broad, branching cluster; leaves 5—15 cm. (2—6 in.) wide.   60.   *Iva xanthifolia*
Heads in a slender spike or scattered; leaves 5—30 mm. (⅕—1⅕ in.) wide.
  Heads in terminal spikes.   61.   *Iva ciliata*
  Heads solitary at leaf bases.   62.   *Iva axillaris*

60. *Iva xanthifolia* Nutt. MARSH ELDER. Coarse, smooth stemmed annual, 1—3 m. (3—10 ft.) high; leaves ovate, 5—15 cm. (2—6 in.) long, toothed, sometimes 3-lobed, velvety hairy; heads 4—5 mm. (⅕ in.) wide in a large, branching top 1—6 dm. (4—24) in.) wide; achenes 2 mm. long. Very common, fields, roadsides, especially old straw stacks and stock yards. Aug. 15—30.

61. *Iva ciliata* Willd. MARSH ELDER. Annual; stems 3—15 dm. high, rough hairy, branching above; leaves ovate or oblong, 3—10 cm. long, toothed or entire; heads 4—5 mm. wide, in spikes 3—5 cm. long at ends of branches; achenes 3—4 mm. long. In 1940, I found a dense stand of slender, crowded plants 3—5 dm. high in a saline marsh at Ardoch, Walsh County, in blossom July 30. It seemed these must be some other species, but Dr. S. F. Blake identified them as *Iva ciliata*. It had not been reported north of Nebraska before. In 1943 the plants were in about the same condition.

62. *Iva axillaris* Pursh. POVERTY WEED. Perennial by spreading roots; stems 2—4 dm. high, often not branched; leaves narrowly ovate to oblong, 1—4 cm. long, thick; heads whitish, 5—8 mm. wide, solitary at leaf bases, drooping; achenes plump, oblong, 2.5—3 mm. long. Frequent on saline clay flats or banks where little else grows; Barnes and Benson Counties, westward. The name poverty weed is applied to *I. ranthifolia* in some regions, but it grows in rich soil. June, July.

Ambrosia   RAGWEED

Rough, native annuals or perennials with opposite leaves; flowers monoecious, staminate heads in terminal spikes, pistillate flowers in small groups at upper leaf bases (fig. 84); pistillate involucre a closed covering, becoming hard and thick in one species.

Key to Species

Plants 1—3 m. (3—10 ft.) high; leaves usually with 3 or 5 palmate
  lobes. 63. *Ambrosia trifida*
Plants 2—10 dm. (8—40 in.) high; leaves pinnately lobed.
  Annual; leaves much divided. 64. *Ambrosia artemisifolia*
  Perennial; leaves pinnately lobed. 65. *Ambrosia coronopifolia*

63. *Ambrosia trifida* L. KINGHEAD. GIANT RAGWEED. Annual; stem rough,
ridged, 1—3 m. (3—10 ft.) high; leaves ovate, 5—25 cm. (2—10 in.) long
sometimes entire but usually deeply 3 or 5-lobed (like fingers of a
hand); staminate heads 4—6 mm. (¼ in.) wide, in dense, terminal
spikes 5—15 cm. long; fruit 6—8 mm. long, top shaped, angular, with 1
central and 3—6 marginal points at top, outer covering thick, woody.
Common in fields, along roadsides and ditches; less common north
central and western parts of State. Begins to bloom July 1, but chiefly
late July-Aug. Wind blown pollen is one of the common causes of
hay fever. The leaves are thin, soft and palatable for livestock. Kinghead is the common name here, referring to the seed, but Giant Ragweed is used in other regions.

64. *Ambrosia artemisifolia* L. COMMON RAGWEED. Annual, 3—10 dm. high,
widely branched; leaves 3—10 cm. long, once or twice divided into
narrow segments; staminate heads 3 mm. wide, in terminal spikes
3—10 cm. long; fruit 3—4 mm. long, rounded, sometimes with small
points, outer covering thin. Fields and roadsides, especially in rich
soil. Late July, Aug. Some authors use *A. elatior* for our plant. Rydberg described two or three additional species.

65. *Ambrosia coronopifolia* T. & G. PERENNIAL RAGWEED. Perennial by long,
shallow, horizontal roots; stem 2—6 dm. high, little branched; leaves
3—10 cm. long, usually pinnately lobed about half way, rather thick;
flowers and fruit similar to No. 64. Common, especially on prairie
and dry soil. Has often been considered the same as *A. psilostachya*
DC.

66. *Franseria acanthicarpa* (Hook.) Coville. FALSE RAGWEED. Annual; stem
widely branched; 3—6 dm. high; leaves 3—10 cm. long, once or
twice divided into narrow segments; staminate heads 3—4 mm.
wide, in terminal clusters 5—10 cm. long; pistillate heads at upper
leaf bases, developing yellowish spines 4—7 mm. long. Records
from Little Missouri River in Billings and McKenzie Counties and
sand dunes in McHenry County. Probably introduced from the west
in recent years. It looks much like common ragweed, but leaf segments are broader and pale green.

67. *Xanthium italicum* Mor. COCKLEBUR. Widely branched rough annual;
stem usually somewhat zigzag and purple spotted, 3—15 dm. (1—5 ft.)
high; leaves alternate, 5—15 cm. (2—6 in.) long, short ovate, coarsely
toothed; flowers monoecious, staminate heads 7—9 mm. (⅓ in.) wide
in a loose or compact cluster; pistillate heads in small clusters at leaf
bases, 2 flowers developing a hard bur 1—2.5 cm. long with hooked
spines (fig. 86). Frequent in fields, waste ground, sand bars, etc. A
troublesome native weed, seedlings poisonous to pigs.

The proper specific name for these plants has long been a puzzle.
They were formerly called *canadense*, which is now considered different. The following forms might be recognized but their relationship is
still in doubt. *X. acerosum* was described from a Fargo specimen and
some of ours have been referred to it but I am unable to see that the
hairs on the spines are different from those of other specimens.

Burs very stout, 2—3 cm. long, densely spiny. *X. speciosum* Kearney
Burs 1—3 cm. long, short and rounded or long and slender.
  Hairs on spines "softly long-pilose"; burs slender.
  X. acerosum Greene
    Hairs on spines stiff, usually hooked.
      Spines hardly as long as width of body of bur.
        Bur 1—2 times as long as wide; spines few. *X echinatum* Murr.
        Bur over twice as long as wide; spines many. *X. commune* Britton
      Spines longer than width of body of bur. *X. italicum* Mor.

68. *Silphium perfoliatum* L. CUP-PLANT. Perennial; stem coarse, 1—2 m.
(3—7 ft.) high, sharply 4-angled; leaves ovate, 1—3 dm. (4—12 in.)
long, coarsely and sharply toothed, opposite leaves grown together
around stem; heads yellow, 5—7 cm. (2—3 in.) wide; only ray flowers
produce achenes, which are very flat, oblong, 1 cm. long, without
pappus. Occasional along wooded streams in 3 southeastern counties
(fig. 19). Aug.

69. *Heliopsis helianthoides* (L.) Sweet. FALSE SUNFLOWER. Perennial; stems
5—15 dm. high; leaves opposite, 5—10 cm. long, broadly ovate, coarsely toothed, rough; petioles 5—10 mm. long; heads orange yellow, 5—7
cm. wide, solitary on long, bare, upper branches; achenes black, 4
mm. long, square in cross section, pappus none. Roadsides, especially
near woods or brush; common north to Barnes and Traill Counties,
and one record for Pembina County. July. Our plants have been considered var. *scabra* (Dunal) Fern.

70. *Rudbeckia serotina* Nutt. BLACK-EYED SUSAN. Rough hairy biennial;
stems 3—6 dm. high, often not branched; leaves narrowly oblong or
lanceolate, the lower 1—2 dm. long, not toothed; heads solitary on
upper branches, 3—5 cm. wide, disk rounded, purplish brown, rays
yellow; achenes black, 2 mm. long, square in cross section, pappus none.
Common in moist grassland east of Missouri River. July, Aug. Fernald
(Rhodora 50:175, 1948) has separated this from the eastern *R. hirta* L.

71. *Rudbeckia laciniata* L. TALL CONEFLOWER. Smooth perennial, 1—2 m.
(3—7 ft.) high; leaves alternate, 1—2 dm. (4—8 in.) long, ovate in
general outline, the upper toothed or entire, the lower pinnately divided into 5—7 somewhat toothed lobes; heads 5—7 cm. (2—3 in.) wide
on spreading upper branches, both rays and disk yellow; achenes as
in last species. Very common in woods or low ground, Red River Valley, also from Turtle Mts. Sept. "Golden Glow", commonly grown as
an ornamental, is a double flowered form (heads with ray flowers
only).

72. *Ratibida columnifera* (Nutt.) Wooton & Standl. LONG-HEADED CONE-
FLOWER. (Fig. 306). Perennial; stem branched, 3—6 dm. high; leaves
gray, rough, 5—10 cm. long, pinnately divided into several linear or
oblanceolate segments; heads solitary on bare, upper branches, 5—8
cm. wide; disk cylindrical, 3—5 cm. long, purplish brown; rays about
5, yellow, sometimes partly or entirely rich, brownish purple (forma
pulcherrima (DC.) Sharp); achenes black, 2 mm. long, flattened, 4-
angled (fig. 309). Common on dry prairies. A striking plant, sometimes grown as an ornamental. The genus name *Lepachys* has been
often used for this plant.

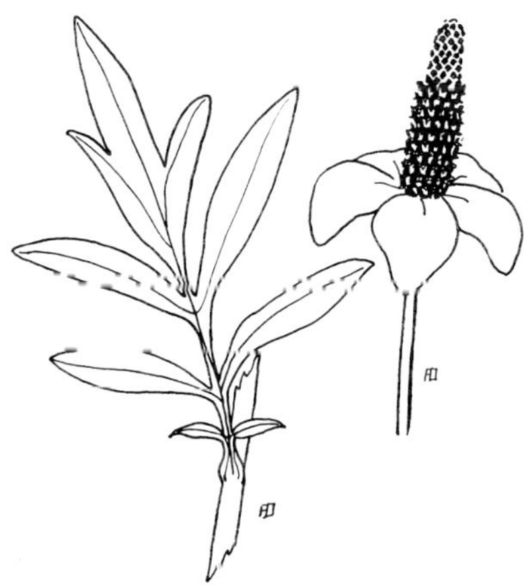

306   Long-headed Coneflower (*Ratibida columnifera*).

73. *Brauneria angustifolia* (DC.) Heller. PURPLE CONEFLOWER. (Fig. 307). Perennial from a thick, black root, stem very rough, 3—6 dm. high; leaves few, elongated lanceolate, 5—20 cm. long; heads 5—8 cm. wide, solitary on a few, bare branches; disk rounded, brown, prickly; rays rose colored to nearly white; achenes yellowish white, 5 mm. long, square in cross section, the angles projecting as sharp points; pappus none (fig. 310). Dry prairie, especially rocky hills, more common westward. July. A striking plant. The heads remain all winter, brown, prickly, 2—3 cm. wide. The roots are used medicinally.

307. Purple Coneflower (*Brauneria angustifolia*).

Helianthus SUNFLOWER

Coarse, rough annuals or perennials with simple leaves and large, yellow heads of disk and ray flowers; achenes wedge shaped, black, spotted or striped, hairy; pappus of 2 thin scales which fall off readily. (Fig. 302.)

Key to Species

Disk of head brown.
  Annual; leaves alternate.
    Leaves toothed, ovate, the lower cordate.   74. *Helianthus annuus*
    Leaves entire, lanceolate or oblong, never cordate.
                        75. *Helianthus petiolaris*
  Perennial by long rhizomes; leaves opposite.   76. *Helianthus rigidus*

Disk of head yellow; leaves mostly alternate.
  Leaves gray, somewhat folded lengthwise, elongated lanceolate.
                                            77. *Helianthus maximiliani*
  Leaves green, flat, ovate to lanceolate.
    Leaves lanceolate, 2—7 cm. (1—3 in.) wide; roots thickened.
      Leaves elongate, narrowly lanceolate.   78. *Helianthus nuttallii*
      Leaves usually short, ovate to lanceolate. 79. *Helianthus rydbergii*
    Leaves ovate or ovate-oblong, 3—10 cm. wide; roots not
      thickened; tubers on long rhizomes.   80. *Helianthus tuberosus*

74. *Helianthus annuus* L. COMMON SUNFLOWER. Very coarse, branching annual, 1—4 m. (3—14 ft.) high; leaves alternate, 1—3 dm. (4—12 in.)
    long, ovate, lower cordate, upper lanceolate; heads 1—1.5 dm. wide.
    Fields and roadsides, especially southern and western parts of State.
    July-Sept. Many of the wild plants may have reverted from plantings.

75. *Helianthus petiolaris* Nutt. SAND SUNFLOWER. More slender and less
    rough than the preceding, 5—15 dm. high; leaves 5—8 cm. long, narrowly ovate, oblong or lanceolate, rather shining; heads 5—10 cm.
    wide. Abundant in very sandy soils. July-Sept.

76. *Helianthus rigidus* (Cass.) Desf. STIFF SUNFLOWER. Perennial by long
    rhizomes forming patches; stems slender, little branched, 8—15 dm.
    high; leaves opposite, 5—15 cm. long, elliptic to lanceolate or ovate,
    thick, stiff; heads 1—5 on slender, bare stem or branch, 4—8 cm. wide,
    disk brown. Common on prairie. July, Aug. Formerly called *H. scaberrimus* Ell.

77. *Helianthus maximiliani* Schrad. NARROW-LEAVED SUNFLOWER. Perennial, growing in dense clumps; stems 1—2 m. tall, branched above;
    leaves elongated, somewhat folded, 5—15 cm. long; heads yellow, 5—8
    cm. wide, usually in a short branched cluster. Roadsides, fields and
    prairies, very common. Aug., Sept. The name "narrow-leaved" is distinctive for our region but might not be elsewhere.

78. *Helianthus nuttallii* T. & G. NUTTALL'S SUNFLOWER. Similar to last except for flat leaves. Mr. E. E. Watson, who examined all of our specimens, placed here 4 specimens from Richland, Cass and Pembina
    Counties. I have not recognized the plant in the field.

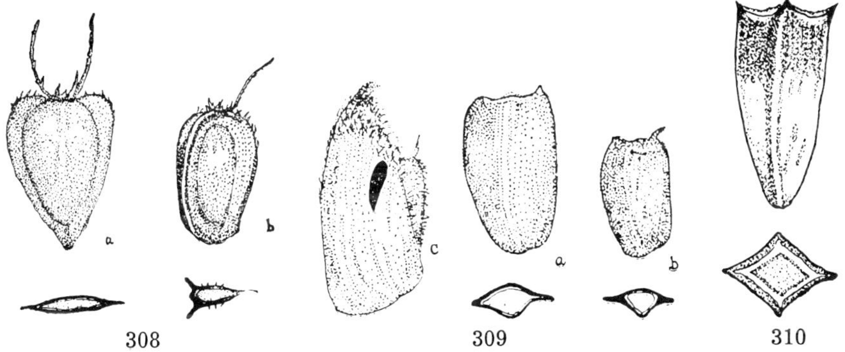

79. *Helianthus rydbergii* Britton. RYDBERG'S SUNFLOWER. Perennial by inconspicuous, short rhizomes, the main roots thickened, tapering gradually to a long, slender tip; stem smooth, 1—2 m. high, branched
    chiefly above; leaves mostly opposite, alternate on upper parts, narrowly ovate to lanceolate, 5—15 cm. long; heads yellow, 4—6 cm. wide.
    Watson's statement (Mich. Acad. Sci., Arts and Letters 9:458) that
    this grows in "sandy, arid soil," is quite incorrect for our area. It is
    characteristic of low meadows or roadside ditches, as compared to
    maximiliani. Specimens referred by Bergman to *H. giganteus* and *H.
    grosse-serratus* are now placed in *nuttallii* and *rydbergii*. It is a difficult group. *H. grosse-serratus*, with tall, smooth stems and large,
    lanceolate, strongly toothed leaves, may occur in Richland County.

80. *Helianthus tuberosus* L. Jerusalem Artichoke. Perennial by tubers on ends of long, slender rhizomes; stem 1.5—3 m. (5—10 ft.) high, widely branched above; leaves mostly opposite, the upper alternate, ovate, 5—30 cm. (2—12 in.) long, 3—10 cm. wide; heads yellow, 5—10 cm. wide. Along streams or other low ground, throughout State but chiefly eastern part. Late Aug. A showy but weedy species. The tubers are not formed until late summer. One specimen from Alice, Cass County, was referred by Watson to *H. canescens* (Gray) Wats., often considered a variety of *tuberosus*. It has smaller, grayer, nearly always opposite leaves.

81. *Coreopsis tinctoria* Nutt. Bugseed. Annual; stem slender, widely branched, 4—6 dm. high; leaves opposite, 5—10 cm. long, pinnately divided into several, narrow segments; heads yellow or purplish, 2—5 cm. wide; outer bracts green, inner yellowish, stiff, thick; achenes black, oblong, flattened, 1.5—2 mm. long. Roadside ditches and shallow water holes on prairie. July, Aug. Rather common western edge of State, rare elsewhere. Also grown as an ornamental.

Bidens   BEGGARTICKS

Fall blooming annuals with opposite, simple or compound leaves; heads yellow, ray flowers often lacking; achenes flat or angular, pappus of 2—4 stiff, barbed awns.

Key to Species

Leaves simple, toothed but not divided.
  Heads nodding after flowering; awns 4.   82. *Bidens cernua*
  Heads erect; awns 3.   83. *Bidens acuta*
Leaves compound, of 3—7 leaflets.
  Leaflets usually 3; achenes black.   84. *Bidens frondosa*
  Leaflets usually 5; achenes brown.   85. *Bidens vulgata*

82. *Bidens cernua* L. Bur Marigold. Stem 2—6 dm. (8—24 in.) high; leaves lanceolate, 3—10 cm. (1—4 in.) long, thin, sharply toothed, sometimes clasping at base; heads nodding, 2—3 cm. wide, usually with well developed rays; achenes brown, 4—5 mm. (⅙ in.) long, 4-angled with 4 awns. Frequent along pond or stream banks.

83. *Bidens acuta* (Wiegand) Britton. Stem stout, little branched, 3—8 dm. high; leaves oblong-lanceolate, 3—10 cm. long, sharply and deeply serrate, no petioles; heads upright, 2—3 cm. wide, rays lacking, outermost bracts leaf-like, 2—5 cm. long; achenes black; flat, 6—7 mm. long,

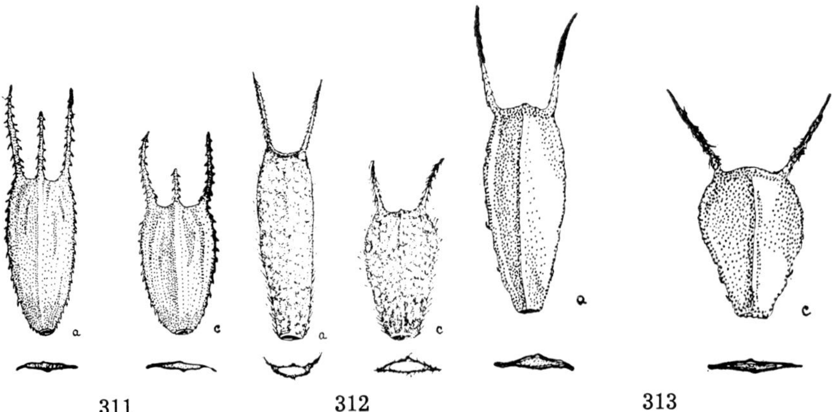

311   312   313

slightly 3-angled with 3 awns (fig. 311). Frequent in low ground, especially ditches. Often called *B. comosa*, var. *acuta* Wiegand. Specimens from Sargent and Cass to Morton County, but it is probably more widely distributed.

84. *Bidens frondosa* L. Stem rather slender, usually purple, widely branched, 3—10 dm. high; leaves long stalked; leaflets lanceolate, 3—8 cm. long, thin, smooth, sharply toothed; heads 6—12 mm. wide, rays usually lacking, outer bracts 1—2 cm. long, narrow; achenes black, 5—8 mm. long, wedge shaped, flat with 2 awns (fig. 312). Frequent along streams, especially wooded areas.

85. *Bidens vulgata* Greene. Stem rather stout, branched above, 1—1.5 m. high; leaflets 3—5, oblong-lanceolate, 5—10 cm. long, sharply toothed; heads 2—3 cm. wide, rays lacking, outer bracts 1—3 cm. long; achenes light brown, 5—10 mm. long, oblong, very thin with 2 awns (fig. 313). Our commonest species, in low spots in fields and other low ground. Our plants seem all or mostly var. *puberula* Wiegand, the leaves finely hairy. One specimen from Hope, Steele County, has the larger leaflets deeply cut at base.

86. *Madia glomerata* Hook. TARWEED. Slender annual; hairy and sticky on flowering parts, 1—4 dm. (4—16 in.) high; leaves alternate, narrowly lanceolate, 2—6 cm. (⅘—2⅖ in.) long; heads yellow, 4—5 mm. (⅕ in.) wide, longer than wide, often in rounded clusters 1—2 cm. wide; rays yellow, small or lacking; bracts in 1 row, somewhat boat shaped, partly enclosing flowers; achenes black, 5—8 mm. long, slightly curved and angled; pappus usually lacking. A common weed farther west. One record from Spring Brook, Williams County, in 1915, where the plants were rather abundant in a coulee near town.

87. *Galinsoga ciliata* (Raf.) Blake. QUICKWEED. Spreading, soft annual, 1—4 dm. high; leaves opposite, ovate, 2—5 cm. long, toothed; heads white, 4—6 mm. wide, on slender, upper branches; rays 4—5, white, very small; achenes black, 1.5 mm. long; pappus a few slender scales or none (fig. 314). Yards and gardens. Found at Fargo in 1911, not again until 1937, since when it has been found there in two or three places. Formerly listed as *G. parviflora* Cav.

88. *Hymenopappus tenuifolius* Pursh. Perennial from a crown; stem rather cottony hairy, 3—6 dm. high, a few almost leafless branches above; leaves mostly near ground, 5—15 cm. long, divided into narrow segments; heads greenish white, 8—12 mm. wide at ends of branches; achenes 1 mm. long, hairy; pappus a few small scales. Hillsides west of Missouri River. Late June. Listed by Bergman as *H. filifolius* Hook., a similar species.

89. *Bahia oppositifolia* (Nutt.) A. Gray. Perennial, gray with short hairs; stem spreading, 1—2 dm. high; leaves opposite, 1—5 cm. long, divided into 5—7 segments; heads yellow, 7—10 mm. wide; achenes 4 mm. long, slender, 4-angled; pappus a few scales. Heavy clay soil. June, July. Records from only Grant, Hettinger and Bowman Counties.

90. *Actinella acaulis* (Pursh) Nutt. BUTTE MARIGOLD. Perennial from a thick crown; leaves linear to narrowly lanceolate, 2—8 cm. long, usually gray hairy, clustered at ground; heads yellow, 2—2.5 cm. wide, each on a stalk 1—2 dm. high; achenes 3 mm. long, very hairy; pappus of several scales. Common on hillsides, Grant to McKenzie County. June, July.

91. *Hymenoxys richardsonii* (Hook.) Ckll. COLORADO RUBBER PLANT. Perennial from a thick crown; stem green, resinous, little branched, 1—2 dm. high; leaves 5—10 cm. long, divided into several narrow segments, finely pitted; heads few, yellow, 1.5—2.5 cm. wide; achenes 4 mm. long, hairy; pappus of several scales. Hills, Morton to Bowman and Golden Valley Counties.

92. *Helenium autumnale* L. SNEEZEWEED. Soft hairy perennial; stem 3—6 dm. (1—2 ft.) high; leaves alternate, oblong to lanceolate, 5—10 cm.

(2—4 in.) long, edges running down stem from base of leaf; heads yellow, 2—3 cm. wide, rays wedge shaped, 3-toothed at tip; achenes 2 mm. ($\frac{1}{12}$ in.) long, hairy; pappus of several lanceolate scales. Locally common in low places; no records west of Missouri River. Aug.

93. *Gaillardia aristata* Pursh. GAILLARDIA. BLANKET FLOWER. Rough perennial; stems with few, slender branches, 3—6 dm. high; leaves oblanceolate, 5—10 cm. long, entire, toothed or coarsely pinnately lobed; heads 3—6 cm. wide, solitary on ends of branches; disk brownish purple, rays wedge shaped, 3-toothed and yellow at tip, purplish at base; achenes 2—3 mm. long, hairy; pappus of several, slender scales. Common on prairie. July, Aug. Often grown as an ornamental.

94. *Dyssodia papposa* (Vent.) Hitchc. FETID MARIGOLD. Ill smelling annual; stems branching, usually erect, 1—3 dm. high; leaves opposite, 2—4 cm. long, divided into narrow segments, green with many orange colored, small dots; heads yellow, 5—8 mm. wide, rays few and very small; achenes black, 3 mm. long, pappus of fringed scales (fig. 315). Dry soil, Missouri River westward (fig. 20). Aug.

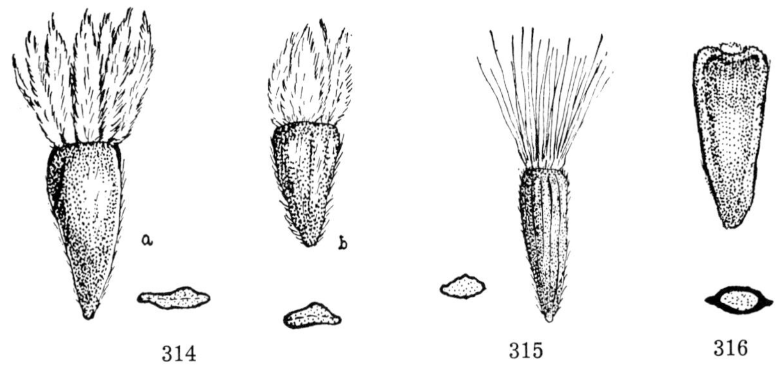

314   315   316

95. *Achillea lanulosa* Nutt. MILFOIL. YARROW. Perennial by short rhizomes; stem 3—6 dm. (1—2 ft.) high; leaves alternate, 5—15 cm. (2—6 in.) long, divided into many, fine segments; heads white, 6—8 mm. (¼ in.) wide, rays about 5, rounded, 2—3 mm. long; achenes white, 2—2.5 mm. long, flat, pappus none (fig. 316). Very common on prairie. June. Strongly aromatic. Related forms, some with pink heads, are grown as ornamentals. Recent work shows that this is the most common American form though very similar to *A. millefolium* L.

96. *Achillea multiflora* Hook. NORTHERN MILFOIL. Similar to last, but coarser and taller, leaves narrow, 5—15 mm. long, deeply fringed but not completely divided into segments. Occasional in woods, Pembina and Turtle Mts. (fig. 21). July, Aug.

97. *Chrysanthemum leucanthemum* L. OX-EYE DAISY. Perennial; stems rather bare, smooth, 3—10 dm. high; leaves pinnately toothed to deeply lobed, obovate to spatulate, upper lanceolate, 3—10 cm. long; heads at ends of branches, 3—6 cm. wide, disk yellow, rays white; achenes 2—3 mm. long, cylindrical, strongly ribbed, without pappus. A troublesome weed in many regions. We have only scattered records from Richland, Cass, Barnes, Cavalier and Stark Counties. One specimen from a field of *Bromus inermis* in Stark County seems to be typical *leucanthemum*, the others var. *pinnatifidum* Lecoq & Lamotte.

Anthemis and **Matricaria** DOG FENNEL. CHAMOMILE

Annuals with finely divided, alternate leaves, often strongly scented; rays white or lacking; achenes small, rounded or angled, pappus none. All introduced weeds.

Key to Species

Heads green, rays none; plants 1—2 dm. (4—8 in.) high, pleasantly
  scented.   99. *Matricaria matricarioides*
Heads with white rays; plants 3—6 dm. high, often ill scented.
  Achenes with small, rounded projections; scales between disk
    flowers.   98. *Anthemis cotula*
  Achenes angled or ribbed; no scales between disk flowers.
    Heads 3—4 cm. (1½ in.) wide, rays 5—10 mm. (⅕—⅖ in.) long.
        100. *Matricaria inodora*
    Heads 1.5—2.5 cm. wide, rays 10—15 mm. long.
        101. *Matricaria chamomila*

98. *Anthemis cotula* L. Dog Fennel. Ill scented annual, 2—6 dm. (8—24
in.) high; leaves 2—5 cm. (1—2 in.) long, finely divided; heads white,
2—3 cm. wide; achenes 1—2 mm. (1/25—$\frac{1}{12}$ in.) long, oblong, roughened
with rounded projections. Occasional around yards or streets. July,
Aug.

99. *Matricaria matricarioides* (Less.) Porter. Pineapple Weed. Stem 1—2
dm. high, spreading; leaves 2—5 cm. long, finely divided; heads green,
rounded, 7—10 mm. wide; achenes slightly angled. Common about
yards and streets, especially railroad yards. June-Sept. Odor suggestive
of pineapple.

100. *Matricaria inodora* L. Scentless Chamomile. Stem 4—10 dm. high,
branched; leaves 2—6 cm. long, dark green, finely divided; heads
white, 3—4 cm. wide; achenes 2 mm. long, 3-angled. Common about
yards and streets in a few localities in Richland, Pembina, Rolette
and Renville Counties.

101. *Matricaria chamomila* L. Wild Chamomile. Stem slender, 2—4 dm.
high; leaves rather narrow, 2—6 cm. long, finely divided; heads white,
1.5—2.5 cm. wide; achenes 1 mm. long, faintly 3—5-ribbed. A specimen
from Wm. A. Hoffman at Carrington, Foster County, in 1935, was
identified as this by Dr. S. F. Blake. Bergman 2399 from Glen Ullin,
Morton County, seems to be the same.

102. *Tanacetum vulgare* L. Tansy. Stout perennial with clumps of stems
5—10 dm. high; leaves alternate, 1—1.5 dm. long, pinnately divided
into several, sharply toothed lobes; heads yellow, 6—10 mm. wide, in
a flat cluster 1—1.5 dm. wide at top of stem; rays not well developed.
Occasional, persisting from old plantings. Aug.

Artemisia WORMWOOD. "SAGE"

Bitter aromatic herbs or shrubs; leaves alternate, entire, toothed
or finely divided, green or gray; heads many, very small, greenish, in a large branching top, rays lacking; achenes very small,
without pappus. Sagebrush is probably best known. Sage of culin-
ary use is a species of *Salvia*.

Key to Species

Biennials or perennials, the base more or less woody.
  Leaves green, not gray hairy.
    Prairie perennial; leaves simple or the lower with a few
      slender segments.   103. *Artemisia glauca*
    Biennials, leaves finely divided or cut.
      Plant of low ground; leaves sharply lobed and toothed.
        104. *Artemisia biennis*
      Plant of dry ground; leaves with narrow, not toothed segments.
        105. *Artemisia caudata*
  Leaves gray with fine hairs.
    Leaves entire or toothed.
      Leaves very narrow; plants tufted.   107. *Artemisia longifolia*
      Leaves narrow to broad; plants with rhizomes.
        106. *Artemisia ludoviciana*

Leaves divided into narrow segments.
  Low perennial, 1—3 dm. (4—12 in.) high.   108. *Artemisia frigida*
  Tall, stout plants, 6—20 dm. high.
    Native biennial; leaf segments about 1 mm. (1/25 in.) wide.
        105. *Artemisia caudata*
    Introduced perennial; leaf segments 2—3 mm. wide.
        109. *Artemisia absinthium*
Woody plants, 4—20 dm. (1¼—7 ft.) high; leaves gray.
  Leaves wedge shaped, 3-toothed at tip.   110. *Artemisia tridentata*
  Leaves linear to lanceolate, not toothed.   111. *Artemisia cana*

103. *Artemisia glauca* Pall. GREEN SAGE. DRAGON SAGEWORT. Dark green
    perennial 5—15 dm. (2—5 ft.) high; upper leaves linear, 2—6 cm.,
    lowest ones 5—10 cm. (2—4 in.) long, often divided into 3 slender
    segments. Common on drier prairie. Aug. This has been called *A.
    dracunculoides* Pursh but *glauca* is older. *A. glauca*, var. *dracunculina*
    (S. Wats.) Fern. is a recent combination.

104. *Artemisia biennis* Willd. BIENNIAL WORMWOOD. (Fig. 317). Biennial;
    stem stout, 1—2 m. high, widely branched in large plants; leaves dark
    green, 5—10 cm. long, 2—4 cm. wide, divided into narrow, sharply
    toothed segments. Common in low places. Late Aug., Sept. Specimens
    are mostly from southern part of State.

317. Wormwood (*Artemisia biennis*).

105. *Artemisia caudata* Michx. Biennial or short lived perennial; stem
    stout, 1—2 m. high, sometimes widely branched; leaves gray or rather
    green, 5—10 cm. long, divided into long, very slender segments. Fre-
    quent on prairie, especially in sandy or gravelly soils. Aug. Some of
    our specimens were referred by Bergman to *A. canadensis* Michx. but
    this does not occur here as I understand it. Rydberg recognized *for-
    woodii* Wats., *camporum* Rydb., and *bourgeuana* Rydb., but these
    are probably not distinct from *caudata*.

106. *Artemisia ludoviciana* Nutt. WHITE SAGE. PASTURE SAGE. Perennial by
    rhizomes, forming patches; entire plant white with cottony hairs;
    stems 3—6 dm. high; leaves linear to oblong, 3—6 cm. long, larger
    ones sometimes pinnately lobed. Very common on prairies (fig. 7).
    Aug., Sept. The typical form of the species has broad, usually lobed
    leaves which become green above. *A. gnaphalodes* Nutt., with narrow-

er, white leaves, and *A. pabularis* A. Nels., with very narrow leaves are often treated as distinct species. Recently Fernald (Rhodora 47:252) has separated several forms including vars. *gnaphalodes* (Nutt.) T. & G., *latifolia* (Bess.) T. & G. and *pabularis* (A. Nels.) Fern. He cites two N. D. specimens of var. *latifolia*. Our material seems to include many of *pabularis* and *gnaphalodes*, but I have difficulty in recognizing *latifolia* or typical *ludoviciana*.

107. *Artemisia longifolia* Nutt. LONG-LEAVED SAGE. Gray perennial; many slender stems 3—10 dm. high in a dense clump; leaves linear or linear-lanceolate, 5—10 cm. long, edges rolled under. Sides of buttes, Missouri River westward.

108. *Artemisia frigida* Willd. LITTLE SAGE. FRINGED SAGE. MOUNTAIN SAGE. Silvery perennial from a branched base; stems tufted, 1—3 dm. (4—12 in.) high, branching above; leaves 1—2 cm. (⅖—⅘ in.) long, divided into very narrow segments; heads 4—6 mm. (⅕ in.) wide, distinctly yellowish when in bloom. Very common, especially on gravelly hills and prairie. July, Aug.

109. *Artemisia absinthium* L. ABSINTH. Stout, branching perennial, 6—15 dm. high; leaves gray, 3—10 cm. long, nearly as wide, divided into narrowly oblong segments. Introduced and well established in pastures, waste ground and roadsides in many parts of State. Aug.

110. *Artemisia tridentata* Nutt. BIG SAGEBRUSH. Much branched shrub, 1—4 m. (3—14 ft.) high; leaves wedge shaped, 1—2 cm. long, 3-toothed at tip (fig. 59). Only locally common, Bowman to McKenzie Counties (fig. 7). Aug. Called "big sagebrush" in western states, but with us is usually less than 1 m. high.

111. *Artemisia cana* Pursh. DWARF SAGEBRUSH. Spreading or erect shrub, 1—2 m. high; leaves silvery, 2—8 cm. long, linear to lanceolate. Common on buttes and especially on creek flats, Missouri River westward; local at Valley City, Jamestown and Pembina Mts. (fig. 7).

112. *Petasites sagittatus* (Pursh) Gray. SWEET COLTSFOOT. Perennial from a thick rhizome, no leafy stem; leaves broad, triangular, 1—3 dm. long, white cottony below; flower stalk 2—3 dm. high, with a few small leaves and branches bearing white heads 7—10 mm. wide; rays small or lacking; achenes small, pappus a conspicuous tuft of hairs. Swamps; very local in Pembina, Benson and Bottineau Counties (fig. 15). Blooms in May before leaves appear.

113. *Arnica fulgens* Pursh. ARNICA. Perennial from a small crown covered with dark brown, wooly scales; stems 2—6 dm. (8—24 in.) high; leaves opposite, mostly basal, 3—10 cm. (1—4 in.) long, usually narrowly oblong; heads 1 or 2, deep yellow, 3—5 cm. wide; achenes 3 mm. (⅛ in.) long, short hairy; pappus a row of unbranched hairs. Local, sometimes abundant, especially westward and northwestward, usually in shallow coulees. June. Apparently it occurred originally through central part of State, but may have been largely destroyed.

Senecio RAGWORT. GROUNDSEL

Annuals or perennials with alternate, green or white hairy leaves; heads yellow, in a branching, flat topped cluster, rays usually well developed, bracts in 1 row, pappus of very fine hairs.

Key to Species

Annual, swamp plants with hollow stems.   114. *Senecio congestus*
Perennial, prairie plants with solid stems.
  Stem leafy to top, leaves 5—15 cm. (2—6 in.) long, deeply.
    pinnatifid.   115. *Senecio eremophilus*
  Upper stem leaves few, small, not much lobed.
    Plants more or less white wooly.
      Basal leaves mostly not toothed; plant very gray.
              116. *Senecio canus*

Basal leaves toothed; stem cottony at base. 117. *Senecio plattensis*
Plants green, with few hairs.
  Stem leaves usually pinnately lobed.
    Basal leaves ovate or some pinnately lobed. 117. *Senecio plattensis*
    Basal leaves rounded, rarely lobed.
      Basal leaves ovate, rounded or cordate.   118. *Senecio aureus*
      Basal leaves elliptic or oblanceolate. 119. *Senecio pauperculus*
  Stem leaves broad, entire or slightly toothed.
                   120. *Senecio integerrimus*

114. *Senecio congestus* (R. Br.) DC. SWAMP RAGWORT. Annual, with coarse, hollow, weak stem, 3—10 dm. (1—3 ft.) high; leaves rather rough hairy, the lower ovate-lanceolate, 1—2 dm. long, toothed or deeply cut; heads 10—15 mm. (½ in.) wide, rays short, 4—5 mm. long. Edges of ponds. Common in Turtle Mts.; elsewhere we have it only from Richland, Ransom and Sargent Counties. July, Aug. Previously known as *S. palustris* L. Fernald (Rhodora 47:256, 1945.) has reduced *palustris* to a variety and described a new var. *tonsus,* to which our material seems to belong.

115. *Senecio eremophilus* Richards. Perennial; stem stout, 5—15 dm. high; leaves dark green, 5—15 cm. long, deeply pinnately lobed and toothed; heads 1—2 cm. wide. Woods; Turtle Mts. only, not common. Aug.

116. *Senecio canus* Hook. GRAY RAGWORT. Gray perennial; stems 1—4 dm. high; leaves oblong or lanceolate, 3—10 cm. long, entire, the upper ones toothed or pinnatifid; heads few, 1.5—2 cm. wide. Hills, chiefly western but also through northern part of State. July.

117. *Senecio plattensis* Nutt. PRAIRIE RAGWORT. Perennial by stolons; stem 3—6 dm. high, green or with some cottony hairs; leaves 1—4 cm. long, the basal rounded, lower stem leaves usually pinnatifid; heads 1.5—2.5 cm. wide. One of the commonest prairie species. Early June. Basal leaves may be dried up in flowering specimens, but can be found on new growth from stolons.

118. *Senecio aureus* L. GOLDEN RAGWORT. Perennial by stolons, green; stem 3—6 dm. high; basal leaves ovate to cordate, 1.5 cm. long; stem leaves pinnatifid; heads 1—2 cm. wide. Frequent in wet meadows, eastern third of State. June.

119. *Senecio pauperculus* Michx. Similar to last and rather difficult to distinguish. Specimens from Richland to LaMoure and Traill Counties. Formerly called *S. balsamitae* Muhl. One specimen from Bottineau (Stevens, July 8, 1917) was identified by Lunell as *S. manitobensis* Greenm., but it seems rather a form of *pauperculus*. Greenman (Ann. Mo. Bot. Gard. 3:180) listed a specimen from McHenry County as *tridenticulatus* Rydb., with *manitobensis* as a synonym. It was distinguished from *pauperculus* by having thick, fleshy leaves.

120. *Senecio integerrimus* Nutt. LAMBSTONGUE RAGWORT. Rather stout, smooth perennial, 3—8 dm. high; leaves broad, 5—15 cm. long, lower oblong, or oblanceolate, upper triangular with clasping base; heads 1.5—2 cm. wide. Fairly common, especially in coulees. June. Specimens from southern half of State.

121. *Arctium minus* (Hill) Bernh. BURDOCK. Stout, widely branched biennial from a thick tap root; stem 1—2 m. (3—7 ft.) high; leaves ovate, 1—3 dm. (4—12 in.) long, softly white hairy below; heads rose purple, 1—1.5 cm. (½ in.) wide, without rays; involucre rounded, the many narrow bracts hook-tipped; achenes 5—6 mm. (¼ in.) long, nearly black. An introduced, troublesome weed in woods and wooded pastures, eastern part of State. Aug. The first year leaves are sometimes called "wild rhubarb," but they are easily distinguished from rhubarb by their narrower shape and white under surface.

Carduus and Cirsium THISTLES

Coarse, prickly, biennial or perennial weeds with alternate, usually pinnatifid leaves and large, rounded, pink heads without ray flowers; achenes nearly cylindrical, smooth, gray, purplish or brown; pappus a large tuft of hairs. The names *Carduus, Cirsium* and *Cnicus* have been variously applied to these plants. As now used, *Carduus* includes the "plume-less" thistles, the pappus hairs of which have no fine branches as in *Cirsium*.

Key to Species

Pappus hairs with slender branches (*Cirsium*).
  Bracts of head all with recurved spines; introduced biennial.
                                122. *Cirsium vulgare*
  Inner bracts without spines, outer with straight spines or none.
    All bracts except inner with well developed spines; heads 2—6
      cm. (⅘—2⅖ in.) wide.
        Leaves quite white, at least below.
          Leaves coarsely toothed, sparsely hairy and green above.
                                123. *Cirsium altissimum*
          Leaves usually pinnatifid, white on both sides.
                                124. *Cirsium undulatum*
        Leaves green on both sides, not white hairy.
                                125. *Cirsium muticum*
    All bracts with spines little developed; heads 1—1.5 cm. wide.
                                126. *Cirsium arvense*
Pappus hairs not branched; bracts not spiny (*Carduus*).
  Heads 1—2 cm. wide, erect.   127. *Carduus crispus*
  Heads 3—5 cm. wide, nodding.   128. *Carduus nutans*

122. *Cirsium vulgare* (Savi) Airy-Shaw. Bull Thistle. Biennial, 1—2 m.
    (3—7 ft.) high, widely branched; leaves somewhat gray, the triangu-
    lar or lanceolate lobes with stout spines 5—15 mm. (⅕—⅗ in.) long;
    heads rose purple, 3—4 cm. (1½ in.) wide, turning down and easily
    breaking off when dry; achenes 3 mm. long, gray, finely purple streaked.
    Chiefly in pastures in wooded areas. Records from Richland, Cass,
    Ransom and LaMoure Counties. July, Aug. A very prickly, introduced
    weed. Formerly included in *C. lanceolatum* (L.) Hill.

123. *Cirsium altissimum* (L.) Spreng. Tall Thistle. Biennial; stem 1—3 m.
    high with few, upright branches; leaves white below, green above,
    coarsely lobed or only toothed, not very prickly; heads 3—4 cm.
    wide, rose purple. Rare in woods and along wooded roadsides, Richland
    and Cass Counties. Late Aug.

124. *Cirsium undulatum* (Nutt.) Spreng. Prairie Thistle. (Fig. 318). Peren-
    nial by roots, forming patches; stem 4—8 dm. high, branches few, up-
    right; leaves white on both sides or greenish above, usually pinnately
    lobed or divided, sometimes lanceolate and only toothed; heads 4—7
    cm. wide, pale purple; achenes yellowish, 4 mm. long. Very common
    on prairie, in pastures, meadows, waste ground or neglected fields.
    July-Sept.

        This is highly variable in size, habit of growth and cut of leaf.
    Bergman separated plants with narrower, more intricately cut leaves
    as *C. flodmanii* (Rydb.) Arthur. An exceptionally stout, little branched
    form, leaves 5—7 cm. wide, with large, triangular teeth, heads 6—7
    cm. wide, is *C. undulatum*, var. *megacephalum* (Gray) Fernald. This is
    locally common in central and western parts of State and sometimes ap-
    pears quite distinct. It seems to grow more as single plants than in
    patches.

125. *Cirsium muticum* Michx. Swamp Thistle. Biennial; stem rather slen-
    der, 1—2 m. high, branches few, upright; leaves green, 1—3 dm. long,
    rather evenly and deeply pinnately divided into 6—10 oblong or lanceo-
    late segments with a few, sharp teeth; heads 3—4 cm. wide, purple.
    Wooded areas, Pembina and Turtle Mts. only.

318   319

318.   Prairie Thistle (*Cirsium undulatum*).

319.   Canada Thistle (*Cirsium arvense*).

126. *Cirsium arvense* (L.) Scop. CANADA THISTLE. Fig. 319. Perennial, spreading by horizontal roots; stems 3—10 dm. (1—3 ft.) high, much branched above; leaves green, a little gray below, much divided into curled, prickly segments; heads numerous, light purple, 1—1.5 cm. (½ in.) wide; achenes brown, 3 mm. long. Common in fields and waste ground, especially low places, chiefly eastern third of State. A very troublesome weed. The plants are dioecious, some patches producing only flowers with stamens, others flowers with pistils. Only the latter bear seed, but pappus hairs develop also on staminate flowers, so that patches of these may appear to be scattering seeds, though none are present.

127. *Carduus crispus* L. CURLED THISTLE. Biennial; stem rather slender, branched above, much covered by jagged wing edges of leaves; leaves deeply pinnately divided into short, remote segments, spines weak; heads 1—2 cm. wide, purple, bracts not spiny; achenes similar to those of Canada Thistle. One plant was found by C. H. Waldron and O. A. Stevens in a wooded, park area at Fargo in 1940 and reported in Bulletin 339 as *Cirsium palustre*, from which it differs by unbranched pappus hairs. In a pasture near Fargo in 1947 (Stevens 1029) it had been abundant for several years but mistaken for *Cirsium vulgare*.

128. *Carduus nutans* L. MUSK THISTLE. Biennial; stem 5—10 dm. high; leaves much divided into spiny segments; heads pale purple, wider than long, 3—5 cm. wide, nodding; bracts lanceolate, more leaf-like than in *Cirsium*, tapering gradually to a point, outer bracts bent back; pappus bristles not branched. This was received from Milton, Cavalier County, in 1904 and was collected by Stevens in 1931 in a farmyard near Park River, Walsh County. It was said to have been established in a nearby field for about 15 years.

Centaurea CORNFLOWER. STAR THISTLE

Annuals or perennials with alternate, usually gray leaves and rounded heads; ray flowers lacking but outer flowers sometimes enlarged, bracts often fringed; achenes smooth, flattened cylindrical or somewhat angled, pappus hairs short and stiff. None are native to N. D. The name "star thistle" refers to species with spiny bracts, none of which are yet found here.

Key to Species

Heads 3—5 cm. (1—2 in.) wide; lower leaves 5—15 cm. long, deeply
  pinnatifid.   129. *Centaurea scabiosa*
Heads 1—3 cm. wide; leaves 3—10 cm. long, entire or somewhat lobed.
  Perennial; heads 1 cm. wide, outer flowers not enlarged.
                          130. *Centaurea repens*
  Annual; heads 2—3 cm. wide, outer flowers much enlarged.
                          131. *Centaurea cyanus*

129. *Centaurea scabiosa* L. SCABIOUS STAR THISTLE. Green perennial; stems 3—6 dm. (1—2 ft.) high; leaves 5—15 cm. (2—6 in.) long, pinnately divided into narrow segments; heads purple, 3—5 cm. wide, at ends of a few, erect branches; bracts black tipped. Specimens received from Ransom, Barnes and Golden Valley Counties, 1922-37.

130. *Centaurea repens* L. RUSSIAN KNAPWEED. Gray perennial from deep and horizontal, black roots; stems 3—8 dm. high, often much branched; leaves 3—8 cm. long, lanceolate, the lower broader and frequently pinnately lobed; heads 1 cm. wide, rose colored to nearly white, at ends of branches; bracts chaffy; achenes 2.5—3 mm. long, chalky white, somewhat angular. Introduced in Turkestan alfalfa seed. First found in Williams County in 1919. Now widely distributed but local. Aug. The name *C. picris* Pall. has been widely used. There is some question about the identity of *C. repens*.

131. *Centaurea cyanus* L. CORNFLOWER. Annual; stem 3—6 dm. high, rather cottony; leaves 3—10 cm. long, lanceolate to linear; heads blue, purple or white, 2—3 cm. wide, outer flowers much enlarged; bracts chaffy; achenes 4 mm. long, smooth, shining, pappus brown. Occasionally escaped from gardens.

132. *Picris echioides* L. OXTONGUE. Prickly biennial, 4—6 dm. high; leaves alternate, oblong, 2—10 cm. long, the upper clasping at base; heads yellow, 1.5—2.5 cm. wide; outer bracts ovate, well separated; achenes yellowish, oblong, 3 mm. long, with fine cross ridges; pappus of soft white hairs. One specimen from a garden at Davenport, Cass County, in 1924 and one more recently (record misplaced).

133. *Tragopogon dubius* Scop. LARGE GOATSBEARD. Smooth, slender leaved biennial, becoming widely branched, 4—8 dm. (16—32 in.) high; leaves alternate, linear-lanceolate, 1—3 dm. long; heads yellow, 3—5 cm. (1—2 in.) wide, their stalks much enlarged just below the head; achenes cylindrical, rough, 2—3 cm. long with a slender "beak" 1—2 cm. long and pappus of branched hairs; fruiting head expanding like that of dandelion, 1—1.5 dm. wide. Very common in prairie and waste ground. Previously reported as *T. pratensis* L., which may also occur. It has smaller heads, broader bracts, flower stalk not enlarged.

134. *Cichorium intybus* L. CHICORY. Perennial; stems 3—6 dm. high, angled, widely spreading, rather bare; leaves oblong or lanceolate, lower 5—15 cm. long, usually pinnatifid, upper quite small; heads blue, 2—3 cm. wide, in small clusters close to stem; achenes nearly black, 2—3 mm. long, angular; pappus of small scales. Once collected at Fargo in 1910.

135. *Lapsana communis* L. NIPPLEWORT. Annual, 1—4 dm. high; upper leaves lanceolate, 3—8 cm. long, somewhat toothed, lower ones obovate, somewhat pinnatifid toward base; heads yellow, 8—10 mm. wide; achenes oblanceolate, 4—5 mm. long, flattened, faintly ribbed. One specimen collected in lawn at Fargo in 1911.

Taraxacum   DANDELION

Perennials from stout tap roots; leaves all basal, oblanceolate, pinnately toothed or lobed; heads yellow, 100—300-flowered; bracts in 2 rows, the outer recurved; achenes short oblong, ribbed, with a few spines at upper end, a slender beak longer than the body and pappus of fine hairs. Mostly introduced weeds.

Key to Species

Achenes greenish brown.   136. *Taraxacum officinale*
Achenes dark red.   137. *Taraxacum laevigatum*

136. *Taraxacum officinale* Weber. COMMON DANDELION. Perennial from a stout taproot; leaves all basal, 1—3 dm. (4—12 in.) long, 1—3 cm. (⅖—1⅕ in.) wide, coarsely or sharply toothed; heads yellow, 2—4 cm. wide, each on a stalk 1—3 dm. high from crown of plant; achenes greenish brown, 3 mm. (⅛ in.) long. Very common, especially eastward. Blooms mainly May 10—30, some plants during summer and late Aug. to Nov.

137. *Taraxacum laevigatum* (Willd.) DC. RED-SEEDED DANDELION Very similar to last but has red achenes; leaves usually finely pinnatifid, heads a little smaller. Appears to be native, occurring chiefly in woods. Specimens from eastern and northern counties, also Slope County (Stevens 888).

Agoseris   PRAIRIE DANDELION

Native perennials; leaves slender, all or mostly basal; heads yellow, on stalks longer than the leaves; achenes cylindrical, pappus of fine, white hairs.

Key to Species

Leaves flat, sometimes toothed; achenes beaked.   138. *Agoseris glauca*
Leaves wavy edged, not toothed; achenes not beaked.
                                                      139. *Agoseris cuspidata*

138. *Agoseris glauca* (Pursh) D. Dietr. Leaves rather waxy, 1—3 dm. (3—12 in.) long; heads 2—5 cm. (1—2 in.) wide; achenes 7 mm. (¼ in.) long, gradually narrowed into a beak 3—5 mm. long. Common in coulees and fairly moist prairie; inclined to grow in patches. June. The leaves vary from linear to broadly lanceolate and sometimes have a few slender teeth or lobes. The bracts are either smooth or hairy.

139. *Agoseris cuspidata* (Pursh) Steud. Leaves 1—3 dm. long, a thick line of hairs along edges; heads 2—5 cm. wide; achenes 10—15 mm. long, only slightly narrowed at upper end. Occasional, chiefly west of Missouri River. One specimen from Stutsman County (Bergman 1721). May, June.

Crepis   HAWKSBEARD

Annuals, biennials or perennials with alternate, pinnatifid leaves; flowers yellow, heads narrow, bracts in 1 row; achenes cylindrical, scarcely beaked, pappus of fine hairs.

Key to Species

Leaves gray; native western plant.   140. *Crepis occidentalis*
Leaves green.
    Leaves mostly basal, 1—3 cm. wide.   141. *Crepis runcinata*
Stems leafy; leaves narrow, entire or with narrow lobes.
                                                      142. *Crepis tectorum*

140. *Crepis occidentalis* Nutt. Perennial; stem 1—3 dm. (3—12 in.) high; leaves gray, mostly basal, 1—2 dm. long, the larger pinnately divided about half way into lanceolate segments; heads yellow, 1.5—2 cm. (¾ in.) wide; 2—6 heads on short branches at top of stem. Hillsides, Bowman to McKenzie Counties (fig. 18). June.

141. *Crepis runcinata* (James) T. & G. Perennial; stems 3—6 dm. high with few, slender, leafless branches; leaves mostly basal, 5—20 cm. long, oblanceolate, entire or irregularly toothed; heads yellow, 1—1.5 cm. wide; achenes brownish, 4—5 mm. long, ribbed. Common in wet coulees. June.

142. *Crepis tectorum* L. Introduced annual; stem 2—4 dm. high, branched from near base; lower leaves 1—1.5 dm. long, toothed or lobed, upper linear, not clasping at base; heads yellow, 8—15 mm. wide; achenes 3—4 mm. long, slender, reddish. This seems well established along roadsides near Walhalla, Pembina County, and we have specimens from Bottineau and Stark Counties.

Lygodesmia SKELETON WEED

Native plants of dry soil; stems nearly bare, with a few, slender or scale-like leaves; heads pink, 5-flowered, at tips of branches; achenes cylindrical, ribbed; pappus of slender hairs.

Key to Species

Widely branched perennial, nearly leafless.   143. *Lygodesmia juncea*
Slender annual; leaves slender, 5—20 cm. long. 144. *Lygodesmia rostrata*

143. *Lygodesmia juncea* (Pursh) D. Don. SKELETON WEED. Perennial by deep and spreading roots; stems branched from base, 2—4 dm. (8—16 in.) high; heads pink, 1—1.5 cm. (½ in.) wide at ends of branches. Common in dry, especially sandy, soils. July, Aug. The lower stems often have clusters of rounded insect galls, 6—8 mm. (¼ in.) wide.

144. *Lygodesmia rostrata* A. Gray. ANNUAL SKELETON WEED. Slender annual, 3—10 dm high; heads scattered along branches. Sand dunes, Richland, Ransom and McHenry Counties (fig. 17).

145. *Hieracium canadense* Michx. HAWKWEED. Native perennial; stem rather stout, branched above; leaves alternate, numerous, oblong-lanceolate, 2—10 cm. long, sharply toothed; heads yellow, 1.5—2.5 cm. wide in a flat topped cluster. Coulees and edges of woods and brush. Aug. Specimens from southeastern and north-central counties.

Prenanthes WHITE LETTUCE. RATTLESNAKE ROOT

Native perennials with broad leaves and slender, nodding, whitish or pinkish heads; achenes cylindrical, ribbed; pappus white or brown.

Key to Species

Woodland plant, very smooth; leaves triangular.   146. *Prenanthes alba*
Prairie plant, hairy above; leaves oblanceolate.

147. *Prenanthes racemosa*

146. *Prenanthes alba* L. WHITE LETTUCE. Perennial; stem very smooth, usually purplish, 1—2 m. (3—7 ft.) high; leaves broadly triangular, the lowest 1—2 dm. (4—8 in.) long, sometimes lobed or coarsely toothed; heads cylindrical, 10—12 mm. (½ in.) long, drooping, in a large, branching, terminal cluster; bracts purplish, flowers white. Woods, eastern counties. Aug., Sept.

147. *Prenanthes racemosa* Michx. Perennial; stem 4—8 dm. high; lower leaves oblanceolate, 1—2 dm. long, usually toothed, upper ovate, clasping; heads cylindrical, 10—12 mm. long, green or purplish, hairy, in a dense, narrow cluster 1—3 dm. long. Frequent in low prairie. Late Aug. Probably widely distributed, records from southeastern counties, also Pierce and Burke.

Sonchus SOW THISTLE

Somewhat spiny, annual or perennial, introduced weeds with yellow heads; achenes oblong, flattened, ridged, without a beak; pappus a tuft of fine hairs.

Key to Species

Heads 2—3 cm. (1 in.) wide; perennials with running roots.
148. *Sonchus arvensis*
Heads 1—1.5 cm. wide; annuals.
Leaves spiny toothed, not pinnatifid. 149. *Sonchus asper*
Leaves not spiny, deeply pinnatifid. 150. *Sonchus oleraceus*

148. *Sonchus arvensis* L. PERENNIAL SOW THISTLE. Vigorous perennial, spreading by roots; stems smooth, 1—2 m. (3—7 ft.) high; leaves light green, 1—2 dm. (4—8 in.) long, irregularly toothed, lobed or pinnatifid, edges spiny; heads deep yellow, 2—3 cm. (1 in.) wide, in a loose, branching top; achenes dark red, 2 mm. ($\frac{1}{12}$ in.) long, slightly flattened, 5—7-ribbed with smaller cross wrinkles. Common in fields and low ground, eastern third of State, rare westward. July-Sept. The prevailing form is var. *glabrescens* Wimm. & Graeb., also called *S. uliginosus* Bieb. The typical form of *arvensis*, with yellowish hairs on stems below heads, has been collected in Dickey and Bottineau Counties but we have no evidence of its spreading.

149. *Sonchus asper* (L.) All. SPINY SOW THISTLE. Prickly annual; stem 3—10 dm. high, not much branched; leaves dark green, the upper 5—15 cm. long with rounded, clasping base; heads pale yellow, 1—1.5 cm. wide, in a rather dense top; achenes 2 mm. long, pale brown, very thin, 3 slender ribs on each side. Frequent, especially about gardens and streets. Aug.—Oct.

150. *Sonchus oleraceus* L. COMMON SOW THISTLE. Annual; stem rather slender, hollow, 6—15 dm. high; leaves 1—3 dm. long, deeply pinnatifid into 5—9 segments, not spiny; heads pale yellow, 1—1.5 cm. wide in a spreading top; achenes yellowish, faintly cross wrinkled. Infrequent about gardens and streets, Richland and Cass Counties. Aug.—Oct. The name "common sow thistle" is misleading here since this is our least common species.

Lactuca LETTUCE

Coarse annuals or perennials with numerous, yellow, blue or purplish flower heads; achenes flattened, usually beaked; pappus of fine hairs.

Key to Species

Flowers yellow.
Leaves usually on edge, strongly spiny. 151. *Lactuca serriola*
Leaves horizontal, scarcely spiny. 152. *Lactuca canadensis*
Flowers blue or purplish.
Heads 15—20 mm. (¾ in.) wide, showy; leaves mostly entire.
153. *Lactuca pulchella*
Heads 8—10 mm. wide; not showy; leaves usually pinnatifid.
Leaves pale green; prairie plants. 154. *Lactuca ludoviciana*
Leaves dark green; woodland plants. 155. *Lactuca biennis*

151. *Lactuca serriola* L. PRICKLY LETTUCE. Prickly annual; stem stout, 1—2 m. (3—7 ft.) high; leaves 1—1.5 dm. (4—6 in.) long, oblong, spiny-edged, toothed or deeply pinnatifid, usually standing edgewise and pointing north or south, midrib spiny below; heads bright yellow, 8—10 mm. (⅓ in.) wide in a large, branching top; achenes greenish, 3—8 mm. long, flattened, elliptical, 7—9-ribbed. Very common, especially on roadsides and in waste ground. Formerly called *L. scariola* L. The en-

tire leaved form is var. *integrata* Gren. & Godr. Rydberg describes the plant as biennial, but it is distinctly annual and winter annual with us, a familiar weed and the only introduced species of *Lactuca.*

152. *Lactuca canadensis* L. WILD LETTUCE. Biennial; stem 6—15 dm. high, rather slender; leaves coarsely pinnatifid, sparsely hairy on midrib below; heads reddish yellow, 8—10 mm. wide; achenes black, 4 mm. long, oblong, faintly 3-ribbed. Rare in woods, all parts of State. Bergman had referred two specimens from Walhalla to *L. hirsuta* Muhl., separated by a few hairs on lower midrib; Mr. L. H. Shinners considers all our specimens var. *latifolia* Kuntze.

153. *Lactuca pulchella* (Pursh) DC. BLUE WILD LETTUCE. Perennial, spreading by roots, forming patches; stems 4—10 dm. high, somewhat branched above; leaves blue green, smooth, 5—15 cm. long, lanceolate and entire or larger ones pinnatifid; heads bluish purple, 1.5—2 cm. wide; achenes gray or reddish, 5—6 mm. long, narrowed into a short beak. Very common along roads, on prairie and especially in neglected fields. July 1—Sept. One of our more persistent native weeds. The flowers remain open until late afternoon, unlike most lettuces which close about noon in bright weather.

154. *Lactuca ludoviciana* (Nutt.) DC. WESTERN LETTUCE. Biennial; stems 5—15 dm. high, widely branched above; leaves pale green, lower oblanceolate, toothed or lobed, upper deeply but coarsely pinnatifid, edges spiny, midrib below with coarse, soft hairs; heads 6—8 mm. wide, bluish purple to nearly white; achenes black, 4 mm. long, flat, oblong, faintly 3-ribbed, beaked. Frequently on prairie. July-Aug. The flowers are usually described as yellow, but I find them usually pale lilac.

155. *Lactuca biennis* (Moench) Fern. BLUE WOOD LETTUCE. Biennial; stem stout, 1—2.5 m. high, flowering branches short; leaves dark green, 1—3 dm. long, coarsely pinnatifid; heads blue, 8—10 mm. wide, in a rather compact, branching top; achenes dark gray, 4—5 mm. long, elliptical, scarcely beaked. Woods, Richland, Ransom, Cass, Pembina and Bottineau Counties. Aug., Sept. Formerly called *L. spicata* (Lam.) Hitchc.

SUMMARY OF FAMILIES WITH NUMBER OF SPECIES

FAMILY		FAMILY	
Ferns and Fern Allies	19	Caltrop	1
Pine	5	Milkwort	3
Cattail	2	Spurge	12
Bur-reed	1	Water Starwort	2
Pondweed	12	Sumac	3
Arrowgrass	2	Staff-tree	2
Water-plantain	3	Maple	4
Waterweed	2	Balsam	2
Grass	145	Buckthorn	2
Sedge	82	Grape	2
Arum	2	Linden	1
Duckweed	3	Mallow	8
Spiderwort	2	St. John's-wort	2
Pickerelweed	1	Waterwort	1
Rush	10	Rock-rose	3
Lily	21	Violet	8
Amaryllis	1	Loasa	2
Iris	2	Cactus	4
Orchid	10	Oleaster	3
Willow	18	Loosestrife	2
Birch	7	Evening Primrose	16
Oak	1	Water Milfoil	3
Elm	3	Ginseng	1
Mulberry	2	Carrot	20
Nettle	4	Dogwood	3
Sandalwood	2	Heath	5
Birthwort	1	Primrose	10
Buckwheat	28	Olive	2
Goosefoot	36	Gentian	6
Pigweed	5	Dogbane	3
Four O'clock	4	Milkweed	9
Carpet-weed	1	Morningglory	10
Purslane	2	Phlox	7
Pink	22	Waterleaf	3
Water Lily	1	Borage	19
Hornwort	1	Vervain	4
Buttercup	30	Mint	28
Barberry	2	Nightshade	10
Moonseed	1	Figwort	32
Poppy	3	Broomrape	3
Fumitory	3	Bladderwort	1
Mustard	54	Lopseed	1
Caper	2	Plantain	7
Sundew	1	Madder	5
Orpine	1	Honeysuckle	8
Saxifrage	10	Teasel	1
Rose	42	Gourd	2
Pea	68	Bellflower	6
Geranium	5	Lobelia	3
Wood Sorrel	3	Aster	155
Flax	4	Total	1143
Rue	1		

DEFINITIONS OF TECHNICAL TERMS

Achene A small, dry, one-seeded fruit.
Acuminate Gradually tapering to a point.
Acute Sharp pointed; ending with a sharp angle.
Alternate Not opposite; with a single leaf at each place of attachment.
Annual Lasting only one growing season.
Anther Upper part of the stamen, containing pollen.
Apex Tip or upper end.
Apical Situated at apex or tip.
Aquatic Growing in water.
Ascending Growing obliquely upward or curving upward.
Auricle An appendage at base of anthers or of leaves.
Awn A slender, bristle-like structure.
Axil Just above base of leaf.
Axillary Borne in axils of leaves.
Barbed With short, fine, stiff points.
Basal Arising from base; lowest leaves, often from stem below ground.
Beak An elongated, tapering tip.
Beard Sometimes used instead of awn, also for a group of stiff hairs.
Berry A fruit in which the seeds are imbedded in a soft or fleshy substance.
Bi-ternate Twice ternate.
Blade Flat, expanded part of a leaf.
Bract A scale or leaf, usually small, sometimes colored, standing below a flower or a flower cluster.
Bristle A stiff hair or any similar outgrowth.
Bulb A short stem with fleshy scales, usually underground.
Bulblets Small bulbs, sometimes formed in place of flowers or at leaf bases.
Bulbous Resembling a bulb; used for fleshy roots, etc.
Capsule A dry fruit which splits open, formed from two or more united carpels.
Carpel A simple pistil, or one member of a compound pistil.
Catkin A slender, dense flower cluster, as in willow or cottonwood.
Calyx All of the sepals, separate or united; usually green, sometimes colorless or bright colored.
Cell Strictly speaking, the individual microscopic part of plant structures; often used for cavity of ovary or anther.
Chaff Thin scales between disk flowers of composite heads; a small thin scale or bract.
Ciliate Provided with marginal hairs.
Cleft Cut about half way to middle.
Compound Composed of two or more similar parts.
Compound leaf Composed of separate leaflets.
Compound ovary Composed of 2 or more united carpels.
Cordate Heart shaped; base broad, each side rounded.
Corm Short, thick, erect, underground stem.
Corolla All of the petals, separate or united; usually showy part of the flower.
Creeping Growing along surface of ground.
Crenate Scalloped; with rounded, short teeth.
Cyme A branching cluster in which the terminal or middle flower blossoms first.

Deciduous Falling off at close of growing period (leaves), or after flowering (flower parts).

Dentate Toothed, with outwardly projecting teeth.

Depressed Vertically flattened, i. e. as if pressed down from above.

Diffuse Widely or loosely spreading.

Digitate Compound, with the members borne in a whorl at apex of support; in leaves, similar to palmate, usually 3 leaflets.

Dioecious Bearing pistils and stamens on different plants.

Disk An enlargement of the axis of a flower around base of pistil; the group of tubular flowers in the aster family.

Dissected Cut or divided into numerous segments.

Distinct Separate; not united.

Divided Lobed to near base or midrib.

Drupe A fleshy or pulpy fruit with inner part stony.

Drupelet A diminutive drupe.

Elliptic Broadest at middle, narrowed equally toward both ends.

Entire Without lobes, divisions or teeth.

Erect Standing upright; vertical.

Exserted Projecting beyond surrounding parts.

Fascicled Borne in dense clusters.

Fertile Bearing seeds, or bearing pollen.

Fibrous Composed of, or resembling fibers.

Filament Stalk of stamen bearing the anther.

Filiform Thread-like, very slender.

Fleshy Thick and soft.

Follicle A dry, simple pistil fruit splitting along ventral side.

Fruit Any ripened ovary, sometimes part of axis including several ovaries or fleshy stem tip.

Glabrous Without hairs; smooth.

Gland A secreting surface or structure; any small appendage or protuberance having the appearance of such an organ.

Glaucous Covered with a white, waxy substance, a bloom.

Glume Scale of spikelets of grasses and sedges; specifically the lowest pair in spikelets of grasses.

Head A dense, rounded cluster of sessile or nearly sessile flowers.

Herb A non-woody plant which dies down to the ground annually.

Indehiscent Applied to fruits that do not open or split to let out the seeds.

Inferior Applied to an organ situated below another one, especially an ovary where the other flower parts are attached to top of it.

Inflorescence The whole flower cluster; spikes, panicles and heads are types of inflorescences.

Internode The part of the stem between two successive nodes.

Involucre A group of leaves or scale-like leaves borne just underneath a flower or a close cluster of flowers.

Irregular Applied to a flower in which the petals are not alike in size and shape.

Keel The two, fused lower petals of the flower of the pea family; also applied to a sharp ridge, as on sepals or bracts.

Keeled Ridged, like the keel of a boat.

Lanceolate Broadest near base, tapering gradually to a slender point.

Legume The dry, simple fruit of the pea family, splitting on both edges when ripe. Also used as a name for the plants of this family.

Lemma The outer scale of a grass flower.

Lenticular Shaped like a biconvex lens.

Ligule A strap shaped flower, as the rays in the aster family; a projection from top of sheath in grasses.

Linear Long and narrow with sides nearly parallel.

Linear-lanceolate Narrowly lance shaped.

Lip Each of the upper and lower divisions where calyx or corolla are so divided; the large, lower petal of an orchid flower.

Lobe One segment of a leaf or corolla.

Membranous Thin, somewhat colorless.

Monoecious Bearing stamens and pistils on the same plant but in different flowers.

Nerve One of the veins (lines or ridges) running through a leaf, scale or sepal.

Net-veined Vein branches running in various directions and connecting with each other.

Nodding Hanging on a bent stalk.

Node Point on stem from which leaf arises; often called "joint."

Nut A dry, one-seeded, indehiscent fruit with a hard shell or covering.

Nutlet A small, hard, one-seeded fruit.

Obcordate Heart shaped but with large end at tip.

Oblanceolate Same as lanceolate but broadest toward tip.

Oblong-lanceolate Intermediate between oblong and lanceolate.

Obovate Ovate but with broad end at tip.

Obtuse Rounded or blunt.

Opposite 2 leaves or flowers at same level on stem but the second on the other side from the first.

Ovary Lower part of pistil in which young seeds are borne.

Ovate With a broad, rounded base and narrower tip.

Ovule Part which develops into seed.

Palet Inner scale of a grass flower.

Palmate Lobes or leaflets extending from or toward a central point.

Panicle A branching flower cluster, the lower branches longest and blossoming first.

Pappus Bristles, hairs, awns or other structures which are borne upon the fruit in the aster family.

Parasitic Growing upon and getting its nourishment from some other living plant.

Parted Cleft nearly but not quite to base.

Pedicel Stalk of a single flower in a flower cluster.

Peduncle Stalk of a flower (if only one flower) or of a flower cluster.

Perennial Stems, underground stems or roots living from year to year.

Perfect A flower having both stamens and pistils.

Perfoliate Leaves appearing to be pierced by the stem.

Perianth Including both sepals and petals, especially where they are not clearly distinct from each other.

Perigynium Sack-like membrane enclosing ovary or achene in genus *Carex*.

Persistent Organs that remain attached after maturity; long continuous.

Petal One of the inner, usually brightly colored, set of flower leaves of a corolla.

Petiole Stalk of a leaf.

Pinnate Leaves divided into leaflets or segments more or less in pairs along a common stalk.

Pinnatifid Pinnately cleft to the middle or beyond

Pistil Central or terminal organ in a flower, containing ovules.

Pistillate With pistils but without stamens.
Pod General term for any dry fruit which opens to release seeds.
Pollen The minute grains found in the anther.
Polygamous A plant having some perfect and some unisexual flowers.
Pubescent With hairs.
Raceme A more or less elongated axis bearing flowers with distinct pedicels of about equal length.
Rachilla Axis of a spikelet.
Rachis Axis of a spike; also applied to axis of a pinnately compound leaf.
Ray One of the strap-shaped flowers in aster family.
Receptacle Tip of a flower stalk or axis bearing the flower parts; in aster family bearing flowers.
Reflexed Abruptly bent or turned backward.
Regular Having the members of each part alike in size and shape.
Reniform Kidney shaped.
Resinous Sticky or with small, yellow particles.
Reticulate Net veined.
Retrorse Turned downward or backward.
Rhombic In outline like a rhombus; obliquely four-sided.
Rootstock An underground stem, often horizontal.
Rotate Flat and circular in outline; wheel shaped.
Scale A minute or much reduced leaf.
Scurfy With small scales on the epidermis.
Segment One of the parts of a leaf or other organ that is cleft or divided.
Sepal One of the outer set of flower parts, usually green.
Serrate With sharp teeth projecting forward.
Sessile Without a stalk.
Sheath The part of a leaf or leaf base which clasps or encloses the stem.
Shrub A woody plant, 1 to 15 ft. high, usually with several stems.
Simple Of one piece, not compound.
Sinuate With strongly wavy margins.
Spathe A leaf-like structure enclosing a flower or a cluster of flowers.
Spatulate Quite narrow with wider tip.
Spike An elongated flower cluster with sessile flowers.
Spikelet A small, few-flowered spike; the flower cluster of grasses and sedges.
Spine A sharp, woody or rigid outgrowth from surface of stem.
Spinulose With small spines or spine-pointed teeth.
Spur A hollow, slender projection from the sepal or petal of a flower.
Stamen The part of a flower which bears the pollen.
Staminate With stamens but without pistils.
Standard Large upper petal of the flower of the pea family.
Stigma Top of pistil to which pollen grains become attached.
Stipule Outgrowths of, or appendages to, the base of a petiole.
Superior Applied to an organ situated above another one; especially the ovary where other parts are attached below it.
Teeth Projections along the margin of a leaf.
Tendril A slender, coiling organ used in climbing.
Ternate Consisting of three leaflets or divisions.
Thorn A stiff, sharp pointed outgrowth; a modified branch.
Trailing Creeping along the ground.
Trifoliate Having three leaflets.

Tubercle A small, usually rounded, projection from surface of leaf, seed, etc.

Twining Winding spirally about a support.

Umbel A flower cluster with all pedicels arising from the same point.

Unarmed Without spines, thorns or prickles.

Unisexual Of one sex, i.e., either staminate or pistillate.

Valve One side of the pod in mustard family; a trapdoor-like opening in the pollen chambers of some anthers.

Verticillate Three or more leaves from same node.

Whorl A group of three or more similar organs, as leaves, radiating from the place of attachment.

Whorled Same as verticillate.

Wing One of the two lateral petals of the flower of the pea family; a thin projecting part, as on a seed or fruit.

CORRECTIONS

The pages of this reprint are identical with those of the first printing.

Additions and revisions in names are not noted here since many of them need further study.

The following errors have come to notice:

Page 11, fig. 9; for Aquilegia, read Delphinium.

29, line 5 from bottom; for Apiaceae, read Ammiaceae.

32, line 14; Elaeagnus has simple, silvery leaves but alternate.

39; Orchidaceae is omitted from the key. Cypripedium would run to Impatiens, others unsatisfactorily in the next three groups.

40a; Campanulaceae is omitted. It would replace Lobelia if the line were changed to read: "corolla tubular, 2-lipped or bell-shaped and regular.

49; delete the period after Eelgrass in No. 11.

63, line 7 from bottom; for 2.5-10 cm., read 5-15 mm. (1/5 - 3/5 in.).

69, line 13; indent two spaces to agree with line 15.

87, in No. 8; for engelmannii, read engelmanni.

116, line 2 and 3 from bottom; for alternate, read opposite.

177; Medicago sativa and M. falcata have 5 leaflets and those of falcata are often not noticeably toothed at tip.

232, line 7; for 5-15 cm., read 5-15 mm.

233, in No. 7; for Navarettia, read Navarretia.

263, in No. 6; for L., read (L.) Blake.
in No. 1; for F. Coult., read T. Coult.

268; delete line 1.

275, line 8; for 3-10 cm., read 3-10 mm.
line 15; after Fig. 303; add 308.

276, line 2 and 3; for 41 and 42, read 42 and 43.

Scientific names of genera and species, both accepted names and synonyms, are printed in italics. Species (and most varieties) are listed except under generic synonyms. Common names are in ordinary type, family and other group names in capitals. Usually only the group or generic common names are indexed, not all of the different kinds of violet, dock, sunflower, etc.

10 MILLI-1
METERS   2   3   4   CENTIMETERS   5   6   7   8   9   10

10 MILLI-1
METERS   2   3   4   CENTIMETERS   5   6   7   8   9   10

10 MILLI-1
METERS   2   3   4   CENTIMETERS   5   6   7   8   9   10

10 MILLI-1
METERS   2   3   4   CENTIMETERS   5   6   7   8   9   10

10 MILLI-1
METERS   2   3   4   CENTIMETERS   5   6   7   8   9   10

10 MILLI-1
METERS   2   3   4   CENTIMETERS   5   6   7   8   9   10

10 MILLI-1
METERS   2   3   4   CENTIMETERS   5   6   7   8   9   10

APPENDIX

Page
11. For *Aquilegia* in Fig. 9, read *Delphinium*.
15. For *purpureum* in Fig. 14, read *maculatum*.
18. In paragraph 3 for *striata*, read *maculata*.
21. In Fig. 21 for *multiflora*, read *siberica*.
32. *Elaeagnus* in Elaeagnaceae has alternate leaves.
33. Fruit dry or splitting open—*Spiraea* has 5 small pods.
39. Orchidaceae omitted. *Cypripedium* would run to *Impatiens*.
40a. *Campanula* was omitted. It would run to *Lobelia* but has symmetrical bell-shaped flowers.
  Verbenaceae (last line) have 4 stamens, flowers in long spikes.
42 *Botrychium lunaria* (L.) Sw. Low prairie, McHenry Co.
43. *Woodsia*. The species is *oregana* D.C. Eaton.
  Onoclea. Also from Pembina Co.
  Marsilea. The species is *mucronata* A.Br.
  Pteretis. Also from Turtle Mts. and Morton Co.
44. *Equisetum variegatum* Schleich. Valley City.
  Equisetum sylvaticum L. Pembina Mts., Cavalier Co.
  Selaginella, has been collected in Eddy and Pembina counties.
45. *Juniperus virginiana* was found at Fargo, presumably an escape.
47. *Sparganium chlorocarpum* Rydb. is reported for McHenry Co. and *S. multipediculatum* (Morong) Rydb. for Morton Co.
46. *Typha angustifolia* is widely established.
48, 49. *Potamogeton perfoliatus* should be *P. richardsonii* (A.Benn.) Rydb.
49. *Potamogeton amplifolius* Tuckerm. Bottineau Co.
  Potamogeton praelongus Wulfen. Turtle Mts.
  Potamogeton friesii Rupr. Burke Co.
53. Key to tribes. Spikelets of *Koeleria* in Aveneae and some species in Hordeae have no awns.
54. Some species of *Festuca* are not awned.
56. *Bromus marginatus* Nees. Slope Co., a native perennial.
  Bromus brizaeformis L. Stutsman Co., an introduced annual.
57. *Puccinellia cusickii* Weatherby. Frequent but hard to distinguish.
58. *Fluminea* should be *Scolochloa festucacea* (Willd.) Link.
61. *Poa fendleriana* (Steud.) Vasey. Bottineau Co.
  Eragrostis diffusa Buckley, Ward Co., and *E. perplexa* L.H. Harvey, Morton Co., are difficult to distinguish from *E. pectinacea*.
63. *Lolium rigidum* should be *L. persicum* Boiss. & Hohen. Line 6 from bottom, for 2.5-10 cm., read 5-15 mm. (⅕-⅗ in).
65. *Hordeum montanense* Beal. Ward Co. Thought to be a hybrid between *H. jubatum* and *Elymus virginicus*.
  Hordeum pusillum Nutt. Bowman and Billings Co., an annual weed.
67. *Avena hookeri* is now *Helictotrichon hookeri* (Scribn.) Henr.
70. *Danthonia spicata* (L). Beauv. Killdeer Mts.
72. *Aristida* is frequent on Sheyenne bluffs at Valley City, Lisbon, etc.; also in sandhills in Eddy County.
74. *Sporobolus airoides* Torr. Bowman Co.
75. *Chloris virgata* Sw. Feather Fingergrass. Annual; several spikes 2-5 cm. long in a group at end of stem. Around grass nursery at Mandan in 1962.
76. *Beckmannia*, read (Steud.) Fern.
  Leptochloa. Abundant in railway yards at Fargo and occasional in other places.
80. *Setaria lutescens* should be *S. glauca* (L.) Beauv.
86. *Hemicarpha micrantha* should be *H. drummondii* Nees.

87. *Cyperus speciosus* should be *C. ferrugineus* Boekl. A second
record is from Sargent Co.
89. **Eleocharis parvula* (R. & S.) Link. Grand Forks and Burleigh
Co. *Eriophorum.* Most specimens are **E. angustifolium* Honck.
90. *Scirpus microcarpus. S. rubrotinctus* Fern. seems preferred.
94. **Carex foena* Willd. Turtle Mts.
**Carex prairea* Dewey. Turtle Mts. and McHenry Co. A bog
plant in large clumps.
95. **Carex aenea* Fern. McHenry and Williams Co. Similar to *C.
brevior.*
**Carex molesta* Mack. Richland Co. Similar to *brevior.*
96. **Carex peckii* Howe. In woods; similar to *C. pensylvanica.*
Carex scirpiformis. Specimens from Cavalier, Bottineau, Dunn
and Williams Co. were identified by F. J. Hermann. The
Williams Co. specimens have 4 or 5 small secondary spikes
and others often one or two.
**Carex garberi* Fern. Mountrail Co.
**Carex richardsonii* A.Br. Sandy prairie, Richland and McHenry
Co. A low, broad-leaved plant.
97. **Carex haleana* Olney. Richland and Benson Co. Similar to *C.
granularis.*
**Carex parrayana* Dewey. Mountrail Co.
Carex torreyi. Bottineau and McKenzie Co. Wooded areas; resembles *C. pensylvanica* but much larger.
**Carex capillaris* L. Collected in Turtle Mts. in 1961.
99. **Calla palustris* L. Wild Calla. Pembina and Rolette Co.
100. **Juncus gerhardi* Loisel. Cass and Richland Co.
105. *Uvularia sessilifolia.* Pembina Mts., Cavalier Co.
107. *Sisyrinchium. S. montanum* Greene is now used for our common
species.
111. **Populus acuminata* Rydb. Narrow-leaved Cottonwood. Billings
and Slope Co.
114. *Ostrya virginiana.* A few trees near Butte, McLean Co.
115. *Betula fontinalis* is now *B. occidentalis* Hook.
**Betula sandbergii* Britton. Turtle Mts. and Richland Co., a
hybrid between paper and dwarf birch.
116. **Ulmus pumila* L. Chinese Elm. Frequent escape.
Humulus has opposite leaves.
118. **Pilea fontana (Lunell)* Rydb. Richland Co.
Laportea is perennial.
121. **Rumex domesticus* Hartm. Common. A little coarser than *R.
crispus,* usually no tubercles on calyx.
**Rumex stenophyllus* Ledeb. Frequent. Like *R. crispus* but calyx
with short teeth.
124. *Polygonum natans.* Should be *P. amphibium* L.
126. **Polygonum douglasii* Greene. McKenzie, Stark and Slope Co. A
slender, hillside annual.
131. **Atriplex glabriuscula* Edmonst. Fargo. Introduced annual.
133. *Corispermum villosum.* Recent authors use *C. orientale* Lam.
134. **Salsola collina* Pall. Common in recent years. More slender and
less spiny than *S. kali.*
Acnida has been considered *Amaranthus tamariscinus* Nutt.
135. *Allionia linearis* Pursh seems better than *albida.*
138. *Gypsophila.* Very abundant in sandy area near Hamar, Eddy Co.
139. For *Spergularia rubra,* read *S. marina* (L.) Griseb.
142. **Delphinium ajacis* L. Rocket Larkspur. Escaped in Stark Co.
143. *Anemone quinquefolia.* Pembina Mts., Cavalier Co.
145. **Myosurus aristatus* Benth. Ward Co.
147. *Ranunculus macounii* is perennial.
154. **Alyssum alyssoides* L. Medora in 1962. Resembles a low peppergrass but pods are divided parallel to sides. Introduced.
157. **Chorispora tenella* (Pallas) DC. Blue Mustard. Williston.
158. *Barbarea vulgaris* was found again at Fargo in 1961, apparently
introduced with grass seed.
159. *Rorippa sylvestris* was found again at Fargo in a different location in 1962.

161. *Berteroa incana.* For pods 5-8 cm. long, read 5-8 mm.
162. *Erysimum parviflorum* should be *E. inconspicuum* (S.Wats.) MacM.
163. *Polanisia.* Our plants are *P. trachysperma* T. & G.
165. **Ribes hirtellum* Michx. Occassional in swampy places.
   **Ribes oxycanthoides* L. Turtle Mts.
168. After 5. *Potentilla plattensis,* insert a line, "Stems upright or spreading, not rooting", and indent next line.
169. *Potentilla nuttallii* is considered a variety of P. gracilis Dougl.
170. *Potentilla atrovirens.* Two more collections are from sandhills in Eddy and Pembina counties. A very robust plant.
173. **Geum rivale* L. Water Avens. Pembina Co. Flowers red.
   **Agrimonia gryposepala* Walr. Richland Co. and Killdeer Mts.
175. *Crataegus rotundifolia* should be *C. chrysocarpa* Ashe.
177. *Medicago falcata* and *M. sativa* have 5 leaflets.
179. *Lupinus plattensis* S.Wats. One specimen from Shell Creek, McLean County. Phillips considered this a sub-species of *L. perennis,* the plant of eastern United States.
181. **Lotus corniculatus* L. Birdsfoot Trefoil. Cass Co.; escaped.
185. *Astragalus purshii.* Dr. S. L. Welsh considers these *A. lotiflorus.*
187. *Oxytropis gracilis.* Now called *O. campestris* (L.) DC. and *O. sericea* dropped. Flowers either purple or yellow.
   **Caragana arborescens* Lam. Much planted and rarely escaped.
190. *Desmodium acuminatum.* Now *D. glutinosum* (Muhl. Wood.
195. *Tribulus terrestris.* A third record is from Mercer Co.
203. *Impatiens biflora. I. capensis* Meerb. is used by recent authors. *Rhamnus* is widely established.
204. *Vitis vulpina.* Our plants are now referred to *V. riparia* Michx.
205. **Althaea officinalis* L. Hollyhock. Escapes locally and persists.
208. **Viola nephrophylla* Greene. Bog Violet. Frequent in wet grassland.
   **Viola bernardii* Greene, a hybrid between *V. papilionacea* and *V. pedatifida,* is occasional.
   Viola eriocarpa. V. pensylvanica Lam. is used by recent authors.
209. **Viola incognita* Brainerd. Swamps in Pembina Co. Small, white flowers.
   **Viola sarmentosa* Dougl. Killdeer Mts.; small, rounded leaves and yellow flowers.
210. *Mentzelia disperma* should be *M. dispersa.*
211. *Opuntia polycantha.* Krueger Lake, Sheridan Co., is an eastern record.
212. *Elaeagnus angustifolia* has become frequent along fences and in low areas.
   Elaeagnus argentea should be *E. commutata* Bernh.
214. *Oenothera rhombipetala* should be dropped.
215. *Oenothera caespitosa.* Collected near Krueger Lake, Sheridan Co.
216. **Myriophyllum verticillatum* Michx. Grant Co.
218. **Sanicula gregaria* Bickn. Richland Co.
   **Osmorhiza claytoni* (Michx.) C. B. Clarke, Richland Co.; had been listed as *O. longistylis,* var. *villicaulis.*
223. *Androsace puberulenta.* Now *A. septentrionalis* L.
   Lysimachia verticillata. Now *L. hybrida* Michx.
   Lysimachia longifolia. Should be *L. quadriflora* Sims.
224. *Dodecatheon media.* A recent study has called our plant *D. pulchellum* (Raf.) Merr.
227. *Asclepias (Acerates) lanuginosa* Nutt. A second record from Stark Co.
228. *Asclepias purpurascens.* Should be *A. sullivantii* Engelm.
231. **Cuscuta glomerata* Choisy. Richland and Ransom Co. Forms a dense band of flowers on aster, goldenrod, etc.
232. In line 7, for 5-15 cm., read 5-15mm.
233. *Naumbergia minima.* Should be *Naumbergia propinqua* Suksd.
236. **Hackelia virginiana* (L.) Johnst. Cass and Barnes Co. Prickles also on back of nutlets; burs more rounded and plant coarser than *americana.*

237. *Lycopsis arvensis* L. Stark Co.
 Cryptantha macounii (Eastw.) Payson. White Earth, Mountrail Co.
238. Line 6 from bottom. For stamens 5, read stamens 4.
241. *Agastache anethiodora.* Should be *A. foeniculum* (Pursh) Ktze.
 Nepeta hederacea. Glechoma hederacea L. seems preferred.
247. *Physalis lanceolata.* These all seem to be *P. virginiana.*
248. *Solanum dulcamara* L. Bittersweet (not to be confused with *Celastrus*).
 Two records, accidentally introduced with garden material.
251. *Linaria canadensis* (L.) Dumort. Grant Co.
 L. dalmatica (L.) Miller. Dalmatian Toadflax. Stark and Steele Co. Escaped from gardens but reported weedy. Leaves ovate, sharply pointed.
255. *Mimulus guttatus* DC. Grand Forks Co. A showy, yellow flowered plant in running water.
 Hydranthelium rotundifolium. Bacopa r. (Michx.) Wettst. is preferred.
258. *Utricularia intermedia* Hayne. McHenry Co.
261. *Galium laboradoricum* Wieg. McHenry Co.
 Sherardia orientalis Boiss & Hohen. Once found at Fargo.
 Linnaea borealis. A second record from Killdeer Mts.
262. *Viburnum affine.* Now *V. rafinesquianum* Schultes.
263. *Sambucus canadensis* L. Black-berried Elder. Occasionally escaped.
265. *Campanula rapunculoides* is proving a most persistent weed.
 Campanula aparinoides. For 50 miles read 15 miles.
268. Strike out all of first line and "involucre woody" above 63-65 middle of page.
272. *Aplopappus. Haplopappus* is the accepted spelling.
275. To Fig. 303, add 308.
276. Lines 2 and 3 refer to 42, 43.
277. *Aster coerulescens. A. simplex* Willd. now seems the best name.
 Aster junceus. Should be *A. junciformis* Rydb.
279. The treatment of *E. glabellus* by Scoggan (Fl. Manitoba) seems better.
 Erigeron asper—upright, rays white, flowering in June.
279. *E. glabellus*—spreading at base, rays pink, flowering in August. *E. asper* is the common prairie plant. *E. glabellus* is usually on steep, north-facing slopes, often near or in wooded areas.
281. *Bellis perennis* L. English Daisy. In lawn at Fargo.
 Iva axillaris has the upper leaves alternate.
282. *Ambrosia trifida* has the pistillate flowers at the upper leaf bases.
 Xanthium. Recent authors use *X. strumarium* L.
284. *Brauneria angustifolia.* Should be *Echinacea angustifolia* DC.
287. *Galinsoga parviflora* Cav. Fargo; in flower beds and yards.
288. *Achillea multiflora.* Should be *A. siberica* Ledeb.
291. *Petasites palmatus* (Ait.) Gray. Turtle Mts.
292. *Senecio vulgaris* L. Common Groundsel. Dickinson and Fargo. A persistent, annual, garden weed.
 Arctium lappa L. Grand Forks and LaMoure Co.
 Arctium tomentosum Mill. Fargo.
293. *Cirsium flodmani* is regarded as the common plant leaving *megacephalum as undulatum.*
295. *Centaurea maculosa* Lam. Spotted Knapweed. Found at Fargo but destroyed.
296. *Taraxacum erythrospermum* Andrz. seems back in favor for *T. laevigatum.*
 Agoseris cuspidata is placed by some authors in *Microseris.*
 Taraxacum koksaghys Robin. Russian Dandelion. Established at Fargo but apparently not aggressive.
297. *Stephanomeria tenuifolia* (Torr.) Hall. Little Bad Lands. Stark Co. Smaller than *Lygodesmia* with a basal tuft of leaves.